DISCARDED
JENKS LRC
GORDON COLLEGE

About Island Press

Island Press is the only nonprofit organization in the United States whose principal purpose is the publication of books on environmental issues and natural resource management. We provide solutions-oriented information to professionals, public officials, business and community leaders, and concerned citizens who are shaping responses to environmental problems.

In 1994, Island Press celebrated its tenth anniversary as the leading provider of timely and practical books that take a multidisciplinary approach to critical environmental concerns. Our growing list of titles reflects our commitment to bringing the best of an expanding body of literature to the environmental community throughout North America and the world.

Support for Island Press is provided by Apple Computer, Inc., The Bullitt Foundation, The Geraldine R. Dodge Foundation, The Energy Foundation, The Ford Foundation, The W. Alton Jones Foundation, The Lyndhurst Foundation, The John D. and Catherine T. MacArthur Foundation, The Andrew W. Mellon Foundation, The Joyce Mertz-Gilmore Foundation, The National Fish and Wildlife Foundation, The Pew Charitable Trusts, The Pew Global Stewardship Initiative, The Rockefeller Philanthropic Collaborative, Inc., and individual donors.

Forest Patches
in Tropical Landscapes

Forest Patches
in Tropical Landscapes

Edited by

John Schelhas

and

Russell Greenberg

JENKS L.R.C.
GORDON COLLEGE
255 GRAPEVINE RD.
WENHAM, MA 01984-1895

ISLAND PRESS

Washington, D.C. • Covelo, California

SD
414
.T76
F67
1996

Copyright © 1996 by Island Press

All rights reserved under International and Pan-American Copyright Conventions. No part of this book may be reproduced in any form or by any means without permission in writing from the publisher: Island Press, 1718 Connecticut Avenue, N.W., Suite 300, Washington, DC 20009.

ISLAND PRESS is a trademark of The Center for Resource Economics.

Library of Congress Cataloging-in-Publication Data

Forest patches in tropical landscapes/[editors] John Schelhas and Russell Greenberg.
p. cm.
Includes bibliographical references and index.
ISBN 1-55963-425-1 (cloth). — ISBN 1-55963-426-x (paper)
1. Forest conservation—Tropics. 2. Forest ecology—Tropics. 3. Forest management—Tropics. 4. Forests and forestry—Environmental aspects—Tropics. I. Schelhas, John. II. Greenberg, Russell.
SD414.T76F67 1996
333.75'16'0913—dc20 96-4289
CIP

Printed on recycled, acid-free paper

Manufactured in the United States of America

10 9 8 7 6 5 4 3 2 1

Contents

Preface

This volume is the culmination of a policy research initiative of the Smithsonian Migratory Bird Center (SMBC) begun in 1991. Over the past five years the SMBC has sponsored a symposium, produced a policy white paper (available in Spanish and English), developed a forest patch project directory, and offered briefings on the topic to organizations interested in conservation and development in the tropics. How is it that an organization dedicated to the conservation of migratory songbirds has become so involved in a topic that involves so many issues seemingly unrelated to birds? The answer to this question can be found in a brief history of the field of migratory bird conservation.

Like many of my colleagues, I have long been interested in the ecology of migratory birds on their tropical wintering grounds. During the 1970s and 1980s, many ornithologists began to realize that the birds we study in North America spend a substantial portion of their lives in habitats and climates quite different from those with which we are most familiar. This realization led to an intense interest in traveling south and discovering the secrets of the tropical lives of migratory birds.

The pioneering work of a number of tropical ornithologists led to another, more ominous, realization. The habitat for migratory birds, indeed for many tropical birds, was being destroyed or severely altered at unimaginable rates. For the past two decades ornithologists have been feverishly trying to document the changes that occur when natural vegetation communities (mostly forest) are transformed. From the beginning the comparisons have often been dichotomous, comparing old-growth forest to a myriad of habitats under the rubric of "disturbed" habitats. This perspective has led to a long debate about the value of disturbed second-growth habitats to migratory birds.

However, when one actually examines the distribution of migratory birds, the dichotomy seems less useful. The seemingly monolithic agricultural landscapes of the tropics are in reality a patchwork quilt. Some patches consist of fields or pastures that are devoid of all but a few species of migratory birds. Other patches, dark green with trees, often sustain very high populations of migratory birds. In fact, these islands of trees (which include remnants of forest, shade coffee plantations, and gallery forest) support the highest densities of many migratory species. Because migratory bird populations spread out over large regions of the tropics, these patches of native arboreous vegetation scattered in the countryside seem to be a necessary component of their habitat, particularly as intact forests become smaller and more isolated.

The SMBC was founded in 1990 to develop creative approaches to the conservation of Neotropical migrant birds. The growing realization of the potential for forest patches to harbor otherwise dwindling populations of migratory birds made the issue of forest patch conservation a natural priority for the nascent center. However, it is the philosophy of the SMBC that conservation actions that benefit migratory birds cannot be accomplished for the love of birds alone.

The purpose of the symposium and this book was to bring together a wide variety of workers in the areas of tropical conservation and rural development to consider what we know about forest patches and to develop the common ground to promote forest patch conservation where it is appropriate.

People love birds. Migratory birds, in particular, catch the imagination of millions. While there are many good reasons to promote the protection of forest patches in agricultural areas, we believe that migratory birds and the interest they engender provide a natural rallying point for promoting programs that focus on these neglected, but increasingly important habitats.

—R. G.

Acknowledgments

I acknowledge the support of the Cornell Program in Ecological and Social Science Challenges of Conservation (NSF Grant # BIR-9113293) during the preparation of this manuscript. Marian Hovencamp, Nancy Bowers, and Carol Rundle provided much needed help with figures, tables, and translating files from diverse word-processing programs. Dialogue with Janis Alcorn on community conservation and comanagement has broadened my view of social and cultural factors in conservation. Discussions with students and faculty at Cornell and with scientists at the Smithsonian Institution have enhanced my interdisciplinary understanding of tropical forest conservation issues. I thank the many farmers in Costa Rica, the Dominican Republic, and Guatemala who have welcomed me into their homes and shown me their farms and forests. Most important, I would like to thank Susie and Robinson for their support throughout this project.

—J. S.

I would like to acknowledge Judith Gradwohl for her continued help and critical thought on issues of tropical conservation. The staff of the

Smithsonian Migratory Bird Center assisted in every phase of the research and preparation of this manuscript. The Smithsonian Institution, particularly Eugene Morton, Bob Hoffman, Ross Simons, and Mike Robinson, has supported the work of the Migratory Bird Center on this and other issues. The Pew environmental scholars program has provided an informal forum for discussing innovative conservation ideas. I particularly thank Janis Alcorn for influencing my thinking about tropical conservation early on in this project.

—*R. G.*

Introduction: The Value of Forest Patches

John Schelhas and Russell Greenberg

Millions of hectares of tropical forest are being converted to agricultural fields and pastures every year, threatening many species with extinction. Efforts to conserve tropical biological diversity have usually concentrated on setting aside large tracts of forest in national parks and in other protected areas. These large, relatively undisturbed forests are essential for the effective conservation of complex tropical ecosystems and many forest species. Yet only 5.9 percent of the world's land area is in designated protected areas, and only 19 out of 131 countries in Asia, Africa, and Latin America have more than 10 percent of their land protected (WRI 1994). Furthermore, these protected areas are concentrated in inaccessible areas and on less economically useful lands, and do not provide representative protection for different ecological habitats. In any event, because of social, economic, and political constraints, there is often little hope of expanding protected area systems. It is becoming increasingly clear that a conservation strategy focusing only on large, protected areas will leave the conservation needs of some organisms and habitats unmet, and that conservation efforts in the tropics must include areas that lie outside large reserves.

While deforestation in the tropics has been extensive over the past few decades, forest clearing is rarely complete and often not permanent. Patches of primary, secondary, and managed forests remain in many agricultural landscapes. Although these forest patches lack some of the biological diversity of large forest tracts, they do contain many species and habitats that are not found in large reserves. Forest patches provide habitat for many organisms with small home ranges, for organisms that range widely over the landscape, and for an array of species that tolerate some human disturbance.

Forest patches, which are widely dispersed through human-occupied landscapes, are owned and controlled by rural people. Because of this, protectionist conservation strategies that would lock up these lands in reserves are neither ethically acceptable nor possible to effectively implement. Forest patches can only be conserved by the people who own and use them. Although some protectionists would have us believe that entrusting the care of forest patches to their rural owners is a prescription for disaster, there is substantial evidence that this course may be the best hope for the conservation of many tropical species and habitats.

Forest patches are not simply left-over and unutilized land areas, they are critical components in the livelihoods of rural people. Forest patches provide products for sale and subsistence, protect watersheds, and have cultural and social benefits. Forest patches can be managed to provide niches for the growing of forest products and crops. As a result, rural people maintain forest patches for their economic, social, and cultural benefits, thereby protecting the diversity of organisms that inhabit them.

Yet forest patches are changing, and in many cases disappearing, under the pressures of human population growth, increased integration of rural households into the world economy, the breakdown of traditional patterns of forest use and conservation, and government policies that create incentives for conversion of forest patches into croplands and pastures. Not only will the loss of forest patches, and the products and environmental services they provide, hurt rural people, it will also have national and global impacts, because of the value of forest patches for watershed protection, biological conservation, and guarding against adverse global environmental changes.

While it is clear that rural people and conservationists have a shared interest in forest patch conservation, there is much that we need to learn and do. We must learn more about which species can survive and thrive in different types and arrangements of forest patches. We must understand the social and economic conditions that promote forest patch conservation and how government policies can promote or harm forest patches. We must develop management schemes that improve the capacity of forest patches to protect biodiversity and that increase their value to local people. And, most important, we must apply this information to develop programs and institutions that support and promote forest patch conservation. This book addresses these concerns by taking a broad, interdisciplinary look at forest patches in tropical landscapes, and presents examples of forest patch conservation strategies that have both social and biological benefits.

Types of Forest Patches

The reader will soon note the diversity of habitats treated by different papers under the rubric "forest patch." This diversity is deliberate—we include everything from patches of trees in pastures and lines of trees along rivers to large fragments of primary forest. Because of the harlequin nature of tropical landscapes, different forest patch types can often be found in close proximity. These forest patch types do not exist in isolation, but together contribute to the resource management system for the region. The interaction of diverse patch types helps to protect biodiversity, since patches may harbor different sets of species, or even the same species, at different times of year. In short, a mosaic of different patch types will provide habitat for the maximum number of species and provide benefits for the most people.

The following is a brief lexicon of the different types of forest patches considered in the different chapters of this book.

Forest remnants are largely unused fragments of old-growth forest, including forest intentionally left uncleared for watershed protection, wildlife conservation, and other benefits, or patches that simply have not been cleared yet because of inaccessibility or recency of human occupation.

Managed forests are patches of forest that are used for timber production and the extraction of nontimber resources.

Natural forest plantations are agricultural systems that integrate the use of native trees with the production of agricultural and other commodities, as well as fuel wood and other household necessities.

Sacred forests are patches of forest protected for their cultural significance.

Gallery or riparian forests are linear strips of forest habitat, a few trees to hundreds of meters wide, that protect streams from erosion and insolation.

In addition, we discuss a number of smaller formations of trees in agricultural settings, including living fences, farm-field border strips, and shade patches for livestock.

Beyond Fragmentation: Biological Considerations for Forest Patches

The value of forest patches to biodiversity conservation is heatedly debated by biologists studying tropical ecosystems. The issue is becoming increasingly important as region after region loses its last remaining tracts of forest and the landscape is increasingly settled. In many areas where natural forest cover has been lost, wooded patches of various kinds are the only remaining habitat for forest flora and fauna. On the other hand, in regions where continuous forest cover remains, it is easy to see that these patches have become highly depauperate. Large wildlife is being hunted out, and even very small organisms are vanishing because their specialized requirements cannot be met in a small and degraded habitat patch.

Observers agree that forest patches contain a highly altered and usually depauperate subset of the original forest fauna. However, assigning a value to these patches is a subjective exercise that results in a wide array of opinions on the matter. One need only read the chapters in this book to gain a sense of this diversity. In general, those authors whose work focuses on rural development and takes place in settled landscapes emphasize the positive contributions that forest patches can make. More skeptical views of the value of forest patches can be found in contributions by ecologists whose work focuses on the conservation of intact tropical forest ecosystems (see Robinson, chapter 6; and Bierregaard and Dale, chapter 10). This has led some to argue that forest patches are "empty forests" and that assigning biological value to them provides justification to developers for further deforestation and fragmentation. There is some truth to this, but it is equally true that in many areas these patches are the only forested habitats left.

To simplify a complex subject, the evaluation of the biodiversity value of forest patches is an intrinsically comparative exercise, and the value placed on forest patch conservation largely depends on the comparison being made. When the comparison is made between intact systems and forest patches, forest patches are invariably shown to be lacking in biodiversity (Robinson, chapter 6; Bierregaard and Dale, chapter 10; Kattan and Alvarez-López, chapter 1; Power, chapter 5). Many of the most common forest organisms are lost at even the most moderate levels of fragmentation and modification of tropical forest habitats. Because of the loss of top predators (Robinson, chapter 6), as well as the agents of pollination (Murcia, chapter 2), seed dispersal (Guindon, chapter 9), nutrient cycling (Kattan and Alvarez-López, chapter 1; Power, chapter 5) and other interspecific interactions, loss of species is usually accompanied by large changes in the entire biological community. However, if the forest patches are compared to alternative land uses in a human-dominated landscape, the patches can easily be shown to increase local diversity and provide refuge for at least some of the species associated with intact forest (Nepstad et al., chapter 7; Power, chapter 5; Kattan and Alvarez-López, chapter 1; Greenberg, chapter 4).

Forest patch conservation and management plays a particularly critical role in certain landscapes and circumstances:

1. As many forest generalists and edge species survive or prosper in forest patches, patches play an important role in increasing local biological diversity in areas where forest cover of any kind is threatened by development. This increase in diversity may be of great importance in regions where most of the intact forest has been lost, and particularly where forests host many endemic species. There are certainly many examples of such regions (Sao Paulo province of Brazil, Haiti, Pacific coastal plain of Central America). Forest patches in these areas may provide the "seed" for the reestablishment of larger tracts of vegetation when ecological restoration becomes feasible.

2. Forest patches may be critical in the buffer zones of large reserves, providing seasonal resources and movement corridors for organisms that spend most of their time in the protected forests. In this regard, streamside vegetation (gallery forests) may play a very important role in regional conservation efforts. It is common for protected forests to occupy a particular elevation in a mountainous region (often the tops of ridges), and clearing up to the boundary of the reserve may eliminate some of the more mobile species (see Guindon, chapter 9; Kattan and Alvarez-López, chapter 1).

3. Forest patches may provide stepping-stone resources for local or long-distance migratory organisms (Greenberg, chapter 4). This is particularly apparent for temperate-zone migratory birds, which often find

refuge in the smallest and most modified of "forest" habitats. But it may be equally true for other groups of migratory organisms that are less well known, such as arthropods that move into patches of woods for dry season refuges.

Having set some general priorities, we hasten to add that even where there is little global significance to the species protected in forest patches, the species that find refuge in forest patches in settled areas provide local human populations access to some remnants of forest flora and fauna that are not locked away in distant and inaccessible reserves. In this regard, rather than focusing on particular regions, forest patch conservation will be most effective if it is woven into the fabric of rural development everywhere.

Forest patches will almost always provide some biological benefit when compared to other land uses, such as pasture and farming. The critical question is, at what point is it worth investing resources to manage forest patches, particularly in light of the often stretched budgets for protecting large tracts of forest? Several authors have pointed out that the value forest patches have is dependent upon human intervention and management (Viana and Tabanez, chapter 8; Lyon and Horwich, chapter 11; Kellman et al., chapter 3). Generating forest patches in agricultural areas may require the initial establishment of small vegetation nuclei (Lyon and Horwich, chapter 11). Established forest patches are particularly sensitive to disturbances, such as blowdowns, fire, and grazing; and invasion of exotic plants and animals, native vines, and aggressive weeds discourages regeneration. Certain important trees must be raised in nurseries and planted, because they lack pollinators and dispersers or because they are already locally extinct. Even strategies that do not necessarily require technical interventions, such as encouraging the conservation of riparian corridors, may require resources for extension programs.

In many cases, there is no necessary trade-off between providing resources for forest patch and natural forest conservation efforts. As Pinedo-Vasquez and Padoch (chaper 16), Alcorn (chapter 12), Lyon and Horwich (chapter 11), and Poffenberger (chapter 18) show, landholders, when given an opportunity and some support, often manage forest patches for both the economic products and the ecological services they provide, and the work can be largely self-supported. In other cases, however, active management of forest patches will require the assistance of outside organizations and institutions. These resources need not be the same as those that would go into wildland management. Because the efforts can be complementary, we believe that it is important that biologists pool their shared interests with the rural development community and work to make forest patch conservation an integral part of long-term, ecologically sound development. Alcorn (chapter 12), Fisher and Bunch (chapter 19), and Poffenberger (chapter 18) outline and provide examples for strategies in which national and international institutions function as partners with local landholders.

In any case, forest patches seldom are isolated and unused fragments of habitat. Rather, with the possible exception of recently colonized frontiers (where forest patches may exist because they are inaccessible), forest patches owe their existence to the value placed on them by local people. Because of this, forest patch conservation is a unique and complex challenge. Many of the decisions affecting forest patch management are made by landholders responding to local or even household concerns. At the same time, even the most isolated patches are currently battlegrounds for conflicts between the different economic forces influencing development in the countryside of most tropical countries. As described by Pinedo-Vasquez and Padoch (chapter 16), traditional forest patch management is part of an overall agricultural system that creates a diverse vegetational mosaic. Small-scale swiddens in these systems may create habitat for forest organisms adapted to small-scale disturbance (Power, chapter 5). The management strategies in these systems are primarily aimed at producing items for the household or local community, or for domestic markets. These locally adapted systems are giving way to ones developed primarily to produce commodities for the global market and provide cash for landowners. Traditional management systems are being replaced or altered by the infusion of external capital into every nook and cranny of the tropical countryside. Schelhas (chapter 13) and Browder (chapter 14) show that economic changes driven by events far away from the land in question have an increasingly important influence.

Given the desire to stay in touch with nature, it is easy to understand the hesitancy of some biologists to become mired in the political and economic complexities of forest patch management. At the very least, however, the realization that forest "fragments" are actually actively managed forest patches has tremendous implications for biologists studying these systems. The island biology and fragmentation paradigm that governed the study of forest patches for the past few decades is giving way to a more holistic approach to forest patch research. Forest patches are seldom just fragments; they are novel habitats that owe their composition and function to a variety of human and natural forces

(Lyon and Horwich, chapter 11). These human influences need to be overlaid on the biological factors controlling the "natural" biological community, including soil type, climate, and natural disturbance regime (lightning fires, hurricanes, flooding).

Fragmentation

Fragmentation refers to the reduction in size and the increased isolation of forest patches. The effect of just these two factors is well described by Bierregaard and Dale (chapter 10). Fragmentation exposes the forest organisms to increased edge effect, bringing with it increased light levels, invasion by open-country species, and desiccating winds. A forest patch may simply be too small to provide resources for animals with large home ranges. More often, a forest patch can support only small populations, which are prone to local extinction from a variety of factors, including stochastic variation in population size. Low-population size introduces a number of potential additional problems related to the loss of genetic diversity. As Bierregaard and Dale point out, sensitivity to fragmentation will vary considerably between species or higher taxonomic groups of species, depending on such things as the degree of ecological specialization, body size, and movement patterns—features that are poorly understood for most tropical organisms. In addition to this within-community variation, the work of Kellman et al. (chapter, 3) suggests that different communities may be comprised of species that vary systematically in how they respond to fragmentation, which may be a result of the different vegetational histories. For example, gallery forests may have provided habitat for forest organisms through dry periods of the Pleistocene, and therefore their trees and organisms are more resistant to extinctions resulting from current isolation in agricultural settings.

Tropical ecosystems not only support a large number of species but consist of a very large number of complex biological interactions as well. Therefore, species loss is not just a numbers game. The loss of large predators and herbivores, for example, has implications for the entire biological community (Robinson, chapter 6). The issue of how fragmentation affects processes that depend on the ecological interplay between species is only beginning to be addressed. For example, Murcia (chapter 2) outlines the potential short- and long-term impacts that fragmentation might have on the pollination of tropical plants, many of which depend on animal vectors and are obligatorily out-crossed. Seed-dispersal systems do not achieve the levels of coevolutionary specialization found in pollination systems; however, Guindon (chapter 9) tackles the problem of maintaining healthy populations of dispersers for specialized fruiting trees of the wild avocado family.

Arrested Succession

Much of the forest present in patches in settled landscapes is secondary, having been cleared or selectively cut at least once or, more likely, numerous times. Eventually, old-growth species may disappear, soil quality will decline, specialized soil symbionts vanish, and the habitat will be retained in a state of continued secondary succession (S. Purata, pers. comm.). In addition, loss of key dispersal or pollination agents may favor plants that can regenerate vegetatively or are pollinated and dispersed by wind (Lyon and Horwich, chapter 11).

Resource Degeneration

Local human populations can systematically harvest certain plants and animals without a sustainable management scheme, thereby reducing the abundance or even causing local extinction of the target organisms in a patch. The effect of hunting on the wildlife of forest patches throughout the tropics provides a graphic example of this type of impact (Robinson, chapter 6). Systematic degradation can apply to patch structure as well as to species composition (Viana and Tabanez, chapter 8). Large trees or palms can be high-graded from the forest, changing the physical structure of the forest. More subtle, but equally as devastating, is pruning, weeding, and vine removal. Such activities, found in silvicultural systems and shade plantation management, will remove epiphytes, vines, mosses, and the dead leaves and detritus captured by these plants—microhabitats that provide niches for many of the most specialized tropical organisms.

Resource Management

Alternatively, people can favor particular plants because of their usefulness or commercial value. (Techniques that favor particularly valuable timber species in forest patches are described by Stanley and Greztinger, chapter 17.) Modern silvicultural systems provide a promising avenue for improving the value of forest patches and thereby their probability of being protected as well. However, these treatments will inevitably change the species composition and the very structure of the forest patch, generally decreasing plant diversity. Pinedo-Vasquez and Padoch (chapter 16) discuss traditional systems where individual rather than tree species selection favors the maintenance of forest diversity.

Resource Augmentation

People can add valued plants to a forest patch. These added plants produce foods or materials used in local households and influence the resources available for wildlife and the functioning of other aspects of the ecosystem. The practice of augmentation provides a tremendous opportunity for improving the biodiversity of human-managed forests. Many common agroforestry species provide fruit or nectar for a wide range of species. However, by decreasing plant diversity and increasing the number of trees of a few species, the resource base for the entire community will be affected. First the presence of a few seasonally abundant tree crops will increase the seasonality of the habitat for animals using these crops. As species decline, the probability of phenological gaps (periods with no fruit or nectar resources available) increases and the needs of only mobile organisms will be met. Second, systematically favoring certain types of plants with particular fruiting or flowering syndromes will alter the types of organisms favored. For example, resources available to frugivorous animals in an agroforestry system will depend on human preferences. A large proportion of trees managed in diverse shade coffee and cacao plantations produce fruit favored by people and other mammals, but bird-dispersed fruit is not favored as often in these systems.

Change in Disturbance Regime

Increased proximity to agricultural fields, change in local climate, and an increase in edge expose forest patches to higher fire frequency (Nepstad et al., chapter 7). In addition, grazing animals often find refuge in forest patches, particularly during the dry season. Because of their isolated and islandlike nature, forest patches may receive high levels of browsing.

Understory Cultivation

As stated earlier, the understory of forest patches is often managed for commodity production or cleared and burned for agricultural, silviculture (Pinedo-Vasquez and Padoch, chapter 16) or livestock use. Cultivation or grazing in forest understory clearly affects understory organisms. For example, coffee and cacao plants appear to support far fewer insectivorous birds, either because of low insect densities or insufficient cover for nesting. The active management of the forest floor will also impact tree regeneration and thus, canopy composition.

Matrix Management

As a number of chapters point out (Bierregaard and Dale, chapter 10; Murcia, chapter 2; Greenberg, chapter 4), the nature of the surrounding vegetation has tremendous implications for patch use. The theory of island biogeography is most useful where the island (patch) is surrounded by an inhospitable barrier that is thoroughly discouraging to travel. In fact, in tropical areas the dichotomy between forest and second growth or field organisms is often quite sharp (Power, chapter 5; Bierregaard and Dale, chapter 10). However, the island model for forest patches is unrealistic for the many organisms that move in and out of, or between, patches if enough trees or shrubs are present in the surrounding fields. Furthermore, it is well known that for birds, the tendency to move through or across clearings varies with the feeding guild or strata used by a species. The permeability of the field–forest border also raises the possibility of influencing the agroecosystem through forest patch management (Power, chapter 5). Preliminary data suggest that forest patches may harbor a high diversity of predatory or parasitic arthropods (although numbers of predators may at times be higher in farm fields). How these shifts affect the dynamics of pest insect populations is an important area of future work.

Creating a Forest Patch Mosaic

Each forest patch type is an altered and often depauperate ecosystem with distinct characteristics. Some of this variation is a result of systematic differences in management, such as that which leads to the creation of diverse shade canopy coffee plantations or stands of thorny acacias in eastern Chiapas (Greenberg, chapter 4). Other differences may result from seemingly idiosyncratic decisions, such as which remnant trees are left standing (Lyon and Horwich, chapter 11). However, the different patch types are usually found intermingled in a complex mosaic of habitats; and as Greenberg emphasizes, there is a need to examine this mosaic. Not only will the different patch types support a different array of forest organisms, but for the mobile species using seasonally available resources in different patches, a proper balance may be required to eliminate regional phenological gaps to sustain their annual activity.

Many factors operate simultaneously to shape the biota of a particular forest patch. This complexity creates a potential epistemological conflict. The best science is often that which examines the effect of individual variables through controlled experiments or, at the very least, comparisons where few factors are varying simultaneously. However, since these systems are the outcome of myriad interrelated inputs, the more controlled experimental result may present an incomplete or even misleading picture. It seems to us that while we pursue standard reductionistic approaches to untangle cause and effect, we also need a basic description of what these different forest patch systems are like from the viewpoint of biology and human activity. More holistic and interdisciplinary studies of the patches, which for better or worse are probably the forest habitats of the future, would be a good first step in understanding what biological diversity will be like in the coming millennium.

Human Dimensions of Forest Patches

Large areas of the tropics have been converted to mosaics of fields, pastures, and forest patches. Although rural people clearly require fields and pastures to survive, the often more subtle values of forest patches should not lead them to be regarded as simply leftover and unutilized lands. Forest patches provide a wide array of benefits to rural people, who in turn value, manage, and protect them. Nevertheless, forest patches have often been overlooked by social scientists, and have rarely been taken into consideration in rural development activities.

Forest patches are used by rural people in many ways, and are an integral part of rural landscapes in the tropics:

Products. Products extracted from forest patches meet diverse household subsistence and income needs. Trees are widely harvested by households for house construction, fence posts, fuelwood, and commercial timber. Wild plants and animals are important sources of protein and other nutrients for many rural dwellers. Forests also provide other products, including medicines, fodder for livestock, thatch, raw materials for handicrafts, leaf litter to maintain soil fertility, and income through nonconsumptive uses such as ecotourism.

Ecological processes. Forest patches provide many ecological services. Forest fallows, which are temporary forest patches, maintain and improve soils in several ways, including restoring fertility by bringing up subsoil and leached nutrients, improving soil aeration and structure, and preventing erosion. Although forest fallows move around the landscape over time, they often remain an important landscape component.

Landholders maintain forests along streams and rivers, around springs, and in other areas to prevent erosion, modulate water runoff, and maintain water quality. In many tropical landscapes, riparian forest strips are both common and highly valued by landholders for their watershed protection benefits. Forest patches may directly benefit agriculture by supporting species that pollinate agricultural crops or prey or parasitize on agricultural pests (Power, chapter 5; Poffenberger, chapter 18).

Forest patches are also managed for forest- and shade-grown crops under a range of intensities. In some cases, predominantly natural forests are subject to low-level management by occasionally either planting useful plants or favoring them by removing competing vegetation. In other cases, management is intensified to produce forests that are a mix of planted and naturally regenerated plants. Management can be intensified further to produce anthropogenic forests such as home gardens and shade-grown crops, which often include native species in the overstory or understory.

Cultural and social values. Rural people maintain forest patches because of their cultural significance, both religious and secular. Sacred groves are found throughout Asia, Africa, and the Americas (Lebbie and Freudenberger, chapter 15). These sacred groves often have multiple uses, including serving as sites for religious rituals and conflict resolution, and providing medicines or other forest products through carefully controlled extraction. Forests also have secular cultural value. Formal private reserves that protect forests' natural heritage value are becoming increasingly common in many tropical countries (Alderman 1991), and rural landholders often informally protect forest patches for their legacy and heritage value (Browder, chapter 14; Schelhas, chapter 13).

Differences in forest patch values. There are many different ways that forest patches are used and valued by rural people. Forest patches are social spaces, shaped by human uses and values (Browder, chapter 14). In some cases, forest patches play important economic and social roles, and hence are valued, managed, and protected by people. In other cases, in which the benefits and uses of forest patches go unrecognized, forest destruction and degradation is high. In yet other cases, forest products are used and forest patch values recognized, but economic, political, or demographic constraints lead to overexploitation or unsustainable use.

Case study examples and efforts by social scientists to draw broad conclusions from research results present a confusing and conflicting array of trends, issues, and proposed strategies to promote forest patch conservation. Many patterns of forest patch conservation and destruction are site specific, and so, too, must be the solutions. But patterns can also be discovered, and understanding the relationships between different variables can help in planning conservation and sustainable development strategies.

Forest Patches in Farming Systems

The chapters in this book not only address many of the ways that rural people use forest patches but also discuss why people do or do not conserve forest patches. Forest patch use and conservation is influenced by factors at the household, community, and national levels. Supporting and enhancing forest patch conservation requires a broad approach that understands the interactions between these multiple levels and the interest groups within them.

Many decisions about forest patches are made by households. Household land-use strategies are often diversified, and forest patches, as a component of diversified land-use systems, meet a variety of household needs (Fisher and Bunch, chapter 19; Browder, chapter 14; Pinedo-Vasquez and Padoch, chapter 16; Schelhas, chapter 13). We have already described the many products extracted from forest patches to meet cash and subsistence needs. Forest patches also reduce economic and subsistence risk, since they can be drawn on during periods of reduced employment opportunities, famines and seasonal food shortages, or to meet both expected and unexpected cash needs. Forest patches are often particularly important to poorer people who lack sufficient agricultural land. The ways in which forest patches are used in a particular place depends on a variety of factors. Rural people choose among alternative land uses by seeking to optimize the returns they receive from their available land, as well as by choosing options that provide attractive returns to labor, make efficient use of household labor resources, and have acceptable levels of risk. Many places, because of forest characteristics and markets, present unique opportunities for forest patch uses. Forest patch use and conservation can only be adequately understood and influenced by viewing it within the broader household and community farming systems and economic strategies within which it occurs (Fisher and Bunch, chapter 19; Schelhas, chapter 13).

Household use of forest patches is also influenced by cultural, political, and macroeconomic contexts. The cultural value of forest patches is often particularly high among indigenous and traditional communities.

Communities that have lived in long association with forest patches often have livelihood systems that are closely tied to them. Traditional village institutions regulate access to forest resources and often protect them from overexploitation (Alcorn, chapter 12; Lebbie and Freudenberger, chapter 15; Pinedo-Vasquez and Padoch, chapter 16). Today, economic and cultural patterns of communities with strong traditional ties to forests are undergoing unprecedented change, and economic pressures resulting from population growth and market involvement, along with the breakdown of cultural-based protection, can lead to forest patch loss (Lebbie and Freudenberger, chapter 15; Schelhas, chapter 13). However, forest patch conservation by indigenous groups today still often compares favorably to that of colonists (Alcorn, chapter 12; Donovan 1994), and it may be possible to combine secular and sacred values of forest patches to reinforce them against the pressures of a changing world (Lebbie and Freudenberger, chapter 15).

Colonists moving to forested frontiers from other regions are fragmenting remaining areas of continuous forest. While government incentives and the logic of shifting cultivation and speculative land markets often lead to initial widespread deforestation in frontier areas (Schelhas, chapter 13), some forest remnants generally remain uncleared. Over time, additional forest patches appear in degraded landscapes as cleared areas that were unsuitable for long-term agriculture, return to forest (Bierregaard and Dale, chapter 10; Browder, chapter 14; Nepstad et al., chapter 7) and as landholders integrate managed forest patches into their farming systems for both human and ecological benefits (Browder, chapter 14; Schelhas, chapter 13).

Government and international agency policies play an important role in influencing the forest patch choices of individual landholders (Alcorn, chapter 12; Browder, chapter 4; Schelhas, chapter 13). Government policies often include incentives for forest clearing as a means to claim and hold land, which promotes expansion of unproductive agricultural lands. Economic subsidies for crops and cattle promote agricultural land uses that would be unprofitable if they were competing with forest land uses on a level playing field. Policies can have positive effects as well, such as programs and incentives that directly promote reforestation and forest conservation. Many countries have laws in place to limit forest clearing, such as those protecting streamside corridors in Brazil and Costa Rica or the Brazilian law requiring forest on 50 percent of landholdings in new settlements to be left uncleared (Bierregaard and Dale, chapter 10; Browder, chapter 14; Schelhas, chapter 13). These laws, however, are often not effective, both because they have not been integrated into landholder and community land management systems and because they are often beyond the enforcement capability of the state.

Although forest patches provide many useful products to rural people, there is a need for new forest management options to increase those benefits. Many of the forest management systems that have been promoted in conservation and development projects provide returns that are too low, to distant, or too uncertain for widespread adoption by landholders. There are many ways that new strategies can be developed. These strategies should draw on both indigenous and scientific knowledge (Pinedo-Vasquez and Padoch, chapter 16; Stanley and Gretzinger, chapter 17) and be developed in collaboration with landholders through on-farm experimentation (Fisher and Bunch, chapter 19; Poffenberger, chapter 18).

The livelihood systems of traditional and indigenous groups often include complex forest management systems that can serve as models of sustainable forest patch use and conservation. Care must be taken to allow for the adaptation of these practices to new social, economic, and political contexts (Schelhas 1994), but marketable crops and cultivation strategies developed over long periods of time for a specific site represent an important knowledge base. Farmer-to-farmer exchange programs, in which newer colonists learn forest patch management strategies from long-term inhabitants, are one possible way to improve forest management by more recent colonists.

Natural forest management for timber has potential to provide economic returns from forests that are relatively diverse biologically and therefore of high conservation value. Many natural forest management systems provide low returns to land, and the high costs of planning and management have led foresters to believe they are only appropriate for large tracts of forest. Stanley and Gretzinger (chapter 17) present an example of a scientific natural forest management system appropriate for use in forest patches. They are implementing this system at sites throughout Central America, where the sustainable timber management potential of forest patches has not been previously recognized. Pinedo-Vasquez and Padoch (chapter 16), on the other hand, provide an example of traditional natural forest management in the Peruvian Amazon. Traditional management is long term, often beginning in an agriculture stage and oriented toward the multiple values of the ribereños, and results in a diverse forest. Not every community will have similar knowledge and tradition of forest management, but traditional forest management models have lessons to teach scientists about the ways forests can be managed both to meet people's needs and to promote biological diversity.

Forest patch management that focuses only on timber ignores many forest resources that are important to local people (Poffenberger, chapter 18). The management of forest patches for nontimber forest products (either alone or in conjunction with management for timber) can provide more immediate and, at times, higher returns than timber management, increasing the attractiveness of forest management to landholders and communities. While management for one product may affect the returns from others (Stanley and Gretzinger, chapter 17), integration of different forest uses into management systems often promotes landholder and community conservation of forest patches. Pinedo-Vasquez and Padoch (chapter 16) provide an example of forest management for a combination of marketable fruits and subsistence products that is carried out under relatively high population density conditions in Kalimantan.

Many rural landholders are interested in reforestation to obtain benefits from forest patches that have been lost on their farms through forest clearing (Browder, chapter 14; Schelhas, chapter 13). Although the more diverse forest that occurs through natural regeneration generally provides better biological habitat than monocultural plantations, it may provide too few economically valuable resources over too long a time period to be attractive to landholders. Landholders often prefer reforestation with fast-growing, high-value timber species—although high plantation establishment costs for any type of tree are often unattractive to landholders unless subsidies are provided. If widespread reforestation is to occur, alternatives with lower costs and higher returns must be developed. There may be many possibilities. Nepstad (in Doyle and Schelhas 1993: 18) suggests taking advantage of natural seed dispersers such as birds and mammals to promote the regeneration of economically valuable forests by, for example, planting lines of disperser-attracting trees out of forest patches into degraded lands. Poffenberger notes that in India natural regeneration through voluntary community protection, with management often oriented toward nontimber forest products, has only 3 to 5 percent of the cost of plantation establishment. In places where the economic value of early successional forest is low, there is a need to develop strategies to enrich regenerating forest with economically valuable species.

More intensively managed forest patch types, such as shade-grown crops and complex agroforestry systems, often have high economic productivity. Where such systems are being used, their biological, watershed, and risk-reduction benefits should be taken into account in agricultural research and extension efforts aimed at "improving" or replacing these systems. Opportunities should also be sought out to enhance these systems, such as through canopy management, to improve biological and economic values.

Forest patches are also conserved as private reserves, often in association with ecotourism (Alderman 1991; Schelhas, chapter 13). In some cases the benefits of this type of forest patch conservation go disproportionately to expatriates and wealthier segments of society, and opportunities need to be enhanced for rural landholders and communities to benefit through employment or small-scale, local tourism operations (see, for example, Lopez Vasquez 1993).

Communities and Forest Patches

Many of the benefits of forest patches are either widely dispersed or very long term, accruing to other landholders, to the community at large, or to later generations. People often make decisions based on short-term, tangible benefits. Long-term occupancy of an area often builds community-level social and cultural incentives for forest patch conservation decisions with dispersed and long-term benefits. Many indigenous and traditional communities have strong cultural values and tenure arrangements supporting the conservation of remnant forests and forest patches (Alcorn, chapter 12). Yet in many places, indigenous and community forest management is under assault from loggers, colonists, and governments. Governments have often supported extractive industries, such as logging, or represented the interests of a narrow and elite segment of the national population to the detriment of community rights and forest management (Alcorn, chapter 12; Peluso 1993). Even government conservation programs may lead to deforestation. State capability to manage forests is often weak, and government expropriation of community forest rights can convert community management and conservation systems to open access situations which encourage resource depletion (Alcorn, chapter 12; Lebbie and Freudenberger, chapter 15). Defending the right of indigenous people and communities to continue to use and manage forest patches against the encroachment of colonists, loggers, or the state may in some areas be the most important action that can be taken in support of forest patch conservation (Alcorn, chapter 12).

Watershed protection is widely recognized by landholders in the tropics as a significant value of forest patches, and landholders often retain or re-establish forests around springs and along streams (Greenberg, chapter 4; Poffenberger, chapter 18; Schelhas, chapter 13). However, many of the benefits of streamside corridors are received by downstream landholders. Maintaining watershed values depends on landscape-level patterns of forest cover, which may not be consistent with the ways in which people have divided up the landscape or with the land-use choices individual landholders have made, and therefore may require community management. Communities will often organize to manage watersheds when faced with clear threats to soil and water resources (Little 1994; White 1992). Riparian forests, which have watershed protection benefits widely recognized by rural people, can form a network over the landscape that connects many different types of forest patches, also providing biological conservation benefits. Community support for biological conservation is likely to be greater when it is combined with programs that address more immediate and direct conservation needs such as soil conservation and watershed protection (Fisher and Bunch, chapter 19). Projects and programs such as the Community Baboon Sanctuary in Belize (Lyon and Horwich, chapter 11), joint forest management in India (Poffenberger, chapter 18), and the Forests on Farms project in Costa Rica (Guindon, chapter 9) provide examples of innovative ways of integrating forest conservation across landscapes in mixed ownership to provide forest benefits to rural people as well as biological conservation benefits.

Collaboration for Forest Patch Use and Conservation

Forest patches are of interest to a diverse and widespread group of people, including local rural people in the tropics, regional and national beneficiaries of forest products and ecological services, and distant people connected to forest patches by migrating species and patterns of global environmental change. Interest groups at different levels—global, national, regional, community, and household—generally receive only a part of the value of forest patches and often ignore other values.

Rural households and communities benefit in many ways from forest patches and manage them for diverse products. This may promote biological diversity in ways that management of forests for a single use, such as timber, does not (Pinedo-Vasquez and Padoch, chapter 16; Poffenberger, chapter 18). But although local landholders and communities often conserve resources that are important to their livelihoods, they may neglect species without recognized local values. Biological conservation, as defined by ecologists, may to some extent be an incidental effect of the management systems of rural people, who may place a higher value on the maintenance of their social and cultural systems than on the ecosystem attributes valued by ecologists (Spooner 1987). In addition, many species valued at national or international levels, such as large mammals, may be undesirable to and actively discouraged by local people (Lebbie and Freudenberger, chapter 15; Poffenberger, chapter 18).

On the other hand, interest groups at national and international levels often make decisions without taking into account the value of forest patches to rural people. Government and private development efforts often ignore the benefits local people receive from forest patches, and instead promote production of a narrow range of commodities through development policies, plans, and projects. Forest patches and rural people suffer adverse consequences from these narrow approaches to rural development and resource management. Governments may also trade off rural values while promoting the interests of urban populations or development at the national level, such as facilitating or subsidizing logging to earn foreign exchange and encouraging colonization of lands claimed and occupied by indigenous people. Conservationists have often sought absolute protection of forest patches, setting conservation in opposition to local interests, rather than looking for mutually beneficial solutions.

Forest patches occur on private and community lands, and if forest patch conservation programs are to be successful they must work with, not against, rural people. Use and conservation interests related to forest patches are often not in conflict, and neither are the interests of local people and broader groups. There is ample evidence that rural people can often be empowered to manage forest patches in ways that also conserve biological diversity (Alcorn, chapter 12; Pinedo-Vasquez and Padoch, chapter 16; Poffenberger, chapter 18). In fact, community institutions often regulate natural resource use to avoid carrying out extreme scenarios of resource degradation (Lebbie and Freudenberger, chapter 15). Alliances between diverse types of organizations, including community groups, NGOs, government agencies, and university researchers, are a critical step in sustaining and promoting forest patch conservation (Fisher and Bunch, chapter 19).

The mutual biological and human benefits from forest patches provide the starting point for sustainable forms of landscape management that considers a broad range of human and ecological benefits. Local households and communities already value forest patches and engage in many practices that protect and conserve them. Although often weighted toward short-term tangible benefits for products and ecological services, these values also include long-term social and cultural benefits. Many household and community forest patch management strategies also have biological diversity benefits. Conservation and rural development agencies, both government and private, can support and supplement rural people's efforts in a variety of ways, including by

(1) recognizing the ways that rural people value and conserve forest patches;

(2) facilitating community and regional planning to link forest fragments and provide benefits beyond those achievable by the actions of individual landholders alone;

(3) enhancing and creating institutions for conflict resolution, both within and between communities;

(4) promoting processes for policy changes that remove disincentives and/or create incentives for conservation;

(5) promoting interchanges between landholders to share existing technologies between communities and fostering collaborative work by researchers and communities to develop, adapt, and promote new technologies for forest patch management;

(6) promoting the development of biologically valuable forest patch management systems that provide economic, social, and cultural benefits to rural people, or are otherwise compatible with rural people's livelihoods;

(7) enhancing understanding of the ecological services and other underrecognized values of forest patches;

(8) assisting communities and governments in strengthening regulations and enforcement mechanism to sanction misuse of forest resources by individuals for short-term gains; and

(9) monitoring biological, watershed, and socioeconomic values in landscapes to evaluate management success and failures and to develop adaptive responses.

References

Alderman, C. L. 1991. "Privately owned lands: Their role in nature tourism, education, and conservation." In *Ecotourism and Resource Conservation,* edited by Jon A. Kusler, pp. 289–323. Berne, NY: Ecotourism and Resource Conservation Project.

Donovan, R. 1994. "BOSCOSA: Forest conservation and management through local institutions (Costa Rica)." In *Natural Connections: Perspectives in Community-Based Conservation,* edited by D. Western, R. M. Wright, and S. Strum, pp. 215–233. Washington, DC: Island Press.

Doyle, J. K., and J. Schelhas. 1993. *Forest Remnants in the Tropical Landscape: Benefits and Policy Implications.* Proceedings of the symposium presented by the Smithsonian Migratory Bird Center. Washington, DC: Smithsonian Institution.

Little, Peter D. 1994. "The link between local participation and improved conservation: A review of issues and experiences." In *Natural Connections: Perspectives in Community-Based Conservation,* edited by D. Western, R. M. Wright, and S. Strum, pp. 347–372. Washington, DC: Island Press.

Lopez Vasquez, T. 1993. "The farmers of ASACODE lead the way: Social forestry development in San Miguel, Costa Rica." *Forest, Trees, and People* 22: 31–35

Peluso, N. L. 1993. "Coercing conservation? The politics of state resource control." *Global Environmental Change* 3(2): 199–217.

Schelhas, J. 1994. "Building sustainable land use on existing practices: Smallholder land-use mosaics in tropical lowland Costa Rica." *Society and Natural Resources* 7(1): 67–84.

Spooner, B. 1987. "Insiders and outsiders in Baluchistan: Western and indigenous perspectives on ecology and development." In *Lands at Risk in the Third World: Local-Level Perspectives,* edited by P. D. Little, M. M. Horowitz, and A. E. Nyerges, pp. 58–68. Boulder, CO: Westview Press.

White, T. A. 1992. "Landholder cooperation for sustainable upland watershed management: A theoretical review of the problems and prospects." University of Wisconsin EPAT/MUCIA working paper 1, Madison, Wisconsin.

WRI/UNEP/UNDP. 1994. *World Resources 1994–95.* New York: Oxford University Press.

Part I

Changing Forests

Chapter 1

Preservation and Management of Biodiversity in Fragmented Landscapes in the Colombian Andes

Gustavo H. Kattan and Humberto Alvarez-López

Introduction

Land use in the Colombian Andes has greatly altered natural ecosystems. Up to 85 percent of the area in premontane and montane forests has been modified to some extent (Orejuela 1985), with the result that in most regions natural vegetation remains only in isolated patches. As a consequence, local extinctions of birds have been documented (Kattan et al., 1994), as well as population declines at the regional and national levels (Hilty 1985; Collar and Andrew 1988). Similar trends at the local level presumably occur in other animal groups (e.g., frogs and dung beetles—Kattan 1993; Escobar 1994), although data are scarce or lacking.

Despite this grim picture, the remaining fragments of natural vegetation still harbor a sizable number of species of plants and animals. While it is unquestionable that large areas need to be protected to preserve the bulk of biological diversity, forest fragments may play an important role in the preservation of a significant fraction of the original regional diversity. At the landscape level, forest fragments of different sizes and ages can be incorporated in a management scheme to maximize the persistence of species. In addition to a sheltering function, these forests would provide important environmental services, such as protection of watersheds.

In this chapter we explore the possibilities for preserving the extant biodiversity in landscape units that include highly fragmented forests in the Colombian Andes. We first describe the natural heterogeneity of Andean landscapes, the impact of land-use practices, and how fragmentation may lead to species extinction. Then we discuss how management at the landscape level may provide tools for the minimization of biodiversity losses. Although we rely mostly on examples from the Colombian Andes, the ideas presented here may be applied to similar situations in the tropical Andes and tropical mountains in general.

The Region

The three ranges of the Colombian Andes extend southwest–north and northeast from the Ecuadorean to the Venezuelan borders, in a latitudinal range from about 1 to 11 degrees north of the equator. Mountaintop elevations range from 1,800 meters at low passes to more than 5,000 meters at snow peaks, encompassing six altitudinal belts (Espinal and Montenegro 1963) from sea level to the highest Andean elevations. The lower belts are essentially continuous along the three cordilleras, but the higher ones, mostly above 3,000 meters, frequently become interrupted by topographical irregularities.

Complex patterns of interactions between altitude, temperature, and rainfall result in a correspondingly complex mosaic of plant communities. Above 1,000 meters at least 15 natural life zones (*sensu* Holdridge 1967; Espinal and Montenegro 1963) have been recognized. These life zones range from premontane thorn woodland in midaltitude inter-Andean valleys affected by rain shadows, to rainforests in the western and eastern foot slopes, to rain tundra adjacent to perpetual snow peaks. Edaphic and topographic factors add to this diversity in the form of an array of transitional life zones and different associations. In addition, as the three ranges of the Colombian Andes are the result of separate orogenic events (González et al. 1989), and the distributions of plants and animals were extensively affected by climatic changes during the Pleistocene (Haffer 1974), historical factors also account for biogeographical features such as complex speciation patterns (Vuilleumier 1986) and high endemism (Terborgh and Winter 1983) and biodiversity (ICBP 1992).

The generally humid mountains generate an extensive, although somewhat irregularly distributed, fluvial network, which in turn provides connectivity among altitudinal belts. Besides, these watercourses, with their associated vegetation, add to the complex interdigitation and irregularity between natural life zones (Espinal and Montenegro 1963).

Although land-use patterns might have initially created a mosaic superimposed on that of natural vegetation, the long-term effect has been one of homogenization of the landscape. While early settlers preferred fertile mountainsides, middle- and high-altitude inter-Andean valleys, and alluvial terraces (Holdridge 1967), demographic growth has forced more intensive land exploitation, and settlement of increasingly steeper slopes and higher and more rainy regions. Therefore, vast regions of the Colombian Andes are presently dominated by pastures and open areas mostly devoid of trees, and sometimes badly eroded. Exceptions to this trend are a few national parks, protected watersheds, some coffee and tree plantations, and the wettest regions of the Pacific and eastern Andean drainages.

Negative Effects of Fragmentation in Andean Ecosystems

Although very few studies have addressed the effects of habitat fragmentation in the tropical Andes, the available evidence indicates that fragmentation and current land management practices may result in the local extinction of large numbers of species. The best evidence available is for birds. The existence of historic faunal inventories dating back to 1911 and 1959 provided a rare opportunity to document avian extinctions in the region of San Antonio, a cloud forest site in the western range of the Colombian Andes (Kattan et al. 1994). The area, which was fragmented mostly during the first half of this century, at present constitutes an isolated archipelago of forest patches ranging in size from 10 to 400 hectares in a matrix of small farms and suburban houses. Forty species, or 31 percent of the original 128 forest bird species, were locally extinct in 1990.

An analysis of patterns of extinction at this site revealed two main trends (Kattan et al. 1994). First, species at the limits of their altitudinal or geographic distributions exhibited a high frequency of extinction. Of 29 species for which San Antonio was either the upper or lower limit of their altitudinal distribution, 19 are locally extinct. Second, large fruit-eating birds, such as quetzals and cotingas, were the trophic group most vulnerable to extinction. Out of 18 large frugivores originally present, 12 went locally extinct. Among 13 extinct bird species for which the study area was well within their altitudinal limits, 7 were large frugivores.

While patterns of extinction reveal groups of organisms that are particularly vulnerable, prevention of biodiversity losses requires knowledge of the processes and mechanisms responsible for such losses. The consequences of fragmentation may occur at several levels. First, there are effects on the physical environment (e.g., light incidence, temperature). Second, there are direct biological effects on the distribution and abundance of organisms, sometimes mediated through the physical effects. Third, there may be complex, indirect biological effects on the interactions among species, such as predation and parasitism (Murcia 1995).

One direct, physical consequence of fragmentation is a reduction in habitat heterogeneity, or the deterioration and disappearance of certain microhabitats. Streams, for example, are particularly vulnerable because of the desiccation process that affects isolated forest patches (Lovejoy et al., 1986; Saunders et al., 1991). In addition, in suburban areas in the Colombian Andes, streams frequently are partially or totally piped to supply water for human consumption. This may have catastrophic consequences, not only in causing the local extinction of populations of stream-dependent organisms, but in altering the fragment's hydric dynamics.

The effect of stream disturbance is more evident on amphibians, because of the dependence of many species on water for breeding. Frogs exhibit a wide diversity of reproductive modes, which refers to a combination of egg deposition site (aquatic or terrestrial) and development mode (varying from free-swimming tadpoles to direct development). Diversity of reproductive modes contributes greatly to species richness in Neotropical anuran assemblages (Crump 1982; Kattan 1987). Thus, diversity of breeding microhabitats is an important factor in determining frog species richness.

In the Brazilian Amazon, diversity of breeding microhabitats was a very important factor determining the persistence of frogs in forest remnants (Zimmerman and Bierregaard 1986). In the San Antonio region, preliminary evaluations of frog populations in sites ranging from undisturbed streams to total piping revealed that species richness depends on disturbance level (G. Kattan and C. Murcia, unpublished data). Seven species of stream-breeding frogs were found in an undisturbed stream. In forest patches where water is partially piped, or where water runs free for a short distance (80–200 meters) and then is piped, only one to four species persisted, depending on the amount of water available. At sites where water is piped right at the stream head, no water-breeding frogs persisted.

In contrast to water-breeding frogs, most terrestrial-breeding species were present in all fragments. Frogs with terrestrial reproduction account for 50 percent or more of the species in Andean frog assemblages, in particular the genus Eleutherodactylus, which is very diverse in the tropical Andes (Lynch 1986). These frogs lay terrestrial eggs that undergo direct development (no aquatic larval stage) and are independent of water for reproduction. Thus, reproductive mode will determine the species assemblages in fragmented habitats. The existence of frogs with terrestrial reproduction means that up to 50 percent of the species may persist in small fragments in the tropical Andes, but diversity at the genera and family levels, as well as diversity in reproductive modes, will be drastically reduced.

In addition to its direct consequences, fragmentation may alter interactions among species. Studies in the temperate zone have shown that processes such as bird nest predation and brood parasitism by brown-headed cowbirds (*Molothrus ater*) are important factors in the decline of bird populations in small forest patches (Robinson 1992; Paton 1994). Rates of nest predation and brood parasitism are frequently elevated in small patches, when compared to large fragments or continuous forest. This frequently occurs as a consequence of an edge effect. In small patches, there is a large perimeter-to-area ratio, which makes the patch vulnerable to invasion by organisms from the matrix or to ecological changes caused by proximity to the edge.

Data on the effects of fragmentation on these ecological processes are sorely lacking in the tropics. Preliminary data suggest that parasitism by shiny cowbirds (*M. bonariensis*) is not a problem in cloud forest fragments (G. Kattan, unpublished data). However, a study on predation of artificial nests revealed that fragmentation may be an important factor in the extinction of some understory birds in Andean cloud forest (Arango 1991). This study, however, did not find support for an edge effect. Instead, circumstantial evidence suggested that the elevated rates of predation may result from the absence of medium-sized mammalian predators in the small fragments and a consequent abundance of small predators (Arango 1991).

Besides affecting predation rates on bird nests, changes in mammal assemblages caused by fragmentation (Malcolm 1988; Fonseca and Robinson 1990) may also have indirect consequences on other organisms. Dung beetles (*Scarabaeidae*) are highly sensitive to deforestation and fragmentation (Klein 1989; Escobar 1994). The species composition and activity patterns in cleared areas are drastically different from those in forests (Escobar 1994), probably reflecting parallel trends in mammals. If large and medium-sized mammals disappear from fragmented areas, dung beetles will follow a similar fate. This, in turn, may have repercussions in the entire ecosystem, as dung beetles may play a major role in secondary seed dispersal and control of pests dispersed in dung (Doube and Moola, 1988; Estrada and Coates-Estrada, 1991).

Frugivory is another potentially vulnerable process that may have important consequences for the preservation of montane forests. Large fruit-eating birds have emerged as vulnerable in several studies (Willis 1979; Kattan 1992; Kattan et al., 1994). There is increasing evidence that many frugivorous birds migrate altitudinally (Loiselle and Blake 1991; Levey and Stiles 1992). Most frugivorous birds may require continuous habitat along altitudinal gradients because fruit availability is variable in time and space, and tracking these resources involves seasonal movements that cover large areas. Forest fragmentation severs the connection between foraging areas and may severely restrict access to a year-round food supply. Because large proportions of plants in tropical montane forests have bird-dispersed seeds, extinction of fruit-eating birds may cause the disruption of ecological processes that pervade the entire ecosystem (Howe 1984; Terborgh 1986).

What Can Be Preserved in Fragmented Forests?

Reduction of continuous forests to small, isolated patches has severe consequences for biological diversity. It is a fact, however, that in the Colombian Andes, except for a few national parks and some remote areas, only fragments remain of the natural vegetation. What can we expect to preserve in such a situation? In this section we evaluate the biological diversity that remains in forest fragments and the contribution that these fragments can make to the preservation of biodiversity in multiple-use landscapes.

We reviewed available studies on bird communities in the Colombian Andes to explore the relationship between forest resident species richness and forest area. Although these studies are highly heterogeneous in terms of types of vegetation sampled and sampling efforts, the analysis revealed some trends. We classified areas in three categories, according to fragment size. Sites with more than 1,000 hectares of forest, with high connectivity and habitat heterogeneity, were defined as large. These were all protected sites contiguous to large extensions of forest (several thousand hectares), and we assumed they contained a full complement of species (Renjifo and Andrade 1987; Orejuela and Cantillo 1990; Naranjo 1994; H. Alvarez-López and G. Kattan, unpublished data). Fragments that ranged between 100 and 600 hectares, isolated in a matrix of agricultural and pasture lands, were classified as medium (Orejuela et al., 1979a; Ridgely and Gaulin 1980; Cuadros 1988; Mondragón 1989; Kattan et al., 1994), while those that ranged between 10 and 50 hectares were defined as small (Orejuela et al., 1979b; Orejuela and Cantillo 1982; Orejuela et al., 1982; Johnels and Cuadros 1986; Corredor 1989).

There were, on average, 144 forest bird species in areas with large expanses of continuous forest and high habitat heterogeneity. Species richness decreased markedly with area (figure 1-1). The mean number of species in small fragments was 25 percent of the mean number of species found in continuous forest areas, while on average 60 percent of the species persisted in medium-sized fragments. The number of species of fruit-eating birds followed a parallel trend. A mean of 16.5 species of large frugivores can be expected in large areas of forest. This number decreases to 7.4 species in medium and 2.4 in small fragments (figure 1-1).

Because there is great concern about the effect that accelerated deforestation of Neotropical habitats may have on migratory birds (Hagan and Johnston 1992), we also looked at the numbers of forest migrants recorded in each of the fragment size categories and found no difference in the number of species as a function of fragment size (figure 1-1). Variability in the number of species, however, was greater in small- and medium-sized fragments. This result suggests that forest migrants are using small- and medium-sized fragments, at least as temporary habitats, as well as large, continuous forests. Some species may even be selecting disturbed habitats, as no migrants were recorded at a continuous forest site in the Pacific lowlands (Orejuela et al., 1980), while six species were recorded at a disturbed site nearby (Hilty 1980).

Preliminary data available for butterflies also suggest that a high diversity may be preserved in fragmented habitats, in particular if they occur in riparian habitats along an altitudinal range, coupled with a diversity of successional stages. In a four-day survey at such a situation on the eastern slope of the western Andes, between 1,550 and 2,100 meters, 85 species of butterflies were recorded (M. D. Heredia and G. Kattan, unpublished data). It is estimated that long-term, more intensive surveys, using complementary capture methods, would increase the number of species by at least 50 percent (M. D. Heredia, pers. com.). The records included sensitive species such as *Pseudohaetera hypaesia,*

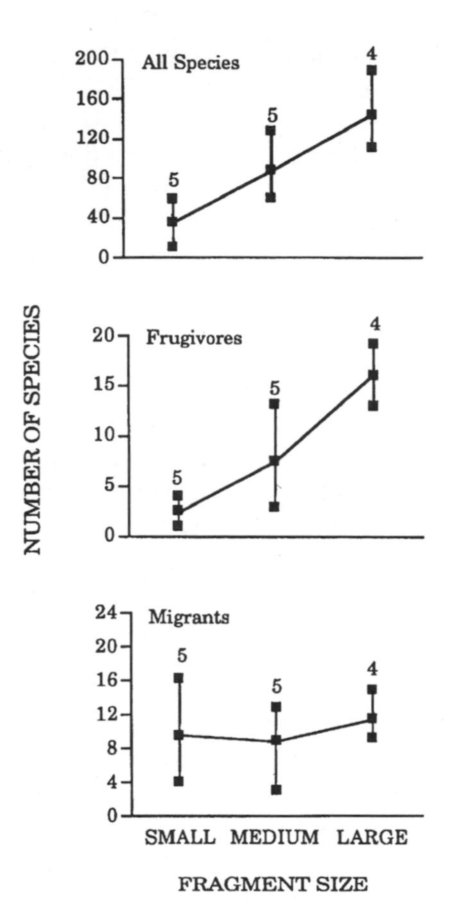

Figure 1-1
Number of bird species as a function of forest fragment size in the Colombian Andes. Fragment-size categories are small (10–50 hectares), medium (100–600 hectares), and large (greater than 1,000 hectares). Symbols indicate the mean, maximum, and minimum number of species. Numbers above the boxes indicate sample size

an understory butterfly that does not tolerate open habitats (Constantino 1992). This butterfly is regularly recorded in forest patches as small as 10 hectares (C. Murcia, pers. com.). It is significant that at a much better preserved site in the central Andes, from 1,800 to 2,600 meters, 92 species of butterflies were recorded in 90 sampling days throughout one year (Andrade 1994).

An alternative way to look at the contribution of forest fragments to local biodiversity is to consider the proportion of forest species that remain at a site, as opposed to open-habitat species. For example, at the San Antonio site (mentioned earlier), 92 out of 151 bird species present are forest dwellers.

An additional justification to preserve forest fragments is that in the long term, these fragments may serve as sources of propagules for the recovery of surrounding lands that are no longer economically productive or that have to be incorporated, for example, into watershed-protection schemes. In this way, even highly degraded landscapes might become important future reserves (Foster 1980). The Río Blanco and Río Otún watersheds in the cities of Manizales and Pereira, respectively, are good examples of the dramatic recovery possible at highly fragmented landscapes. After about 30 years of protection and reforestation, both sites exhibit high levels of biodiversity (Uribe 1986; Rangel 1994; pers. obs.).

Possibilities for the Preservation of Biodiversity in Fragmented Landscapes

The above results suggest that, by including at least some fragments in the range of several hundred hectares, it may be possible to manage multiple-use landscapes in the Colombian Andes to preserve a significant part of the original biodiversity. Landscape ecology may provide a framework for preserving and managing biodiversity in such situations. The main variables in landscape ecology are patch area, connectivity, and habitat heterogeneity (Forman and Godron 1986). By manipulating these variables in a scheme of multiple land use, a substantial portion of biological diversity could be preserved in forest fragments. For example, creating connections among fragments could allow for the maintenance of metapopulations in situations where small fragments will not maintain a viable population if managed individually (Saunders and de Rebeira 1991). By manipulating connectivity and increasing habitat heterogeneity in the matrix, it may even be possible to increase the number of species close to the values expected for well-preserved areas.

One important factor that contributes to high diversity in Andean communities is the extent of the altitudinal gradient, because of the replacement of species assemblages that usually occur along elevational ranges. Along a transect of 1,000 meters (1,600–2,600 meters) at Ucumarí Regional Park in the central Andes, three different bird assemblages have been identified: one restricted to the lower half of the range, another restricted to the upper half, and a third common to the whole altitudinal range (Naranjo 1994). Thus, habitat preservation along the altitudinal range results in the preservation of more species than would be accomplished by preserving a similar area but restricted to a single altitudinal belt. Conversely, fragmentation along the altitudinal gradient would result in the local extinction of populations isolated at their altitudinal limits (Kattan et al., 1994).

Another factor contributing to high diversity is habitat heterogeneity, in particular the presence of patches in different successional stages (Levey and Stiles 1994; Loiselle and Blake 1994). Because it may not be possible to maintain the natural patch dynamics of tropical forests in small areas (Pickett and Thompson 1978), maintenance of habitat heterogeneity will probably require intensive manipulation. In this respect, some land-use schemes such as agroforestry may prove valuable to increase regional habitat diversity, as opposed to land-use practices that produce a homogeneous and simplified matrix. For example, although very few Neotropical birds nest in conifer plantations, up to 44 percent of forest species may at least move through them (Mondragón 1989). Patches and corridors of natural vegetation are increasingly being preserved and managed as integral parts of conifer plantations, because of their value for maintaining organisms useful for the biological control of pests, and as fire breaks (Madrigal and Sierra 1975; Mondragón 1989). Similar patterns have been reported for other types of plantations. Shaded coffee plantations support a wide variety of forest birds, including most Neotropical migrants common in cloud forests (Corredor 1989). Ash (*Fraxinus chinensis*) plantations are intensively used by birds in Ucumarí, including sensitive species such as the endemic Cauca guan (*Penelope perspicax*) and the red-ruffed fruitcrow (*Pyroderus scutatus*) (Naranjo 1994; Serrano 1994).

A feasible management strategy in Andean landscapes would be to maintain native vegetation at least along rivers and streams, as well as on steep slopes. This would accomplish two objectives. First, these forests would provide important environmental services such as protection of watersheds and erodable slopes and buffering of local weather. Second, this strategy would create a network of highly connected forest patches along extended altitudinal gradients, which could preserve an important fraction of the regional biodiversity. Besides, riparian habitats are important as seasonal refuges and as natural altitudinal corridors for wildlife migrations (Foster 1980; Naiman et al., 1993). It is noteworthy that riparian corridors could perform similar services even through urban areas.

Therefore, as opportunities to preserve extensive blocks of habitat containing a full complement of species are restricted in the densely populated Colombian Andes, preservation of the extant diversity in fragmented landscapes, which represents a significant portion of total biodiversity, becomes one of the most urgent tasks for conservation biologists. Integrated management of multiple-use landscapes containing successionally diverse and interconnected forest fragments is not only desirable for long-term goals of conservation but is also feasible and necessary from the viewpoint of watershed protection, soil stability, and other environmental services that support economic development.

Coda: Research Needs

Conservation biology has placed most of its emphasis on the preservation of pristine habitats and on the negative consequences that will result from human influence. Forest fragments, and second-growth and "degraded" habitats in human-dominated landscapes, however, can offer many opportunities for biological preservation. Furthermore, the biological diversity preserved in such areas represents, for most of the people, the only opportunity to have any contact with nature. Therefore, an important research effort should be devoted to asses the potential for biological preservation in such areas. One limitation we found in writing this article was the lack of information on even very simple aspects, such as the biodiversity contained in fragmented areas. Here we present some questions we think are crucial for the effective management of forest fragments.

1. What is the species content of habitat fragments of different sizes? How does it compare with the diversity of large expanses of continuous habitat? These inventories should be done for as many taxa as possible.
2. What is the contribution of landscape elements to total biodiversity? What species (e.g., special-interest species) are contributed by each habitat type?
3. What is the potential for small fragments to sustain viable populations? Demographic studies of species in different-sized fragments are required.
4. What is the role of corridors (potential positive and negative effects) and their value in maintaining metapopulations? What organisms disperse along corridors?
5. What is the role of corridors at the landscape level (e.g., altitudinal corridors) in maintaining patterns of regional migrations?

6. What are the spatial and temporal patterns of habitat "degradation" in small patches (e.g., tree fall caused by senescence or edge effects)? Are natural regeneration rates enough to counteract degradation rates? If not, then intensive management will be required.
7. How do fragment size and isolation affect ecological processes and species interactions, such as frugivory and seed dispersal, predation, and parasitism, among others?

Acknowledgments

We thank Carolina Murcia and Maria Dolores Heredia for discussion of ideas and for sharing unpublished data. Our work has been supported by Universidad del Valle, Corporación Autónoma Regional del Valle del Cauca (CVC), Corporación Autónoma Regional de Risaralda (CARDER), Wildlife Conservation Society, Fundación para la Promoción de la Investigación y la Tecnología (Banco de la República), and the Biodiversity Support Program, a consortium of World Wildlife Fund, The Nature Conservancy, and the World Resources Institute, with funding by the U.S. Agency for International Development. The opinions expressed herein are those of the authors and do not necessarily reflect the views of the U.S. Agency for International Development.

References

Andrade, M. G. 1994. "Las mariposas del Parque Regional Natural Ucumarí: Distribución local y estacional de rhopalocera." In *Ucumarí, Un Caso Típico de la Diversidad Biótica Andina,* edited by J. O. Rangel, pp. 247–274. Pereira, Colombia: Corporación Autónoma Regional de Risaralda.

Arango, N. 1991. "La depredación de nido y su relación con la fragmentación del hábitat en un bosque Nublado Tropical." Thesis, Universidad del Valle, Cali, Colombia.

Collar, N. J., and P. Andrew. 1988. "Birds to watch: The ICBP World Checklist of Threatened Birds." International Council for Bird Preservation Technical Publication No. 8. Washington, DC: Smithsonian Institution Press.

Constantino, L. M. 1992. "*Paradulcedo,* a new genus of Satyrinae (Nymphalidae) from western Colombia." *Journal of the Lepidopterists' Society* 46: 44–53.

Corredor, G. A. 1989. "Estudio comparativo entre la avifauna de un bosque natural y un cafetal tradicional en el Quindío." Thesis, Universidad del Valle, Cali, Colombia.

Crump, M. L. 1982. "Amphibian reproductive ecology on the community level." In *Herpetological Communities,* edited by N. J. Scott, pp. 21–36. U.S. Fish and Wildlife Service Research Report No. 13.

Cuadros, T. 1988. "Aspectos ecológicos de la comunidad de aves en un bosque nativo en la Cordillera Central en Antioquia (Colombia)." *Hornero* 13: 8–20.

Escobar, F. 1994. "Excremento, coprófagos, y deforestación en bosques de montaña al suroccidente de Colombia." Thesis, Universidad del Valle, Cali, Colombia.

Espinal, L. S., and E. Montenegro. 1963. "Formaciones vegetales de Colombia: Memoria explicativa sobre el mapa ecológico." Instituto Geográfico Agustín Codazzi, Bogotá.

Fonseca, G. A. B., and J. G. Robinson. 1990. "Forest size and structure: Competitive and predatory effects on small mammal communities." *Biological Conservation* 53: 265–294.

Forman, R. T. T., and M. Godron. 1986. *Landscape Ecology.* New York: Wiley.

Foster, R. B. 1980. "Heterogeneity and disturbance in tropical vegetation." In *Conservation Biology: An Evolutionary-Ecological Perspective,* edited by M. E. Soulé and B. A. Wilcox, pp. 75–92. Sunderland, MA: Sinauer.

González, E., G. Guillot, N. Miranda, and D. Pombo (eds.). 1989. *Perfil Ambiental de Colombia.* Bogotá: ESCALA.

Haffer, J. 1974. "Avian speciation in tropical South America." *Publication Nuttall Ornithology Club* 14: 1–390.

Hagan, J. M., and D. W. Johnston (eds.). 1992. *Ecology and Conservation of Neotropical Migrant Landbirds.* Washington, DC: Smithsonian Institution Press.

Hilty, S. L. 1980. "Relative abundance of north temperate zone breeding migrants in western Colombia and their impact on fruiting trees." In *Migrant Birds in the Neotropics: Ecology, Behavior, Distribution, and Conservation,* edited by A. Keast and E. S. Morton, pp. 265–271. Washington, DC: Smithsonian Institution Press.

Hilty, S. L. 1985. "Distributional changes in the Colombian avifauna: A preliminary blue list." In *Neotropical Ornithology,* edited by P. A. Buckley, M. S. Foster, E. S. Morton, R. S. Ridgely, and F. G. Buckley, pp. 1000–1012. Ornithological Monograph 36. Washington, DC: American Ornithologists' Union.

Holdridge, L. R. 1967. *Life Zone Ecology.* San José, Costa Rica: Tropical Science Center.

Howe, H. F. 1984. "Implications of seed dispersal by animals for tropical reserve management." *Biological Conservation* 30: 261–281.

ICBP. 1992. *Putting Biodiversity on the Map: Priority Areas for Global Conservation.* Cambridge, UK: International Council for Bird Preservation.

Johnels, S. A., and T. C. Cuadros. 1986. "Species composition and abundance of bird fauna in a disturbed forest in the central Andes of Colombia." *Hornero* 12: 235–241.

Kattan, G. H. 1987. "Patrones de composición taxonómica y modos reproductivos en comunidades de ranas en el Valle del Cauca." *Cespedesia* 15: 75–83.

Kattan, G. H. 1992. "Rarity and vulnerability: The birds of the Cordillera Central of Colombia." *Conservation Biology* 6: 64–70.

Kattan, G. H. 1993. "The effects of forest fragmentation on frogs and birds in the Andes of Colombia: Implications for watershed management." In *Proceedings of the Symposium on Forest Remnants in the Tropical Landscape: Benefits and Policy Implications,* edited by J. K. Doyle and J. Schelhas, pp. 11–13. Washington, DC: Smithsonian Institution Press.

Kattan, G. H., H. Alvarez-López, and M. Giraldo. 1994. "Forest fragmentation and bird extinctions: San Antonio eighty years later." *Conservation Biology* 8: 138–146.

Klein, B. C. 1989. "Effects of forest fragmentation on dung and carrion beetle communities in Central Amazonia." *Ecology* 70: 1715–1725.

Levey, D. J, and F. G. Stiles. 1992. "Evolutionary precursors of long-distance migration: Resource availability and movement patterns in neotropical landbirds." *American Naturalist* 140: 447–476.

Levey, D. J., and F. G. Stiles. 1994. "Birds: Ecology, behavior, and taxonomic affinities." In *La Selva: Ecology and Natural History of a Neotropical Rain Forest,* edited by L. A. McDade, K. S. Bawa, H. A. Hespenheide, and G. S. Hartshorn, pp. 217–228. Chicago: University of Chicago Press.

Loiselle, B. A. , and J. G. Blake. 1991. "Resource abundance and temporal variation in fruit-eating birds along a wet forest elevational gradient in Costa Rica." *Ecology* 72: 180–193.

Loiselle, B. A., and J. G. Blake. 1994. "Annual variation in birds and plants of a tropical second-growth woodland." *Condor* 96: 368–380.

Lovejoy, T. E., R. O. Bierregaard, A. B. Rylands, J. R. Malcolm, C. E. Quintela, L. H. Harper, K. S. Brown, A. H. Powell, G. V. N. Powell, H. O. R. Schubart, and M. B. Hays. 1986. "Edge and other effects of isolation on Amazon forest fragments." In *Conservation Biology: The Science of Scarcity and Diversity,* edited by M. E. Soulé, pp. 257–285. Sunderland, MA: Sinauer.

Lynch, J. D. 1986. "Origins of the high Andean herpetological fauna." In *High Altitude Tropical Biogeography,* edited by F. Vuilleumier and M. Monasterio, pp. 478–499. New York: Oxford University Press.

Madrigal, J. A., and G. Sierra. 1975. "Inventario de fauna benéfica de plantaciones de Ciprés en Caldas, Antioquia." Thesis, Universidad Nacional de Colombia, Medellín, Colombia.

Malcolm, J. R. 1988. "Small mammal abundances in isolated and non-isolated primary forest reserves near Manaus, Brazil." *Acta Amazonica* 18: 67–83.

Mondragón, M. L. 1989. "Estructura de la comunidad aviaria en bosques de coníferas y en bosques aledaños de vegetación nativa." Thesis, Universidad del Valle, Cali, Colombia.

Murcia, C. 1995. "Edge effects in fragmented forests: Implications for conservation." *Trends in Ecology and Evolution* 10: 58–62.

Naiman, R. J., H. Décamps, and M. Pollock. 1993. "The role of riparian corridors in maintaining regional biodiversity." *Ecological Applications* 3: 209–212.

Naranjo, L. G. 1994. "Composición y estructura de la avifauna del Parque Regional Natural Ucumarí." In *Ucumarí, Un Caso Típico de la Diversidad Biótica Andina,* edited by J. O. Rangel, pp. 305–325. Pereira, Colombia: Corporación Autónoma Regional de Risaralda.

Orejuela, J. E. 1985. "Tropical forest birds of Colombia: A survey of problems and a plan for their conservation." In *Conservation of Tropical Forest Birds,* edited by A. W. Diamond and T. E. Lovejoy, pp. 95–115. Cambridge, England: International Council for Bird Preservation Technical Publication No. 4.

Orejuela, J. E., and G. Cantillo. 1982. "Estructura de las comunidades aviarias en tres areas seleccionadas como posibles refugios ecológicos en el Departamento del Valle del Cauca." *Cespedesia* 11: 121–140.

Orejuela, J. E., and G. Cantillo. 1990. *Aves de la Reserva Natural la Planada.* Cali, Colombia: Fundación para la Educación Superior.

Orejuela, J. E., R. J. Raitt, and H. Alvarez-López. 1979a. "Relaciones ecológicas de las aves en la Reserva Forestal de Yotoco, Valle del Cauca." *Cespedesia* 8: 7–28

Orejuela, J. E., R. J. Raitt, H. Alvarez-López, C. Benalcázar, and F. Silva de Benalcázar. 1979b. "Poblaciones de aves en un bosque relictual en el Valle del Río Cauca, cerca a Jamundí, Valle, Colombia." *Cespedesia* 8: 29–42.

Orejuela, J. E., R. J. Raitt, and H. Alvarez-López. 1980. "Differential use by North American migrants of three types of Colombian forest." In *Migrant Birds in the Neotropics: Ecology, Behavior, Distribution, and Conservation,* edited by A. Keast, and E. S. Morton, pp. 253–264. Washington, DC: Smithsonian Institution Press.

Orejuela, J. E., G. Cantillo, J. E. Morales, and H. Romero. 1982. "Estudio de la comunidad aviaria en una Pequeña Isla de hábitat de bosque premontano húmedo cerca a Argelia, Valle, Colombia." *Cespedesia* 11: 103–119.

Paton, P. W. 1994. "The effect of edge on avian nest success: How strong is the evidence?" *Conservation Biology* 8: 17–26.

Pickett, S. T. A., and J. N. Thompson. 1978. "Patch dynamics and the design of nature reserves." *Biological Conservation* 13: 27–37.

Rangel, J. O. (ed.). 1994. *Ucumarí, Un Caso Típico de la Diversidad Biótica Andina.* Pereira, Colombia: Corporación Autónoma Regional de Risaralda.

Renjifo, L. M., and G. I. Andrade. 1987. "Estudio comparativo de la avifauna entre un area de bosque andino primario y un crecimiento secundario en el Quindío, Colombia." In *Memorias del III Congreso de Ornitología Neotropical,* edited by H. Alvarez-López, G. Kattan, and C. Murcia, 121–127. Cali, Colombia: Sociedad Valledaucana ole Ornitología.

Ridgely, R. S., and S. J. C. Gaulin. 1980. "The birds of Finca Merenberg, Huila Department, Colombia." *Condor* 82: 379–391.

Robinson, S. K. 1992. "Population dynamics of breeding neotropical migrants in a fragmented Illinois landscape." In *Ecology and Conservation of Neotropical Migrant Landbirds,* edited by J. M. Hagan and D. W. Johnston, pp. 408–418. Washington, DC: Smithsonian Institution Press.

Saunders, D. A., and C. P. de Rebeira. 1991. "Values of corridors to avian populations in a fragmented landscape." In *Nature Conservation 2: The Role of Corridors,* edited by D. A. Saunders and R. J. Hobbs, pp. 221–240. Chipping Norton, NSW, Australia: Surrey Beatty & Sons.

Saunders, D. A., R. J. Hobbs, and C. R. Margules. 1991. "Biological consequences of ecosystem fragmentation: A review." *Conservation Biology* 5: 18–32.

Serrano, V. H. 1994. "Generalidades sobre la selección de hábitat, el ciclo reproductivo y el sistema lek de apareamiento de *Pyroderus scutatus* (Toro de Monte)." In *Ucumarí, Un Caso Típico de la Diversidad Biótica Andina,* edited by J. O. Rangel, pp. 343–357. Pereira, Colombia: Corporación Autónoma Regional de Risaralda.

Terborgh, J. 1986. "Keystone plant resources in the tropical forest." In *Conservation Biology: The Science of Scarcity and Diversity,* edited by M. E. Soulé, pp. 330–344. Sunderland, MA: Sinauer.

Terborgh, J., B. Winter. 1983. "A method for siting parks and reserves with special reference to Colombia and Ecuador." *Biological Conservation* 27: 45–58.

Uribe, D. A. 1986. "Contribución al conocimiento de la avifauna del bosque muy húmedo montano bajo en cercanías de Manizales." Thesis, Universidad de Caldas, Manizales, Colombia.

Vuilleumier, F. 1986. "Origin of the tropical avifaunas of the high Andes." In *High Altitude Tropical Biogeography,* edited by F. Vuilleumier and M. Monasterio, pp. 586–622. New York: Oxford University Press.

Willis, E. O. 1979. "The composition of avian communities in remanescent woodlots in southern Brazil." *Papeis Avulsos Zoologia* 33: 1–25.

Zimmerman, B. L., and R. O. Bierregaard. 1986. "Relevance of the equilibrium theory of island biogeography and species–area relations to conservation with a case from Amazonia." *Journal of Biogeography* 13: 133–143.

Chapter 2

Forest Fragmentation and the Pollination of Neotropical Plants

Carolina Murcia

Assessing the consequences of the reduction and isolation of populations is fundamental for predicting the fate of species in forest patches and for designing effective management programs for small forest areas. This is a complex task, however, because the effects of fragmentation may spread indirectly through a network of interspecific interactions (Lovejoy et al., 1986; Murcia 1995). The most critical interactions are, perhaps, those that directly affect the reproductive success.

Reproduction in plants involves several steps: pollination, fruit and seed set, seed dispersal, and seed germination. In tropical plants almost every step involves direct interactions with animals. Thus, processes such as pollination, seed output, and seed dispersal could be susceptible to fragmentation through the effects of fragmentation on the plants, on the animals, or on both. Because these processes differ in their nature and in the organisms involved, each may be affected in a different way by forest fragmentation, and each would require different management strategies.

Here, I will focus on the effects of forest fragmentation on pollination. There are several reasons why we should be concerned about the pollination of tropical plants in small and isolated populations. First, pollination is one of the first steps in the reproduction of plants, and any effect at this level may be magnified at each subsequent step. For example, the amount of pollen delivered to the stigmas partially determines fertilization rates, which in turn affect the rates of seed production and fruit abortion. Second, in many tropical plants, pollination involves interactions with animals, making this process doubly susceptible to the effects of fragmentation. For instance, a fragmentation-induced change in pollinator activity could affect the quantity or quality of pollen received by the plants, resulting in changes in seed production or seed germination. Third, this is perhaps one of the reproductive processes less amenable to manipulation; for example, seeds can be collected and manually dispersed or germinated in nurseries and the seedlings planted. Hand-pollinating plants, in contrast, would require intensive work at a large scale, making manipulation of this process next to impossible.

I will begin by summarizing the available information on pollination in fragmented areas. Then I will discuss how factors such as breeding systems, life form, and pollinator type may affect the susceptibility of species to forest fragmentation. In the last section, I will suggest management strategies for enhancing the chances of continued plant reproduction and survival in forest patches.

Current Knowledge on the Effect of Fragmentation on Pollination

Very little is known about the effects of forest fragmentation on plant pollination. Only two studies have been conducted in the Neotropics. The first one was carried out in dry forest (*Chaco serrano*) in Argentina (Aizen and Feinsinger 1994a). In this community-level study, pollination levels and seed and fruit set were measured in small (less than 1 hectare) and large (less than 10 hectares plus one 20.5 hectares) remnants, and in continuous forest, using four replicate sites. Ten of 16 species exhibited a change in pollination levels in association with remnant size. Nine of these species showed a decline in pollination with a reduction in remnant size, while 1 showed the opposite trend. Fruit set was affected in 7 of 15 species, but while 5 species showed a decrease in fruit set in small remnants, the other 2 showed an increase. Among 14 multiseeded species, 3 showed a reduction in seed set in small fragments. In 2 other species, seed set was associated with remnant size but showed a significant interaction with the site, suggesting that the direction of the effects was inconsistent. The identity of flower visitors for 2 plant species also varied with fragment size (Aizen and Feinsinger 1994b). Plants in smaller fragments received a higher proportion of visits from nonnative bees than plants in larger fragments. Thus, small populations were probably subject to gene flow patterns different from those in continuous forest.

In contrast, a study carried out in cloud forest remnants in the Colombian southwestern Andes found no consistent effect of remnant size on pollination levels (Murcia and Arango, unpublished data). The study took place in two small (10–19 hectares) and three medium-sized remnants (75, 300, and 700 hectares), and a 150,000-hectare forest (Farallones National Park) located 10 kilometers away. The remnants formed part of an archipelago of forest patches scattered along one ridge of the western Andes and were originally connected to the national park (see Murcia 1993 for a more detailed description of the site). For the 14 species measured, there were no significant differences in pollination between small and medium-sized fragments. For the 3 species sampled in the fragments and the Farallones National Park, there were no differences associated with remnant size.

The effects of fragmentation that result from edge effects have also been assessed. Because the conditions near the forest edge are influenced by the climatic and biotic conditions of the surrounding matrix, it is expected that, either directly or indirectly, the plant–pollinator interactions would be altered and the pollination levels of plants affected. Edge effects have been found in many structural variables and in the dynamics of species interactions in forest edges (Lovejoy et al., 1986; Murcia 1995; Saunders et al., 1991). At the same field site in the Colombian Andes, the pollination levels of 16 plant species were measured at different distances from the edge. Three medium-sized remnants were used. No species (including 2 that were introduced experimentally) showed consistent changes in pollination levels in association with exposure to the forest edge (Murcia 1993).

Two other studies conducted in temperate zones did find reductions in pollination levels in plants from small, isolated populations. In Sweden, individuals of two small, isolated populations of *Dianthus deltoides*, a perennial caryophyllaceous herb, exhibited a reduced seed set relative to conspecifics in a nonfragmented site (Jennersten 1988). In this case, the reduction in seed set in one of the remnants was associated with low pollinator visitation, reflecting the low numbers and diversity of pollinators in the remnant relative to the continuous habitat.

Another study conducted in the United States compared pollination and seed set in an insular setting for two species: *Centrosema virgini-*

anum (Fabaceae) and *Opuntia stricta* (Cactaceae)(Spears 1987). *C. virginianum* had populations in Seahorse Key (approximately 112.5 hectares, 8 kilometers off the western coast of Florida) and Cedar Key (approximately 225 hectares, midway between Seahorse Key and the mainland), as well as a mainland population located 2 kilometers inland. *O. stricta* had populations on the two islands but not on the mainland. Individuals of both species had lower pollinator visitation and fruit set on the smallest and farthest island than their conspecifics in the larger and closer island or on the mainland. Experiments with fluorescent dye on both species showed a progressive increase in pollen dispersal (many more flowers and farther away from a pollen donor) from the small to the medium-sized island to the mainland.

It is not possible to extract patterns that could explain why some studies have found strong effects of fragmentation in the pollination of plants while others have not. In the two community-level studies, plants of several life forms, breeding systems, and pollinators were sampled. Aizen and Feinsinger (1994a) did not find any particular group of species (by breeding system or life form) with a particular sensitivity to fragmentation. Small sample size and correlation among some variables, however, prevented a strict analysis. Nevertheless, differences in results between the two studies do not seem related to differences in the kinds of plants sampled (very sensitive versus very insensitive).

Perhaps the major difference between the results of Murcia and Arango and the results of other studies lies in the size of the remnants. Effects of fragmentation have been evident on remnants smaller than 10 hectares (Aizen and Feinsinger 1994a) and strongest in remnants smaller than 1 hectare (Aizen and Feinsinger 1994a; Jennersten 1988). It is likely that pollination is affected by fragmentation only when populations are reaching an extremely small size. Remnants smaller than 10 hectares in the Colombian site were so disturbed by human intervention that they had a completely different species composition and could not be included in the study.

It is likely that, at least in some sites, remnants smaller than 10 hectares face other problems that are far more urgent than changes in pollination levels. Aizen and Feinsinger (1994a) observed that cattle grazing and trampling inside the remnants in the Argentinean Chaco constituted perhaps a larger threat to the survival of the plant species than a reduction in pollination. In the face of this evidence, albeit circumstantial, the management of forest patches that are very small (less than 10 hectares) should focus on problems different from pollination.

Many remnants are larger than 10 hectares, or have little interference from humans and cattle. However, they may be facing reproduction-related problems that, if appropriately managed, could be diminished. To manage such remnants, we must consider the mechanisms by which fragmentation could affect the pollination of plants in quantity and quality, and the possible subtle effects that would be noticeable only on a long-term scale.

Potential Effects of Fragmentation on Pollination

Forest fragmentation may disrupt the process of pollination through direct and indirect effects on plants and pollinators. Direct effects of fragmentation on the pollinators include changes in the number of flower visitors or in their identity. Direct effects on the plants may include changes in the reproductive population structure, either through changes in flower density or in the spatial distribution of sexually compatible individuals. Due to the interdependence of plants and pollinators, any effect of fragmentation on one party is likely to cause indirect effects on the other, ultimately influencing the pollination process as a whole.

Consequences of Changes in Pollinator Abundance and Identity

Factors such as a reduction in floral rewards, microclimatic changes inside the fragment, pesticides that diffuse from the matrix into the fragments, or invasion of competitors or predators from the matrix may reduce the populations of pollinators in fragmented forests. A reduction in the number of pollinators may result in a decline in the amount of pollen that is delivered to the stigmas. It is currently unknown to what extent natural levels of pollination (i.e., number of pollen grains delivered to the stigmas) limit seed production Yet, it is very likely that reductions in pollination associated with forest fragmentation will negatively affect seed production and reproductive success.

Fragmentation can also cause changes in the pollinator assemblage (Aizen and Feinsinger 1994b). Unlike an equivalent area in a continuous forest, small fragments may not have sufficient numbers of flowers to sustain the same number of pollinator species. Because many plant species are pollinated by several animal species, the local extinction of one pollinator may be compensated by increased visitation from other species (Aizen and Feinsinger 1994b). Not all flower visitors, however, are equivalent in their effectiveness as pollinators (Murcia 1990; Roubik 1989; Schemske and Horvitz 1984; Schmitt 1980; Spears 1983,

p. 352). Flower visitors differ in their flight distances and in the amount of pollen that adheres to their bodies (Feinsinger 1983; Murcia 1990; Schmitt 1980; but see Waser 1982). Consequently, animals differ in their capacity to disperse pollen from donors (Schmitt 1980; Waser 1982), affecting the patterns of gene flow in a plant population. The extinction of a pollinator species from a fragment may not affect significantly the pollination levels of the plants, if other species still visit the plant and deliver sufficient pollen for seed set. Long-term cryptic effects, however, could result from a shift in the identity of the pollinators. Pollinators that fly short distances tend to deposit more genetically related pollen and increase the level of inbreeding or self-fertilization. Conversely, pollinators that tend to fly long distances in a single foraging bout could bring pollen from genetically distant populations. Changes in the degree of relatedness between pollen donors and receptors may result in a reduction in fitness through either inbreeding or outbreeding depression (Waser and Price 1989, and references therein).

Consequences of Changes in Flower Number and Density

The consequences of changes in flower number and density are likely to differ among species, depending on the specificity of their interaction with pollinators. In interactions that are highly species specific, pollinators may be capable of persisting in very low densities in association with naturally rare plants; or else, pollinators may have life cycles that allow them to be temporarily independent from the floral resource during the plant's off flowering season. Forest fragmentation, however, may cause a decline in the floral density to a point where the plants fail to provide sufficient floral rewards to maintain a pollinator population. The final consequence is the loss of the pollinator by extinction or emigration, followed by the loss of the plant population as a result of a lack of reproduction.

Plant–pollinator interactions in the Neotropics, however, often involve several plant species sharing several pollinator species. Although reciprocal adaptation and diffuse coevolution may occur among plants and their pollinators (Feinsinger 1983), they often involve groups of species rather than plant–animal pairs. Consequently, plants usually have more than one animal species, occasionally from various taxa, visiting the flowers (Murcia 1990; Ramirez 1989; Schemske and Horvitz 1984; Stiles 1985; Vaughton 1992). Individual animals in turn feed from more than one flower species (Feinsinger 1976; Ramirez 1989; Thomson 1983).

Interactions with generalist flower visitors may have positive and negative effects. Species that depend on highly energy-demanding pollinators (such as hummingbirds) may not provide sufficient floral rewards at the beginning or end of their flowering seasons. During those times of low flower density, the plants may be unattractive to pollinators. Yet, if two or more plant species are flowering (albeit sparingly) in an area, pollinators may be attracted to the patch and provide adequate pollination service to all species. The effect of collectively attracting pollinators by several intermixed species is also known as *facilitation* (Busby 1987). Conversely, because generalist flower visitors forage in a wide spectrum of species, they may be able to persist longer on a forest patch, despite reductions in the flower density of particular species and changes in species composition of the floral offer.

Sharing generalist flower visitors also has negative consequences for some, if not all, plants involved. As pollinators feed from several species in a single foraging bout, pollen from one species could be brushed on the structures of heterospecific flowers, becoming unavailable for fertilization. In addition, pollen from heterospecific flowers could be deposited on the stigmas, using up space in the stigmatic surface (stigma clogging) and precluding conspecific pollen grains from germinating. Consequently, plant species sharing pollinator individuals may experience reduced pollination levels, lowered seed production, and limited pollen dispersal when growing intermingled (Campbell 1985; Campbell and Motten 1985; Feinsinger et al., 1988; Feinsinger and Tiebout 1991; Murcia and Feinsinger in press).

In unfragmented forests, the positive and negative effects of pollinator sharing may balance each other. In a forest remnant, however, plant populations are reduced and so are their flower numbers. Faced with an extremely low availability of floral rewards (pollen or nectar) in forest fragments, flower visitors may resort to feeding from many more species in each foraging bout. As a consequence, the negative effects of intermixing heterospecific flowers may be exacerbated. Those species that flower out of synchrony with others, while not suffering from the negative effects of intermixing with heterospecific flowers, may not provide sufficient floral rewards by themselves to attract pollinators.

Consequences of Changes in the Structure of Reproductive Populations

Besides changing flower density, fragmentation is also likely to affect the spatial distribution of sexually compatible individuals. The consequences of the changes in spatial distribution of plants to their pollination depend on the plant's breeding system and life form. Flowering plants have a diverse array of breeding systems that vary from having male and female sexual organs in the same flower and the capacity of self-fertilizing in the absence of flower visitors to having individuals that produce only either male or female flowers and depend exclusively on pollen vectors for pollination. Between the two extremes, there is a wide range of reproductive systems that vary in the disposition of sexes in the individuals and in the degree of genetic compatibility (Bawa and Beach 1981).

Given the wide variety of breeding systems, an analysis of how fragmentation affects each one is out of the scope of this chapter. Therefore I have chosen the four breeding systems most commonly found in Neotropical forests: dioecy, distyly, and self-incompatible and self-compatible hermaphroditism. Of all breeding systems, dioecy is perhaps the one most sensitive to fragmentation. In dioecious plants, individuals produce only male or female flowers, and it is necessary that individuals from both sexes be present in the same fragment, unless there is adequate pollen flow between fragments. Moreover, pollen should flow from male to female individuals, which reduces the number of effective interplant movements to one-half. In these species, only female individuals bear fruit.

Distyly imposes some of the same limitations as dioecy. In distylous species, all individuals produce hermaphroditic flowers (with male and female parts), but individuals differ in the floral morphology. There are two morphs in the population, and fertilization does not occur between gametes of the same flower morph (Bawa and Beach 1981). Consequently, it is necessary that at least two plants from opposite morphs be present in a fragment. Because all individuals have hermaphroditic flowers, they all set seed, but interplant pollen movement will be effective only if it takes place between plants of opposite morphs. Thus given a 1:1 morph ratio in the population, and random interplant movement, half of the movements between plants will not have any potential for fertilization.

One example of how fragmentation may affect pollination through changes in the spatial distribution of distylous plants is *Palicourea obesifolia,* a subcanopy tree that occurs in the western Andes of Colombia. During a study on the effects of fragmentation on pollination of cloud forest plants, Murcia and Arango (unpublished data) sampled the two individuals extant in a 15-hectare forest fragment. Both plants produced flowers of the same morph. Through the flowering season no flowers from the two individuals received compatible pollen, eliminating the chances of seed production. Consequently, unless there is an influx of seeds from neighboring areas, the species will disappear from the fragment. In contrast, in larger adjacent remnants where populations reached a few dozen individuals of both morphs, individuals received sufficient pollination for seed production in at least half of their flowers (Murcia 1993).

The largest group among flowering plants are those with hermaphroditic flowers. A significant proportion of this group is self-incompatible (i.e., the pollen cannot fertilize ovules in the same flower or plant). There are several self-incompatibility mechanisms, determined by one to several loci (Kress 1984). Fertilization cannot take place between plants that share the same incompatibility allele (Bawa and Beach 1981). Because determining the number of alleles involved in each species requires extensive manipulation, it is commonly assumed that plants are only incompatible to their own pollen, and that pollen from any other individual should fertilize the ovules. This is an oversimplification that must be carefully considered in cases where populations are extremely small. If all individuals in a small population are closely related, the chances of sharing incompatibility alleles increases, as does the probability of incompatibility among them. Therefore, although hypothetically a self-incompatible species would require only two individuals in a fragment to produce seeds, the genetic relatedness of the two could profoundly affect the outcome.

The co-occurrence of male and female organs in the flowers, and the absence of genetic incompatibility, makes hermaphroditic self-compatible species the group of plants least susceptible to fragmentation. As long as there is flower visitation, and pollen delivery to the stigmas, a single individual would suffice for fruit set.

In all four breeding systems discussed, I have made gross generalizations about the minimum number of individuals required in a fragment for any seeds to be produced. These generalizations, however, serve only for illustrative purposes and do not take into account the long-term genetic consequences of limited pollen flow and inbreeding. Therefore, they ought to be taken as relative measures of sensitivity to fragmentation and not as suggested population sizes for management programs.

A plant's life form also determines the structure of its reproductive population. In general, the size and growth form of a plant determine its population density and flower number. Life form and breeding systems are not completely independent characters. Community-wide studies have found a higher incidence of some breeding systems among particular life forms than expected by chance (Bullock 1985; Flores and Schemske 1984; but see Bawa et al., 1985b). Therefore, the sensitivity of a plant species to fragmentation may depend on both breeding system and life form. An additional modulating factor of the sensitivity of pollination to fragmentation is the pollinator type. Pollinators differ in their sensitivity to fragmentation and to changes in the floral offer, depending on their physiological requirements, flight capacities, and social and ecological constraints.

Trees in moist, tropical ecosystems often have very low densities. In Barro Colorado Island, for example, only 22 of 114 overstory tree species had more than 100 individuals per square kilometer, and 25 species had fewer than 10 individuals in a 0.5-kilometer plot (Hubbell and Foster 1986). Thus, medium-sized remnants (50–500 hectares or 0.5–5 square kilometers) may not support a viable population for most tree species. This very rough estimate is complicated by the fact that a proportion of tropical canopy trees are obligate outcrossers, either dioecious or self-incompatible. Twenty-two percent of the tree species in semideciduous forest in Costa Rica are dioecious (Bawa and Opler 1975). Assuming a 1:1 sex ratio, in dioecious species only half of the adults bear fruit. In addition, 80 percent of 28 hermaphroditic species studied at a tropical lowland rainforest in Costa Rica are self-incompatible (Bawa et al., 1985b). Palms, many of which form the canopy or subcanopy of tropical forests, also exhibit a high frequency of obligate outcrossing. Of the species for which the breeding system is known, 58 percent are obligate outcrossers, of which 45 percent are dioecious (calculated after Henderson 1986). In a montane tropical forest, five of eleven hermaphroditic trees were self-incompatible (Sobrevila and Arroyo 1982), and in a tropical dry forest in Argentina, all five tree species sampled were self-incompatible (Aizen and Feinsinger 1994a). With such levels of obligate outcrossing and low population densities, trees would be the group of plants most susceptible to forest fragmentation. It is unlikely that most remnants contain a large enough number of unrelated individuals (for self-incompatible species), or of both sex morphs (for dioecious species), for a population to be self-sustaining.

Despite this grim scenario, trees could be managed to maintain reasonably sized populations. In most areas, increasing the area covered by forest might be unrealistic if the matrix surrounding the fragments is under heavy human use. The key for the conservation of many tree species lies on the landscape approach, where the remnants are units in the landscape and the whole area is managed in a more integral way. Each species, however, must be managed independently, outside the context of the ecosystem in which it belongs. Species could be maintained as extended populations, with some individuals in the remnants but many more as part of live fences, riparian forests, or shade trees in pastures and near houses.

The spacing of trees should take into account the pollinators and their foraging flight behavior, so that pollinators are able to move pollen between individuals. Medium- and large-bodied bees, hummingbirds, bats, and sphyngid moths are strong flyers that can cover long distances while foraging. In contrast, beetles, wasps, butterflies, and noctuiid moths travel short distances. Trees pollinated by the latter group would then be more susceptible to spatial isolation of individuals.

At a lowland rainforest in Costa Rica, medium- to large-sized bees constitute by far the largest pollinator group of canopy trees. Nearly half of the 33 canopy tree species are pollinated by this group (Bawa et al., 1985a). Although there is much variation among species in their flight capacity, bees can easily forage within 1 kilometer from their nest (Roubik 1989, pp. 85–87 and references therein). Marked individuals of the euglossine bee *Eufriesea surinamensis* were able to return to their nest from a distance of 23 kilometers (Janzen 1971). Because bees are central-place foragers (i.e., they must return often to the same point—their nest), the location of trees, especially for dioecious species, is critical. The ideal distribution for pollen to flow from male to female trees requires not only a well-balanced sex ratio, but also that male trees are close to female trees. Moreover, factors such as prevailing winds, slope, and the spatial distribution of food sources (linear versus scattered) can affect the bees' flight distance and directionality (Roubik 1989, p. 86). Therefore, for each situation, it becomes necessary to determine how these variables affect pollinator movement among trees to design effectively a management program. The changes in pollinator type that could occur in those trees outside the fragments should also be considered. As discussed above, changes in pollinators may affect pollen flow and genetic structuring of the population. Therefore, management of seed dispersal and location of seedlings would be necessary to correct for any population substructuring that may result from some trees being pollinated by one species in the fragment and others being pollinated by a different species outside the fragment.

Bat- and hummingbird-pollinated trees may also tolerate wide spacing among conspecifics. Bats and hummingbirds are strong flyers, capable of flying long distances and crossing open spaces. The long-nosed bat (*Glossophaga soricina*), for example, has wide home ranges that vary between 1.5 and 51 hectares (Lemke 1984). Many tropical hummingbirds are capable not only of regional migrations (Levey and Stiles 1992) but, on a daily basis, of covering long distances to foraging sites

(Gill 1988) or along presumably preestablished foraging paths (Stiles and Wolf 1979). Although the number of tree species that are bat- and hummingbird-pollinated is low in the humid lowlands (Bawa et al., 1985a), their relative importance increases with elevation (Arroyo et al., 1983) and in some dry habitats (Howell 1983). In contrast, trees pollinated by poor flyers, such as small insects, beetles, and some butterflies, may require higher plant densities.

The smaller size of subcanopy trees relative to canopy trees permits higher densities and larger populations to be contained in forest fragments. An analysis of population densities of woody species in Barro Colorado Island showed a particular absence of rare species among midstory trees (Hubbell and Foster 1986). Among subcanopy trees, dioecious and distylous species may require special management. Staminate and pistillate morphs of dioecious tree species may be intermingled in a random fashion at a large scale (Bawa and Opler 1977); but at a smaller scale, individuals of one sex/morph could have a clumped distribution. Therefore it is critical to avoid having all individuals in a forest patch by chance belong to a single-morph clump. The same precautions should be taken with distylous species. As discussed above for *P. obesifolia* in the Colombian Andes, the spatial isolation of individuals from genetically compatible conspecifics may lead to significant reductions in ovule fertilization. The case of *Palicourea lehmannii* is another example of the potential effects of fragmentation on distylous species. In a cloud forest fragment in southwestern Colombia, individuals of *P. lehmannii,* located some 100 meters inside the forest, exhibited a significant reduction in pollination levels relative to conspecifics near the edge (Murcia 1993). The plants in the interior, however, were all the same flower morph and circumscribed to an area of 400 square meters, while those near the edge had a flower morph ratio closer to one. The low numbers of this species in other fragments precluded separating the two factors. Although differences in pollination between the two groups could be attributed to an edge effect, it is also likely that their spatial distribution, regardless of distance to the edge, influenced the pollination.

The breeding systems and spatial requirements of herbaceous and woody understory plants are very poorly known. It is possible, however, that these plants experience the lowest effects of fragmentation on pollination. At least some groups exhibit high levels of self-compatibility (Kress 1984), and many cloud forest species are capable of clonal growth and adventitious rooting (personal observation). Therefore, in species that are obligate outcrossers, asexual reproduction may compensate for deficiencies in seed set caused by a negative impact of fragmentation on their pollination. Additionally, because of the small body size of herbs and shrubs, forest remnants may contain populations sufficiently large to be self-sustaining. Particular problems may arise with rare species and those with very species-specific plant–pollinator relationships. The vagueness of this analysis is a reflection of the great void in our knowledge on plant reproduction of tropical shrubs and herbs. A few isolated species have been the focus of pollination studies (e.g., McDade and Davidar 1984), but no community-level studies exist.

Information on breeding systems and plant–pollinator interactions of epiphytic angiosperms is very scant. Bawa and Opler (1975) found no dioecious species among vines or epiphytes in a tropical lowland, semideciduous forest in Costa Rica. Except for Bawa and Opler's study, no community-level studies on reproduction of orchids or other vascular epiphytes exist. This lack of information prevents any attempt to draw projections on possible effects of fragmentation on the pollination of epiphytes.

Potential Long-Term Consequences on Fragmented Populations

Besides the obvious loss of species that results from reducing and isolating plant populations, a loss of genetic variability is to be expected. It is likely that self-incompatible systems disappear over time. Among hermaphroditic, self-incompatible species, the collapse of the self-incompatibility could result from selection of self-compatible phenotypes. Species may contain at least a few individuals that have limited or total self-compatibility (Bawa 1974; Bawa et al., 1985b; Sobrevila and Arroyo 1982); therefore, the potential for natural selection exists. In heterostylous species, the incompatibility barriers could disappear through random recombination and selection in favor of self-compatible individuals (see Sobrevila et al., 1983, for a discussion of the mechanisms).

The increase in inbreeding, caused by the shift to self-compatibility, may be associated with inbreeding depression as the deleterious alleles are expressed in recessive homozygotes. However, given the right circumstances for the deleterious alleles to be purged from the population (ample time, population size, no catastrophes), the inbreeding depression could be drastically reduced over time (Simberloff 1988). The results of a recent study suggest this possibility. In a study comparing germination of seeds collected from populations of different sizes, those collected from populations smaller than 150 individuals showed a significantly reduced germination relative to seeds from larger populations

(Menges 1991). The results confirmed the prediction that small populations are subject to inbreeding depression. Yet, in Menges' results, a subset of the samples from small populations had germination rates equivalent to those from large populations and low, within-sample variation. These results suggest that those small populations with no inbreeding depression are perhaps the ones that had remained isolated the longest and therefore have already been purged from deleterious alleles. The low variation of those samples also supports my interpretation, because long-term inbreeding may have an associated loss of genetic variability (Levin and Kerster 1974). The potential shift toward higher self-compatibility in the forest fragments would be associated not only with a loss of genetic diversity (Levin and Kerster 1974) but with a loss of diversity in breeding systems, such as heterostyly and andromonoecy.

Certainly, there are no magic recipes to prevent extinction from occurring in forest fragments. Loss of species and genetic diversity is inherent to fragmentation. Although some species may be impossible to preserve in small forest patches, other species could, with some management, persist in fragmented landscapes. To determine how to manage those species, however, we must be aware of the factors that make them more susceptible to fragmentation. Until recently, the connection between pollination ecology and conservation biology did not exist. Increasing awareness of the importance of pollination as a process that could affect the persistence of species in fragments is opening many possibilities for research with immediate application in conservation.

Acknowledgments

The ideas presented in this paper are the result of lengthy discussions with Peter Feinsinger and Gustavo Kattan. I thank them for their patience and interest. My work on forest fragmentation has been supported by the Department of Zoology, University of Florida, Underhill Foundation, and Fundación para la Promoción de la Investigación y la Tecnología–Banco de la República de Colombia.

References

Aizen, M. A., and P. Feinsinger. 1994a. "Forest fragmentation, pollination, and plant reproduction in a chaco dry forest, Argentina." *Ecology* 75: 330–351.

Aizen, M. A., and P. Feinsinger. 1994b. "Habitat fragmentation, native insect pollinators, and feral honey bees in Argentine "Chaco Serrano." *Ecological Applications* 4: 378–392.

Arroyo, M. T. K., J. Armesto, and R. Primack. 1983. "Tendencias altitudinales y latitudinales en mecanismos de polinización en la zona andina de los Andes templados de Sudamérica." *Revista Chilena de Historia Natural* 56: 159–180.

Bawa, K. S., 1974. "Breeding systems of tree species of a lowland tropical community." *Evolution* 28: 85–92.

Bawa, K. S., and J. H. Beach. 1981. "Evolution of sexual systems in flowering plants." *Annals of the Missouri Botanical Gardens* 68: 254–274.

Bawa, K. S., and P. A. Opler. 1975. "Dioecism in tropical forest trees." *Evolution* 29: 167–179.

Bawa, K. S., and P. A. Opler. 1977. "Spatial relationships between staminate and pistillate plants of dioecious tropical forest trees." *Evolution* 31: 64–68.

Bawa, K. S., S. H. Bullock, D. R. Perry, R. E. Coville, and M. H. Grayum. 1985a. "Reproductive biology of tropical lowland rainforest trees. II. Pollination systems." *American Journal of Botany* 72: 346–356.

Bawa, K. S., D. R. Perry, and J. H. Beach. 1985b. "Reproductive biology of tropical lowland rain forest trees. I. Sexual systems and incompatibility mechanisms." *American Journal of Botany* 72: 331–345.

Bullock, S. H. 1985. "Breeding systems in the flora of a tropical deciduous forest in Mexico." *Biotropica* 17: 287–301.

Busby, W. H. 1987. "Flowering phenology and density dependent pollination success in *Cephaelis elata* (Rubiaceae)." Ph.D. Dissertation, University of Florida.

Campbell, D. R. 1985. "Pollinator sharing and seed set in *Stellaria pubera:* Competition for pollination." *Ecology* 66: 544–553.

Campbell, D. R., and A. F. Motten. 1985. "The mechanisms of competition for pollination between two forest herbs." *Ecology* 66: 554–563.

Doube, D. M., and F. Moola. 1988. "The effect of the activity of the African dung beetle, *Catharsius tricornutus* De Geer (*Coleoptera: Scarabaeidae*) on the survival and size of the African buffalo fly, *Haematobia thirouxi potans* (Bezzi) (Diptera: Muscidae), in bovine dung in the laboratory. *Bulletin of Entomological Research* 78: 63–73.

Estrada, A., and R. Coates-Estrada. 1991. "Howler monkeys (*Alouatta palliata*), dung beetles (*Scarabaeidae*) and seed dispersal: Ecological interactions in the tropical rain forest of the Los Tuxtlas, Mexico." *Journal of Tropical Ecology* 7: 459–474.

Feinsinger, P. 1976. "Organization of a tropical guild of nectarivorous birds." *Ecological Monographs* 46: 257–291.

Feinsinger, P. 1983. "Coevolution and pollination." In *Coevolution,* edited by D. Futuyma and M. Slatkin, pp. 282–311. Boston, MA: Sinauer.

Feinsinger, P., and H. M. Tiebout III. 1991. "Competition among plants sharing hummingbird pollinators: Laboratory experiments on a mechanism." *Ecology* 72: 1946–1952.

Feinsinger, P., W. H. Busby, and H. M. Tiebout III. 1988. "Effects of indiscriminate foraging by tropical hummingbirds on pollination and plant reproductive success: Experiments with two tropical treelets (Rubiaceae)." *Oecologia (Berlin)* 76: 471–474.

Flores, S., and D. W. Schemske. 1984. "Dioecy and monoecy in the flora of Puerto Rico and the Virgin Islands: Ecological correlates." *Biotropica* 16: 132–139.

Gill, F. B. 1988. "Trapline foraging by hermit hummingbirds: Competition for an undefended, renewable resource." *Ecology* 69: 1933–1942.

Henderson, A. 1986. "A review of pollination studies in the palmae." *The Botanical Review* 52: 221–259.

Howell, D. J. 1983. "*Glossophaga soricina.*" In *Costa Rican Natural History,* edited by D. H. Janzen, pp. 472–474. Chicago: The University of Chicago Press.

Hubbell, S. P., and R. B. Foster. 1986. "Commonness and rarity in a Neotropical Forest: Implications for Tropical Tree Conservation." In *Conservation Biology: The Science of Scarcity and Diversity,* edited by M. E. Soulé, pp. 205–231. Sunderland, MA: Sinauer.

Janzen, D. H. 1971. "Euglossine bees as long-distance pollinators of tropical Plants." *Science* 171: 203–205

Jennersten, O. 1988. "Pollination in *Dianthus deltoides* (Caryophilaceae): Effects of habitat fragmentation on visitation and seed set." *Conservation Biology* 2: 359–366.

Karr, J. R. 1982. "Avian extinction on Barro Colorado Island, Panama: A reassesment." *American Naturalist* 119: 220–239

Kress, W. J. 1984. "Self-incompatibility in Central American Heliconia." *Evolution* 37: 735–744.

Lemke, T. O. 1984. "Foraging ecology of the long-nosed bat, *Glossophaga soricina,* with respect to resource availability." *Ecology* 65: 538–548.

Levey, D. J., and F. G. Stiles. 1992. "Evolutionary precursors of long-distance migration: Resource availability and movement patterns in Neotropical landbirds." *American Naturalist* 140: 447–476.

Levin, D. A., and H. W. Kerster. 1974. "Gene flow in seed plants." *Evolutionary Biology* 7: 139–220.

Lovejoy, T. E., R. O. Bierregaarrd Jr., A. B. Rylands, J. R. Malcolm, C. E. Quintela, L. H. Harper, K. S. Brown, A. H. Powell, G. N. V. Powell, O. R.

Schubart, and M. B. Hays. 1986. "Edge and other effects of isolation on Amazon forest fragments." In *Conservation Biology: The Science of Scarcity and Diversity,* edited by M. E. Soulé, pp. 257–285. Sunderland, MA: Sinauer.

McDade, L. A., and P. Davidar. 1984. "Determinants of fruit and seed set in *Pavonia dasypetala* (Malvaceae)." *Oecologia* 64: 61–67.

Menges, E. S. 1991. "Seed germination percentage increases with population size in a fragmented prairie species." *Conservation Biology* 5: 158–164.

Murcia, C. 1990. "Effect of floral morphology on pollen receipt and removal in *Ipomoea trichocarpa.*" *Ecology* 71: 1098–1109.

Murcia, C. 1993. "Edge effects on the pollination of tropical cloud forest plants." Ph.D. dissertation, University of Florida.

Murcia, C., and P. Feinsinger, in press. "Interspecific pollen loss by hummingbirds visiting flower mixtures: effects of floral architecture." *Ecology.*

Murcia, C. 1995. "Edge effects in fragmented forests: Implications for conservation." *Trends in Ecology and Evolution* 10: 58–62.

Ramirez, N. 1989. "Biología de polinización en una comunidad arbustiva tropical de la alta guayana Venezolana." *Biotropica* 21: 319–330.

Roubik, D. W. 1989. *Ecology and Natural History of Tropical Bees.* Cambridge, UK: Cambridge University Press.

Saunders, D. A., R. J. Hobbs, and C. R. Margules. 1991. "Biological consequences of ecosystem fragmentation: A review." *Conservation Biology* 5: 18–32.

Schemske, D. W., and C. Horvitz. 1984. "Variation among floral visitors in pollination ability: A precondition for mutualism specialization." *Science* 225: 519–521.

Schmitt, J. 1980. "Pollinator foraging behavior and gene dispersal in *Senecio* (Compositae)" *Evolution* 34: 934–943.

Simberloff, D. 1988. "The contribution of population and community biology to conservation science." *Annual Review of Ecology and Systematics* 19: 473–511.

Sobrevila, C., and M. T. K. Arroyo. 1982. "Breeding systems in a montane tropical cloud forest in Venezuela." *Plant Systematics and Evolution* 140: 19–37.

Sobrevila, C., N. Ramirez, and N. X. d. Enrech. 1983. "Reproductive biology of *Palicourea fendleri* and *P. petiolaris* (Rubiaceae), heterostylous shrubs of a tropical cloud forest in Venezuela." *Biotropica* 15: 161–169.

Spears, E. E., Jr. 1983. "A direct measure of pollinator effectiveness." *Oecologia* 57: 196–199.

Spears, E. E., Jr. 1987. "Island and mainland pollination ecology of *Centrosema virginianum* and *Opuntia stricta.*" *Journal of Ecology* 75: 351–362.

Stiles, F. G. 1985. "Seasonal patterns and coevolution in the hummingbird–flower community of a Costa Rican subtropical forest." In *Neotropical Ornithology,* edited by P. A. Buckley, M. S. Foster, E. S. Morton, R. S. Ridgely, and F. G. Buckley, pp. 757–785. Washington DC: The American Ornithologists' Union.

Stiles, F. G., and L. L. Wolf. 1979. "Ecology and evolution of a lek mating system in the long tailed hermit hummingbird." *Ornithological Monographs* No. 27. Washington DC: American Ornithologists' Union.

Thomson, J. D. 1983. "Component analysis of community-level interactions in pollination systems." In *Handbook of Experimental Pollination Biology,* edited by C. E. Jones and R. J. Little, pp. 451–460. New York: Van Nostrand Reinhold.

Vaughton, G. 1992. "Effectiveness of nectarivorous birds and honeybees as pollinators of *Banksia spinulosa* (Proteaceae)." *Australian Journal of Ecology* 17: 43–50.

Waser, N. M. 1982. "A comparison of distances flown by different visitors to flowers of the same species." *Oecologia* 55: 251–257.

Waser, N. M., and M. V. Price. 1989. "Optimal outcrossing in *Ipomopsis aggregata*: Seed set and offspring fitness." *Evolution* 43: 1097–1109.

Chapter 3

The Consequences of Prolonged Fragmentation: Lessons from Tropical Gallery Forests

Martin Kellman, Rosanne Tackaberry, and Jorge Meave

Introduction

The universality of tropical forest fragmentation and the possibility that systems of small forest patches will comprise the only forest surviving in many future tropical landscapes makes an ability to predict future conditions in these systems a matter of some urgency. The capacity of these systems to conserve regional biodiversity and withstand prolonged human intervention is of particular concern. The recent origin of most tropical forest fragments means that few are likely to be in equilibrium with their new biogeographic context (*sensu* Pickett et al., 1992), and they are more likely to be dominated by transient, rather than by equilibrial, conditions. Thus, while studies have provided important information on the short-term adjustment processes operating within recently created fragments (e.g, Bierregarrd et al., 1992), they are unlikely to be very useful at indicating the long-term consequences of fragmentation. For example, the average tropical forest canopy tree appears to live for approximately one century (Rankin-de-Merona et al., 1990; Swaine et al., 1987). However, demographic adjustment of most biological populations to new habitat conditions is likely to require several generations, making it improbable that the full impact of fragmentation on these organisms will become evident before several centuries have elapsed.

Most long-term projections about conditions in community isolates have drawn on analogies between these and oceanic islands (MacArthur and Wilson 1967 and the extensive literature flowing from it), and generally have predicted a loss of diversity because local immigration rates are assumed to fall after isolation while local extinctions continue. However, oceanic islands may provide poor analogs of habitat fragments as the two have very different histories and are embedded in different matrices. Moreover, serious questions have been raised about the generality of the theory upon which these predictions are based (Gilbert 1980; Williamson 1989).

In this chapter we propose that the gallery forests that occur along the banks of rivers in many tropical savanna landscapes represent naturally occurring patch systems of probable long isolation and provide a means of making tentative predictions about the long-term consequences of fragmenting tropical forest. This proposal assumes that gallery forest patches are of great antiquity. Long-term paleoecological data are of scattered occurrence and normally provide only for regional, rather than local, reconstructions of past vegetation; so it is generally not possible to confirm unequivocally that gallery forests are very old. However, we believe that certain conditions make this probable. Gallery forests are by far the most frequent form of natural forest patch in subhumid tropical landscapes, suggesting that riparian habitats are especially favored for forest persistence. Streams are enduring landscape features, and the improved moisture supply and reduced fire incidence associated with them may be expected to be equally persistent and favorable to forest preservation. The assumption is also strengthened when, as is the case in our study sites, paleoecological data indicate a prolonged regional absence of a forest cover. While it could be argued that riparian habitats are specialized and generally not representative of the habitats of recent forest patches, many of the latter also occur in riparian locations and, for these, natural gallery forests provide very close analogs.

Our discussion is confined to trees. While restrictive, we believe that these provide the basic structural framework of forest patches without which few other life forms could persist. We begin by summarizing the results that we have obtained in an examination of several gallery forest systems, then go on to discuss the processes that seem to be important to their form and function. Finally we outline the changes that we anticipate will develop gradually in recently created patches.

Gallery Forests and Their Composition

Tropical gallery forests have received frequent mention in the literature (e.g., Beard 1955), but little systematic examination. There has been some attention paid recently to the flooded variants of these that occur along major rivers (e.g., Balslev et al., 1987; Campbell et al., 1986; Junk 1989), but these are of little use as analogs of forest fragments as they occupy an atypical habitat and contain a small and very specialized forest flora. While some documentation of nonflooded tropical gallery forest has taken place (e.g. de Oliveira-Filho et al., 1990; Gibbs and Leitao Filho 1978), nonstandardized sampling procedures and the absence of comparative data from continuous forests of the same region have made it difficult to use these data to answer questions about the internal structure of the forests and the preservation of regional floras within them.

Our own data have come from three gallery forest systems in differing biogeographic milieux. Most data have been collected in gallery forests of the Mountain Pine Ridge savannas of Belize, Central America, an area of hilly terrain at approximately 500 meters elevation. Here gallery forests comprise discontinuous elongate patches, generally less than 100 meters in width, that occupy steep-sided valleys embedded in a plateau covered by pine savanna. The fire recurrence interval for areas in the savanna not subject to fire control has been estimated to be 18 years (Hutchinson 1977). Less complete data have also been collected in the Apure savannas of southwestern Venezuela, an area of flat, generally treeless savanna at elevations less than 100 meters above sea level, in which gallery forests form highly continuous systems usually less than 200 meters in width. In both areas, paleoecological data suggest prolonged persistence of savanna since at least Late Glacial time (Kellman 1975; Vaz and Miragaya 1989). More recently, we have begun to collect comparative data on gallery forest fragments at elevations between 500 and 1,500 meters above sea level in the Gran Sabana of southeastern Venezuela, an area of variable physiography and exceptionally infertile soils, where forest fragmentation appears to be a more recent phenomenon (Dezzio 1994; Rull 1991). The results achieved to date have been published in detail elsewhere (Kellman et al., 1994; Kellman and Tackaberry 1993; MacDougall and Kellman 1992; Meave and Kellman 1994) and only a summary will be presented here.

All gallery forests examined tend to have low canopies and low stem basal areas relative to continuous tropical forest, but are otherwise variable in structure. In Belize they tend to be characterized by high densities of small stems, but in Apure a more typical diameter-size class distribution prevails. The number of species of trees equal or larger than 10 centimeters diameter at breast height (dbh) in sample areas up to one hectare in size are comparable to those in similar areas of continuous forest in the same region (Kellman et al., 1994). In Belize, 50 to 60 tree species per hectare were found, while in Apure, floristic richness is approximately 50 percent greater at this scale. A detailed analysis of the floristic composition of woody taxa in the Belize gallery forests has shown that approximately 75 percent of the common species are recorded as present in continuous tropical forests elsewhere in the region (Meave and Kellman 1994). Mammal-dispersed species and those with separate male and female plants are somewhat underrepresented in this flora relative to those of continuous forests, while bird-dispersed species are somewhat more prominent. Light-demanding pioneer species are rare, but some taxonomic groups, such as the family Melastomataceae, are more abundant than is usual in continuous forests. A comparable floristic analysis has not been performed in the other gallery forests that we have studied, but the presence in these of tree species common in continuous forests suggests that equivalent conditions prevail there. While some subtle differences appear to exist between the gallery forest floras and those of continuous forests, these results do not support the major prediction of island biogeographic theory—that biotic diversity will fall as a result of insularization of habitats.

Internal Organization and Edge Effects

Continuous tropical forests usually show weak internal spatial structure, with few distinct changes in community composition that are closely correlated with local habitat variability. Most internal structure is associated with irregular canopy gap formation or past disturbances (Whitmore 1989). If gallery forests are simply small spatial segregates of these larger systems, we would expect, but have not found, comparably weak internal structuring. Rather, most gallery forests possess consistent spatial organization, with many species showing monotonic changes in abundance relative to distance from the creek, the savanna–forest boundary, or both. In the Belize system, an analysis of the distributions of 51 species revealed that only 10 did not show significant correlation of abundance with one or both ends of this gradient (Meave 1991). Surprisingly, few consistent changes in soil properties were found along the gradient, suggesting that other processes whose intensity varied in this dimension were more important contributors to the pattern. Of these, the processes known collectively as *edge effects* seemed most likely to be contributing to the pattern; and so far we have examined two of these: light gradients and fire intrusions.

Light Gradients

Intrusion of external atmospheric conditions into the microenvironment of forest patches is frequently seen as the edge effect most threatening to the coherence of patch communities, and deleterious effects associated with this intrusion have been documented by Kapos (1989) and Laurance (1991). Many microclimatic conditions are correlated with solar radiation loading, and the seedlings of most tropical forest plants are sensitive to light variability, making light levels a useful single indicator of atmospheric edge effects in forest patches.

An analysis of light penetration into the Belize gallery forests (MacDougall and Kellman 1992) has shown this to be surprisingly limited. In contrast to the results at recent forest boundaries (Kapos 1989), changes were found to be generally limited to the outer 10 meters of forest edge, beyond which light levels characteristic of closed, continuous forest (1–2 percent of external illumination) were found. We attribute this primarily to the closed forest edge canopy that exists in these forests (figure 3-1) which is a product of both lateral branching by canopy trees and an aggregation of tree saplings and other small woody forest plants at the boundary. Comparably limited light intrusion has been documented in temperate forest fragments by Brothers and Springarn (1992). An analysis of the light sensitivity of six common tree species in gallery forests showed that two were significantly more abundant in better-illuminated microsites and three significantly more abundant in poorly illuminated microsites than random expectation (MacDougall and Kellman 1992). The seedlings of only one species showed no sensitivity to differences in prevailing light levels. These re-

Figure 3-1
A "closed" gallery forest boundary, Apure savannas, Venezuela. Lateral branching by canopy trees, and a concentration of smaller woody plants, provides a deep peripheral canopy that effectively insulates the forest interior from high light levels. (Photo by M. Kellman.)

sults suggest that the light gradient associated with gallery forest patch edges provides a habitat dimension to parts of which some forest species are specialized.

Fire Intrusions

Fire is normally a rare event in tropical evergreen forests (Uhl et al., 1988), and its effects in these places are often lethal to tree species (Uhl and Buschbacher 1985). However, tropical gallery forests exist in a matrix of frequently burned vegetation, as do most forest patches deriving from human activities. While most savanna fires that we have observed at gallery forest boundaries failed to enter the forests, we have also found widespread signs of deeper fire intrusions in many gallery forests examined. An analysis of the distribution of fire signs in the gallery forests of Belize showed a decreasing frequency with distance from the boundary, suggesting a gradient in fire-disturbance frequency (Meave

1991). An analysis of the distribution of 42 tree species relative to signs of past fires in one gallery forest patch in that area showed 8 species positively associated and 2 negatively associated with past fire signs (Kellman and Tackaberry 1993).

Severe fires during the 1991 dry season in Belize resulted in some deep fire penetration of gallery forests and allowed a more direct analysis of the responses by trees. Fire intrusions were found to be patchily distributed and of a variety of intensities. At some intrusions, soil-surface light levels increased slightly relative to those in adjacent unburned sites as a result of seedling and sapling mortality, but three years later had undergone little further change (table 3-1). In contrast, at other intrusions light levels had continued to increase over this time period, suggesting longer-term canopy dieback. Thus, while all fire intrusions create a specialized seedbed for germination, a variety of light conditions subsequently develop over these places. Two years after the fire, the distributions of seedlings of eight species that had established in sufficient numbers to permit an analysis showed five to be significantly more abundant at burned sites than in adjacent unburned areas, and two significantly less so (table 3-2). One species, *Vochysia hondurensis,* showed converse abundances at two sites of differing illumination, suggesting sensitivity to subtle seedbed–light interactions. Survivorship of

Table 3-1

Light levels measured over one 24-h period at five fire-intrusion sites in Belize gallery forests, two months and three years after the fire.[a]

	Two months			Three years		
Site	Burned	Unburned	Ratio (B/U)	Burned	Unburned	Ratio (B/U)
A	4.78 ± 3.54	1.18 ± 0.19	4.1	5.20 ± 1.00	2.97 ± 0.68	1.8
B	3.70 ± 2.07	2.69 ± 1.20	1.4	9.48 ± 3.20	4.31 ± 0.50	2.2
C	9.93 ± 1.05	3.12 ± 1.51	3.2	65.82 ± 22.13	1.90 ± 0.39	34.6
D	7.47 ± 4.25	1.79 ± 1.21	4.2	33.80 ± 9.70	3.79 ± 0.73	8.9
E	2.54 ± 0.67	1.28 ± 0.35	2.0	12.46 ± 3.55	7.71 ± 4.79	1.6

[a]Light levels measured concurrently in adjacent unburned control plots are also shown; burned/unburned (B/U) ratios express the degree of increased illumination at burned sites relative to controls. All light measurements made ca. 10 cm above the soil surface, except in the three-year burned plots, where measurements were made above vegetation that had established after the fire. Data expressed as percentage of external illumination measured concurrently. $\bar{x} \pm 1SD$ shown, $N = 5$.

Table 3-2

Frequencies of seedlings ($N \geq 10$) in 16-m^2 plots at sites of fire intrusions in Belize gallery forests and in adjacent unburned control plots, two years after the fire.

Species	Plot[a]	Burned	Unburned	*P*[b]
Alchornea latifolia	A	30	0	< 0.001
	D	61	0	< 0.001
Luehea sp. (#JM74)	A	19	1	< 0.001
Meliaceae (#RT15)	D	4	15	< 0.01
	E	5	6	NS
Myrsine coriaceae	A	23	0	< 0.001
	C	30	0	< 0.001
	E	25	0	< 0.001
Photinia sp. (#JM1061)	B	424	83	< 0.001
Symphonia globulifera	E	1	10	< 0.01
Vochysia hondurensis	B	49	0	< 0.001
	D	17	48	< 0.001
Xylopia frutescens	B	9	1	< 0.05

[a]Plot identifications as in Table 3-1.

[b]Probabilities of the samples being drawn from the same population (chi-square analysis).

seedlings on burned plots over the ensuing year was comparable to that of a subset of seedlings of comparable size in adjacent unburned control plots. Three species, *Myrsine coriaceae, Photinia* sp., and *Xylopia frutescens,* which were significantly more abundant as seedlings in burned plots (table 3-2), have also been documented as occurring in significantly higher frequencies as mature trees in areas of old fire intrusions than in unburned areas (Kellman and Tackaberry 1993).

Collectively, these data indicate that occasional fire intrusions in gallery forests provide a unique class of microsites for seedling establishment, from which at least a subset of tree species may benefit. Although some species are sensitive to fire, the patchy nature of fire intrusions makes it improbable that these species will be eliminated from a fragment.

A comparison of species identified as either "edge-concentrated" or "core-concentrated" (Meave 1991) with those found to be sensitive to fire intrusions (Kellman and Tackaberry 1993) or light levels in the forest understory (MacDougall and Kellman 1992) shows that 8 of 22 edge-concentrated species were associated with fire or higher light levels, while 3 of 17 core-concentrated species showed the reverse pattern. In only two instances were contradictions found between the two data sets: *Calophyllum brasiliense* was identified as a species insensitive to location by Meave (1991) but found to be shade-requiring by MacDougall and Kellman (1992), while *Alchornea latifolia* was identified as a core species by Meave (1991) but found to be abundant at fire intrusion sites. However, populations of trees equal to or greater then 10 centimeters diameter of the latter species did not show any significant difference in distribution between burned and unburned zones (Kellman and Tackaberry 1993).

It would thus appear that the distribution of an appreciable proportion of the species showing segregation along creek–savanna gradients in gallery forests can be attributed to associated changes in light levels and fire frequencies. A more complete sampling of these effects would almost certainly increase the proportion of species showing interpretable patterns, and a contribution by other unexamined edge effects cannot be excluded.

Thus, while edge effects are commonly assumed to be threatening to the integrity of a patch and its ability to sustain biodiversity, we find that in gallery forests these edge effects contribute to tree species coexistence by providing a complex microenvironmental to disturbance gradient along which many species have segregated.

Interpatch Dispersal

Numerical modeling has repeatedly indicated that interpatch dispersal is important in preventing global extinction of species in fragmented systems (e.g., Goodman 1987), and a decrease in immigration rate due to isolation is one of the major predictions of island biogeographic theory (McArthur and Wilson 1967). This leads to the further prediction of eventual diversity reduction in fragmented systems. While we have no data available with which to test the first prediction directly, there exist scattered empirical data and theoretical reasons leading one to suppose that reductions in immigration rates to patches may be less extreme than anticipated.

The maintenance of species richness in gallery forests at levels comparable to those in continuous forests in itself suggests that no dramatic drop in immigration rate has occurred (if in fact the dynamic species equilibrium of island biogeographic theory prevails). Moreover, the tendency of gallery forest species richness to mirror that of continuous forests in the same region suggests floristic connection between the two systems. In one completely enumerated gallery forest patch in Belize, 8 of the 61 tree species (equal to or greater than 5 centimeters diameter) occurred only as single stems. A reenumeration five years later has shown that although two of these species have disappeared, two new ones have entered the measured-size class. It is possible that the long life spans of some rare tree species allow their persistence at low densities by requiring only one successful establishment from each individual over several decades. However, the data presented above indicate that at least some species turnover takes place.

Bird-dispersed species were found to be slightly more abundant in these forests than in continuous forests in the region (Meave and Kellman 1994), implying floristic change toward a more dispersable assemblage. In theory, restriction of foraging areas of frugivores in a fragmented system should induce movement over greater linear distances within or between patches than would occur in continuous forest, which could help offset the effects of the loss of some animal seed dispersers. However, this depends on the survival of a sufficient suite of frugivores capable of interpatch movement. The high levels of plant diversity in the gallery forest study sites suggest that despite the isolation of patches and reduced habitat area, it is biologically possible to maintain populations of dispersers in a system of this type over the long term. However, it is likely that such disperser population maintenance will be more problematic in areas where forest patches are set in a matrix of higher densities of human populations, because the fauna will be much more vulnerable to pressure from hunting and other disturbances. In planning patch systems, it is clearly desirable to take all measures possible to ensure the survival of animal seed dispersers and facilitate interpatch mobility. However, we are still a distressingly long way from being able to go beyond such simple qualitative statements.

Thus, while the concerns expressed by Terborgh (1992) and others about the eventual extinction of some plant species as a consequence of the loss of their animal seed dispersers in fragmented systems cannot be ignored, there appears to be a subset of relatively mobile tree species that exists ephemerally in the natural patch systems of gallery forests. These mobile species may well be superimposed on a set of less mobile species to form a bifunctional system analogous to that described for orb spiders on islands by Schoener and Spiller (1987). While the less mobile set of species may ultimately be more vulnerable to global extinction, the existence of the more mobile subset offers promise for both the persistence of some form of forest patch after isolation and for the reconstitution of patches in forest rehabilitation efforts.

Tropical Forest Patches and Biological Conservation

Preservation of a significant proportion of a previously forested region's biota in a system of forest fragments requires that diversity be "compressed" at some spatial scale. In theory, either of two forms of compression could achieve this: (1) Increased local species densities, with each subregion's biota packed into one or a few patches, or (2) increased turnover in species composition between patches, with each patch having a distinctive biota and a high degree of dissimilarity with other patches.

Data collected so far indicate that, although at local scales (i.e., areas, approximately one hectare in size) gallery forests have not lost tree species diversity, this diversity has not been augmented above that which is found in continuous forests. This precludes the first alternative for species conservation. Our data at present are insufficient to evaluate the second alternative. Bell et al., (1993) have recently shown that habitat similarity is distance dependent in a variety of ecological systems, implying that a widely dispersed patch system is likely to contain much more habitat heterogeneity than one local block of similar area. In theory then, patch systems should retain a proportion of habitat heterogeneity that is greater than the proportion of total area they represent, thereby providing for considerable habitat dissimilarity between patches. This would be expected to contribute to interpatch species turnover. If compositional heterogeneity in the preexisting continuous forest were to be largely forced by habitat heterogeneity, no increase in floristic dissimilarity would necessarily follow from fragmentation. However, were composition in continuous forest to be affected by other processes that are sensitive to fragmentation, an increase in postfragmentation biotic differences between patches would be anticipated. The most likely change of this sort is a reduction in the tendency for competitive exclusion to take place in isolated patches. Exclusion of less effective competitors by more aggressive species frequently results in gradual biotic impoverishment at local scales, unless the process is interrupted by extraneous disturbances (Connell 1978; Huston 1979), and patch isolation may provide a suitable refuge from more aggressive species (cf. Quinn and Robinson 1987).

The data that we have so far collected on species richness beyond local scales have focused on the rate of new species accumulation along linear gallery forest patches. This is the main dimension of new species acquisition in these systems and can be taken as equivalent to interpatch species accumulation. Our research (Tackaberry 1992) has shown that at local scales (i.e., distances of over hundreds of meters) new tree species accumulate more rapidly at forest edges than in cores, presumably reflecting the enrichment effects of such processes as fire intrusions. However, over longer distances (i.e., kilometers) the reverse pattern prevails, with the edge floras stabilizing at moderate floristic richness, while the cores continue to accumulate new species. This suggests that the greatest regional floristic richness is likely to be preserved in core areas that are less frequently and severely affected by such edge effects as fire intrusions and that are enclosed within a less diverse, but highly protective boundary assemblage of "edge-tolerant" trees.

Much ecological theory assumes that communities are species saturated and effectively closed to further enrichment. However, increasing evidence suggests that tropical forest communities are loosely organized and capable of sustaining large numbers of coexisting, nonspecialized tree species almost indefinitely (Hubbell and Foster 1986). The existence of local species richness in the gallery forests of Apure that is approximately 50 percent larger than that in the Belize gallery forests strongly suggests a lack of species saturation, at least in the latter system. This raises the possibility that artificial enrichment of patches by transplantation may be a viable management technique for conservation of biological diversity.

Post-Fragmentation Change in Forest Patches

The differences that exist between recently created forest patches (e.g., those described by Lovejoy et al., 1986) and long-isolated gallery forests suggest that long-term adjustments take place in patches following isolation. However, in the absence of a well-dated sequence of patches of differing age, details of the adjustments can only be inferred.

Recently created patches are characterized by an unspecialized edge zone (figure 3-2), deep and often destructive edge effects (Kapos 1989), possible edge retreat (figure 3-3), temporary supersaturation by mobile organisms, and loss of species requiring large home ranges (Lovejoy et

Figure 3-2
An "open" forest patch boundary, probably of recent origin, in the Gran Sabana, Venezuela. Boundaries of this sort offer little protection from external atmospheric conditions and are vulnerable to frequent fire intrusions. (Photo by M. Kellman.)

al., 1986) or specialized symbionts (Terborgh 1992). Presumably internal spatial organization in these patches reflects the canopy regeneration mosaic at the time of isolation.

We believe that the subsequent change most critical to patch integrity is the development of a boundary community composed of trees that are both tolerant of fire and other edge conditions, and capable of forming a closed peripheral canopy (figure 3-1). The speed with which this change develops is likely to vary a great deal, depending on the tree flora available locally. Many tropical tree floras presumably contain some tree species preadapted to fire (e.g. Uhl and Kauffman 1990) and capable of lateral branching; and the rapid development of a closed canopy at recent tropical forest boundaries has been documented by Williams-Linera (1990). Because many of the tree species prominent at gallery forest boundaries are also present in the surrounding savanna (e.g., *Clethra hondurensis* and *Quercus oleoides* in Belize and *Caraipa llanorum* in Apure), access to such a flora would greatly increase the speed

Figure 3-3
Fire damage to a remnant forest patch, Gran Sabana, Venezuela. The absence of well-developed edge communities, and the deep organic mats in which most trees are rooted, makes forest patches in this area especially vulnerable to destructive fire intrusions. (Photo by M. Kellman.)

with which an insulative boundary would develop. In contrast, in areas such as those covered naturally by continuous cloud forest, absence of a fire-tolerant tree flora may delay the development of a protective forest boundary and leave patches vulnerable for prolonged time periods (figures 3-2, 3-3).

The appearance at gallery forest boundaries of a new group of edge-tolerant species rarely found in continuous forests represents the major floristic addition to these systems. In the Neotropics, members of the family Melastomataceae are particularly common at gallery forest boundaries. The size of this group is sufficient, at local scales, to compensate for any extinction of other tree species and result in the equivalence in floristic richness between gallery and continuous forests in the region. An internal reorganization of the community structure along the core–edge gradient probably proceeds gradually and coincidentally with the development of a distinctive edge community. Less clear is whether significant between-patch reorganization of floras takes place during the period of adjustment, which would result in increases in interpatch floristic differences. However, if transformation from continuous forest to a system of patches were a gradual process and involved temporary supersaturation by both plant and animal species, such a reassortment could derive from selective extinction within individual patches.

It seems inevitable that some loss of plant species that are obligate symbionts of extinguished frugivores will take place within systems of small forest fragments; these species are likely to be preserved only in large forest reserves. However, the gallery forest data indicate that a proportionate increase in more readily dispersed species, plus those tolerant of edge conditions, compensates for this loss at local scales. Two more specific questions may be asked about post fragmentation change: (1) Can biodiversity be maintained over exceptionally long time periods in fragmented systems? and (2) Are human activities within patches or in the surrounding matrix likely to cause an ongoing change in patch composition?

Very Long-Term Fragmentation

The paleoecological record for some (but by no means all) tropical locations indicates a virtually complete loss of moist forest cover during periods of Pleistocene drought (e.g., Hopkins et al., 1993; Kershaw 1974, 1976; Leyden 1984). The rapid return of a rich forest flora during interglacial periods attests to its survivability in scattered local refugia (Kellman et al., 1994). This implies that systems of forest patches can act as effective reservoirs of forest diversity for periods in the order of 10,000–20,000 years. However, the depauperate flora of the patch system of monsoon forests in northern Australia (Russell-Smith 1991; David Bowman, pers. comm.) also indicates that maintenance of biodiversity may not be indefinite. These forest patches have probably been isolated since the late Tertiary (i.e., approximately 20 million years) as a consequence of intensified continent-wide aridity (Adam 1992). The existence of such patches testifies to the remarkable persistence of some form of forest in a fragmented state throughout geological time. However, their depauperate flora indicates that some loss in diversity must be expected over very long time periods, in the absence of human intervention to prevent this.

Human Activities

Most gallery forests have probably developed in the absence of intense human activities. Some hunting and gathering within these forests, and increased fire frequencies in the surrounding savannas, are likely to have been the major consequences of the appearance of human populations. While gallery forests have persisted in the face of these activities, the ability of these and other forest patches to withstand more intense human intervention remains unknown.

The small size of most forest patches makes it most unlikely that structural integrity could be sustained in the face of the removal of large trees, but harvesting of small timber and other forest products for local uses could be sustainable. The boundaries of gallery forest patches appear able to remain intact at low fire frequencies, but may be unable to sustain increased frequency. For example, in the Apure savannas, annual savanna fires set by ranchers appear to be causing a gradual retreat of forest boundaries (Kellman et al., 1994), and selectively logged forest remnants in Amazonia seem to be especially vulnerable to intrusion of pasture fires (Uhl and Buschbacher 1985). Similarly, shifting cultivators forced by overpopulation to use small forest patches may establish fields that are contiguous with surrounding savanna (figure 3-4), which

Figure 3-4

A shifting cultivator's field in a small remnant forest patch, Gran Sabana, Venezuela. The positioning of the field contiguous with the surrounding savanna leaves it vulnerable to penetration by savanna fires after abandonment and lowers the probability of successful forest regeneration. (Photo by M. Kellman.)

facilitates fire penetration and a conversion of forest to savanna. Successful preservation of forest patches would therefore seem to require fire regimes in the surrounding matrix that are much longer than one year. Preservation would also seem to require that the use of small patches be limited to nonagricultural activities and concentrated in patch interiors, preserving an intact and resilient boundary zone.

Natural gallery forests, which are now coming under intense human usage in some areas while in others remaining minimally affected, provide ideal places in which to examine the sustainability of biodiversity in forest patch systems in the face of human activities.

Conclusions

The gallery forest data that we have gathered offer some cause for cautious optimism in forest conservation efforts, suggesting that loss of plant diversity is not an inevitable consequence of fragmentation and that diverse forest patches can form stable components of tropical landscapes. Similar conclusions have been drawn for some ancient temperate forest fragments (Dzwonko and Loster 1989; Peterken and Game 1984). In contrast to the situation among many animal species, populations of perennial plants do not normally undergo rapid local extinctions. Rather, they drift gradually to this condition, which, in natural systems, provides an extended period for dispersal and establishment elsewhere. In managed systems, this gradualness provides more time to document the process and, if necessary, to take corrective measures.

Our data also indicate that edge effects can be of very limited extent in stabilized tropical forest patches but can play a significant role in promoting tree species diversity in patches. This contrasts with the depauperizing effect that is frequently ascribed to edge effects. Limited edge effects also imply that the establishment of narrow forest corridors in riparian habitats is likely to be a realistic endeavor; light measurements in the Belize system indicate that corridors as narrow as 30 meters could, in theory, preserve a shady forest core, although widths greater than this would clearly be desirable. Many gallery forests comprise systems of corridors of irregular width, with wide floodplain sectors connected by relatively narrow links through which seed dispersion appears to occur freely.

We emphasize that diverse gallery forest patches do not represent forest microcosms identical to those of continuous forests within the same region. They contain a flora that, while dominated by tropical forest taxa, has undergone some change; certain functional groups of species may have been lost and other new ones may have appeared, notably at forest boundaries. The relative proportions of species have almost certainly changed as has the internal organization of the community.

We do not believe that it is realistic to attempt to preserve intact analogs of continuous forest in systems of small forest patches; this is best done in large forest reserves. Rather, we believe that conservation of biodiversity in small patch systems will be maximized by seeking to move recent fragments as rapidly as possible to a condition that is sustainable in the new biogeographic milieu created by fragmentation. In this new milieu, some species may be lost and others gained, while many others may play roles that differ from those played in continuous forests. Rapid achievement of a resistant boundary community seems to be especially important in ensuring patch persistence, and intervention that even extends to transplantation of suitable taxa to achieve this end would seem to be an appropriate conservation strategy. Gallery forests provide a useful source of information for the planning of such management efforts, as well as a potential seed source for forest species well fitted to the fragmented milieu.

Acknowledgments

We are grateful to the Belize Forestry Department for permission to work in the Mountain Pine Ridge Forest Reserve; the assistance of Earl Green, Chief Forest Officer, has been especially appreciated. Augustine Howe has been our continuous and invaluable field assistant in Belize. Our field work in Belize and Venezuela has benefited from the assistance and companionship of many people, among whom we would especially like to mention Alberto Briceño, Andrew MacDougall, Lisandro Pacheco, Lesley Rigg, Judith Rosales, and Ramón Silva. The research has been supported by grants (to M. K.) from the National Geographic Society and the Natural Sciences and Engineering Research Council of Canada.

References

Adam, P. 1992. *Australian Rainforests.* Oxford, UK: Clarendon Press.

Balslev, H., J. Luteyn, B. Ollgaard, and L. B. Holm-Nielsen. 1987. "Composition and structure of adjacent unflooded and floodplain forest in Amazonian Ecuador." *Opera Botanica* 92: 37–57.

Beard, V. S. 1955. "A note on gallery forest." *Ecology* 36: 339–340.

Bell, G., M. J. Lechowicz, A. Appenzeller, M. Chandler, E. DeBlois, L. Jackson, B. Mackenzie, R. Preziosi, M. Schallenberg, and N. Tinker. 1993. "The spatial structure of the physical environment." *Oecologia* 96: 114–121.

Bierregaard, R. O., T. E. Lovejoy, V. Kapos, A. A. dos Santos, and R. W. Hutchings. 1992. "The biological dynamics of tropical rainforest fragments." *BioScience* 42: 859–866.

Brothers, T. S., and A. Springarn. 1992. "Forest fragmentation and alien plant invasion of Central Indiana old-growth forests." *Conservation Biology* 6: 91–100.

Campbell, D. G., D. C. Daly, G. T. Prance, and U. N. Maciel. 1986. "Quantitative ecological inventory of terra firma and varzea tropical forest on the Rio Xingu, Brazilian Amazon." *Brittonia* 38: 369–393.

Connell, J. H. 1978. "Diversity in tropical rain forest and coral reefs." *Science* 235: 167–171.

de Oliveira-Filho, A. T., J. A. Ratter, and G. J. Shepherd. 1990. "Floristic composition and community structure of a Central Brazilian gallery forest." *Flora* 184: 103–117.

Dezzio, N. M. (ed.) 1994. *Ecología de la Altiplanice de la Gran Sabana (Guyana Venezolana). I Investigaciones Sobre la Dinamica Bosque-Sabana en el Sector SE: Subcuencas de los Rios Yuruani, Arabopo y Alao Kukenan.* Scientia Guaianae Vol. 4, *Caracas.*

Dzwonko, Z., and S. Loster. 1989. "Distribution of vascular plant species in small woodlands of the Western Carpathian foothills." *Oikos* 56: 77–86.

Gibbs, P. E., and H. F. Leitao Filho. 1978. "Floristic composition of an area of gallery forest near Mogi Guacu, State of Sao Paulo, S. E. Brazil." *Revista Brasiliera de Botanica* 1: 151–156.

Gilbert, F. S. 1980. "The equilibrium theory of island biogeography: Fact or fiction?" *Journal of Biogeography* 7: 299–335.

Goodman, D. 1987. "Consideration of stochastic demography in the design and management of biological reserves." *Natural Resources Modelling* 1: 205–234.

Hopkins, M. S., J. Ash, A. W. Graham, J. Head, and R. K. Hewett. 1993. "Charcoal evidence of the spatial extent of the *Eucalyptus* woodland expansions and rainforest contractions in North Queensland during the late Pleistocene." *Journal of Biogeography* 20: 357–372.

Hubbell, S. P., and R. Foster. 1986. "Biology, chance and history and the structure of tropical rain forest communities." In *Community Ecology,* edited by T. J. Chase and J. Diamond, pp. 314–329. New York: Harper and Row.

Huston, M. 1979. "A general hypothesis of species diversity." *American Naturalist* 113: 81–101.

Hutchinson, I. 1977. *Ecological Modelling and the Stand Dynamics of* Pinus caribaea *in Mountain Pine Ridge, Belize.* Ph.D. thesis, Simon Fraser University, Burnaby B.C.

Junk, W. J. 1989. "Flood tolerance and tree distribution in Central Amazonian floodplains." In *Tropical Forests: Botanical Dynamics, Speciation and Diversity,* edited by L. B. Holm-Nielson, I. C. Nielson, and H. Balslev, pp. 47–64. London: Academic Press.

Kapos, V. 1989. "Effect of isolation on the water status of forest patches in the Brazilian Amazon." *Journal of Tropical Ecology* 5: 173–185.

Kellman, M. 1975. "Evidence for Late-Glacial age fire in a tropical montane savanna." *Journal of Biogeography* 2: 57–63.

Kellman, M., and R. Tackaberry. 1993. "Disturbance and tree species coexistence in tropical riparian forest fragments." *Global Ecology and Biogeography Letters* 3: 1–9.

Kellman, M., R. Tackaberry. N. Brokaw, and J. Meave. 1994. "Tropical gallery forests." *National Geographic Research and Exploration* 10: 92–103.

Kershaw, A. P. 1974. "A long continuous pollen sequence from northeastern Australia." *Nature* 251: 222–223.

Kershaw, A. P. 1976. "A Late Pleistocene and Holocene pollen diagram from Lynch's Crater, northeastern Queensland, Australia." *New Phytologist* 77: 469–498.

Laurance, W. F. 1991. "Edge effects in tropical forest fragments: Application of a model for the design of nature reserves." *Biological Conservation* 57: 205–219.

Leyden, B. W. 1984. "Guatemalan forest synthesis after Pleistocene aridity." *Proceedings of the National Academy of the USA* 81: 4856–4859.

Lovejoy, T. E., R. O. Bierregaard, A. B. Rylands, J. R. Malcolm, C. E. Quintela, L. H. Harper, K. S. Brown, A. H. Powell, G. V. N. Powell, H. O. R. Schubart, and M. B. Hays. 1986. "Edge and other effects of isolation on Amazon forest fragments." In *Conservation Biology, the Science of Scarcity and Diversity,* edited by M. Soule, pp. 257–285. Sunderland, MA: Sinauer.

MacArthur, R. H., and E. O. Wilson. 1967. *The Theory of Island Biogeography.* Princeton, NJ: Princeton University Press.

MacDougall, A., and M. Kellman. 1992. "The understory light regime and patterns of tree seedlings in tropical riparian forest patches." *Journal of Biogeography* 19: 667–675.

Meave, J. 1991. *Maintenance of Tropical Rain Forest Plant Diversity in Riparian Forests of Tropical Savannas.* Ph.D. Thesis, York University, Toronto.

Meave, J., and M. Kellman. 1994. "Maintenance of rain forest diversity in riparian forests of tropical savannas: Implications for species conservation during Pleistocene drought." *Journal of Biogeography* 21: 121–135.

Peterken, G. F., and M. Game. 1984. "Historical factors affecting the number and distribution of vascular plant species in the woodlands of Central Lincolnshire." *Journal of Ecology* 72: 155–182.

Pickett, S. T. A., V. T. Parker, and P. L. Fiedler. 1992. "The new paradigm in ecology: Implications for conservation biology above the species level." In *Conservation Biology, the Theory and Practice of Nature Conservation, Preservation and Management,* edited by P. L. Fiedler and S. K. Jain, pp. 65–88. New York: Chapman and Hall.

Quinn, J. F., and G. R. Robinson. 1987. "The effects of experimental subdivision on flowering plant diversity in a California annual grassland." *Journal of Ecology* 75: 837–856.

Rankin-de-Merona, J. M., R. W. Hutchings, and T. E. Lovejoy. 1990. "Tree mortality and recruitment over a five-year period in undisturbed upland rainforest of the Central Amazon." In *Four Neotropical Rainforests,* edited by A. H. Gentry, pp. 573–584. New Haven, CT: Yale University Press.

Rull, V. 1991. *Contribución a la Paleoecología de Pantepui y la Gran Sabana (Guyana Venezolana): Clima, Biogeografía y Ecología.* Scientia Guaianae Vol. 2, *Caracas.*

Russell-Smith, J. 1991. "Classification, species richness and environmental relations of monsoon rainforest vegetation in the Northern Territory, Australia." *Journal of Vegetation Science* 2: 259–278.

Schoener, T. W., and D. A. Spiller. 1987. High population persistence in a system with high turnover." *Nature* 330: 474–477.

Swaine, M. D., D. Lieberman, and F. E. Putz. 1987. "The dynamics of tree populations in tropical forest: A review." *Journal of Tropical Ecology* 3:359–366.

Tackaberry, R. 1992. *Patterns of Tree Species Richness in Fragmented Riparian Forests of Neotropical Savannas.* M.Sc. Thesis, York University, Toronto.

Terborgh, J. 1992. "Maintenance of diversity in tropical forests." *Biotropica* 24: 283–292.

Uhl, C., and R. Buschbacher. 1985. "A disturbing synergism between cattle ranch burning practices and selective tree harvesting in the eastern Amazon." *Biotropica* 17: 265–268.

Uhl, C., and J. B. Kauffman, 1990. "Deforestation, fire susceptibility, and potential tree responses to fire in the Eastern Amazon." *Ecology* 71: 437–449.

Uhl, C., J. B. Kauffman, and D. L. Cummings. 1988. "Fire in the Venezuelan Amazon 2: Environmental conditions necessary for forest fires in the evergreen rainforest of Venezuela." *Oikos* 53: 176–184.

Vaz, J. E., and J. G. Miragaya. 1989. "Thermoluminesence dating of fossil sand dunes in Apure, Venezuela." *Acta Cientifica Venezolana* 40: 81–82.

Whitmore, T. C. 1989. "Canopy gaps and the two major groups of forest trees." *Ecology* 70: 536–538.

Williams-Linera, G. 1990. "Origin and development of forest edge vegetation in Panama." *Biotropica* 22: 235–241.

Williamson, M. 1989. "The MacArthur and Wilson theory today: True but trivial." *Journal of Biogeography* 16: 3–4.

Chapter 4

Managed Forest Patches and the Diversity of Birds in Southern Mexico

Russell Greenberg

Introduction

Anyone who has traveled much in the tropics knows that most areas that have been cleared of their native forest are not expanses of treeless farm fields, but complex mosaics of field, shrub, and wooded vegetation. In recently occupied areas, fragments of the original forest can be found in increasingly inaccessible sites. As sites become settled longer, the haphazard patches of degraded woods give way to a more thoroughly managed landscape. In these settled landscapes the continued existence of wooded vegetation owes itself to the value placed on it by local farmers and ranchers. Because wooded vegetation is used, the structure and composition of the landscapes are partially the result of selective removal or deliberate planting of particular species of trees.

Many aspects of the floristics and vegetative structure of these habitats are affected by human decision making. Managed forest patches (MFPs) in settled agricultural areas can no longer be viewed as fragments of the original forest ecosystem, but as unique and usually simplified forest ecosystems. Although individually these habitats are depauperate, taken together they may contribute significantly to the protection of biodiversity in a developed region. A number of recent studies have examined bird use of individual managed forest patch types, such as shade coffee or cacao plantation (Alguilar-Ortiz 1982; Canaday 1991; Robbins et al., 1992), riparian or agricultural border strips, or pasture shade patches (Greenberg 1992; Greenberg and Salgado 1994; Hutto 1980; Powell et al., 1992; Saab and Petit 1992; Salgado 1993; Villasenor 1993; Warkentin et al., in press), acacia woodlots, (Greenberg et al., in press), degraded forests (Johns 1992; Rappole and Morton 1985; Thiollay 1992), and forest "islands" (Bierregaard 1986; Robbins et al., 1987). However, there has been little by way of comprehensive overview of the different MFPs.

This chapter will examine the diversity of different MFPs and how they may combine to provide habitat for a higher diversity of birds than any would protect individually. I will focus on eastern Chiapas, an area where we have been conducting research over the past four years. This regional focus should not obscure the fact that most of the land uses and associated MFPs are typical of the entire northern Latin American region and can be found in other tropical areas as well. I will define the major forest patches and then compare the different MFPs for several community attributes, including overall diversity, similarity with one another as well as agricultural and forest habitats, abundance and diversity of migratory versus resident species, and seasonality of use (particularly during the dry season).

Methods

The methodology of this project has been described in detail in other publications (Greenberg 1992; Greenberg et al., in press; Warkentin et al., in press). I will present a brief overview to make the following discussion more understandable.

We employed two different types of surveys to address different aspects of habitat use. The first survey types were standard fixed-radius-point counts, which are conducted for ten minutes and all birds detected within 25 meters tallied. Points were generally surveyed once—during the dry season—and located along transects through appropriate habitat at a minimum interpoint distance of 200 meters. Because the focus of the research was on small habitat patches, a small radius circle was selected to ensure that detected birds were within the area considered. Point counts were conducted in the major natural vegetation types (table 4-1) as well as in the major agricultural habitats and managed for-

Table 4-1

The number of individual and species of migrants per point and overall species richness in acacia groves and other habitats in Chiapas based on point counts.[a]

Habitat	N^b	Migrant species[c]	Migrant Individuals per point	Total Individuals per point	Total Diversity[d]	Trend[e]
Acacia woodlot	73	18	9.0	11.8	50.8	0
Low-elevation pasture	70	3	0.7	4.1	38.0	0
Low-elevation 2nd growth	100	6	1.3	6.2	68.7	0
Low-elevation gallery	187	13	4.7	12.1	65.0	
Low-elevation forest	102	5	1.2	10.0	81.6	0
Midelevation pasture	70	5	0.7	3.2	46.4	0
Midelevation 2nd growth	100	4	2.1	3.6	49.9	0
Midelevation milpa	70	8	1.2	4.4	56.1	–
Shade coffee[f]	212	14	4.9	12.3	66.2	+
Midelevation gallery	52	13	5.1	7.7	56.8	0
Midelevation pine–oak liq.	100	3	1.3	2.6	56.0	
Midelevation pine–oak	82	7	3.2	13.2	68.0	0
Midelevation pine	70	3	1.1	3.0	37.0	
Midelevation tropical forest	80	3	1.1	3.2	75.6	
High-elevation milpa	70	3	1.1	2.6	29.8	
High-elevation 2nd growth	100	2	1.3	2.5	39.5	
High-elevation pine	70	4	1.8	4.1	27.0	
High-elevation pine–oak	50	4	1.3	2.4	40.0	

[a]Significant seasonal trends in overall bird numbers based on repeated transect censuses of same habitats.

[b]Number of points.

[c]Includes only those species with mean > 0.10 per point.

[d]Diversity is based on rarefaction analysis for randomly selected 225 individuals per habitat.

[e]Seasonality is represented by the results of a regression analysis of bird number versus week from late wet season to mid dry season. Results are based, in most cases, on two-year averages for repeated transect census: + denotes significant positive trend, 0 denotes no significant trend, and – indicates significant negative trend.

[f]Rustic and planted-canopy plantations are pooled.

est patches. Diversity was compared using rarefaction (James and Rathbun 1981), which calculates the expected number of species from randomly selected resamples of the census data for comparable numbers of individuals. For the second survey technique, we conducted weekly surveys along 1-kilometer by 40-meter transects, most of which were censused over two temperate zone winters. (The transects allow us to examine the seasonality of bird population changes, as well as construct a more detailed map of habitat use, by accumulating observations in transect grid units over time.) During these surveys we recorded other information about the birds, including the perch height and plant height.

The Eastern Chiapas Landscape

The study was conducted in eastern Chiapas, between the Ucimacinta and Lacantun Rivers (2–300 meters above sea level) and the Central Mesita (San Cristobal de las Casas and vicinity, 2,000–2,500 meters above sea level). The natural vegetation of the region encompasses a complex mosaic of tropical and temperate ecosystems including lowland moist tropical forest, pine–oak woodland, montane tropical forest, and pine–oak liquidamber (Breedlove 1981). Superimposed on this ecological complexity is a diversity of human land-use patterns. The midelevation and highland areas are heavily settled with dominant land use in cattle, milpa agriculture, and coffee production. Much of the land is under the control of indigenous *ejidos,* (community-managed lands) but large estates can be found in the valley floors. Lowland areas have mostly been converted to cattle pasture (Calleros and Brauer 1983), with a large tract of forest remaining in the Montes Azules Biosphere Reserve. Within this protected region are small tracts managed in traditional manner by Lacandon Indians (De Vos 1988). Lowland areas of Chiapas have been colonized in successive waves described by De Vos (1988). Therefore, this region encompasses an array of human-influenced landscapes (see figure 4-1), ranging from the indigenous management patterns described by Alcorn (chapter 12, this volume), to recently colonized areas and intensively managed valley land discussed by Schelhas (chapter 13, this volume).

Scattered Trees, Shade Patches, and Living Fences

Before discussing the role of MFPs, I would like to explore the importance of trees in the tropical agricultural landscape at a fine scale. Scat-

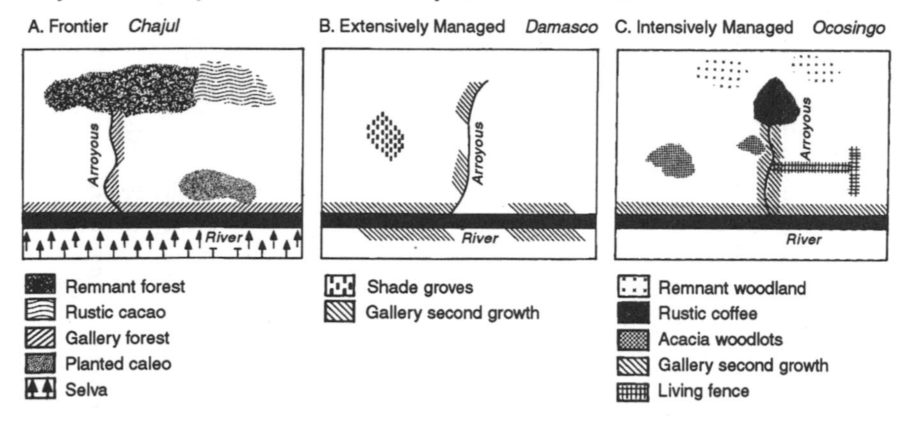

Figure 4-1

Schematic diagram of managed forest patches in three different landscapes of eastern Chiapas.

tered trees or small patches are commonly found in pastures and croplands throughout the region. In some areas, trees are managed in living fences (Sauer 1977) or as border strips to fields. The contributions of hedgerows and other managed patches of trees and shrubs have received some attention in the temperate zone but little research in tropical areas, where such features are particularly common (but see Greenberg 1992; Saab and Petit 1992; Villasenor 1993). In our research in the Selva Lacandona, we censused 4 kilometers of pasture and milpa on a weekly basis over two winters. During these censuses we recorded whether the birds we detected were found in shrubby vegetation (defined as plants shorter than 5 meters) or trees (plants taller than five meters). At the same time we recorded vegetation height every 10 meters (N = 400) and determined that tree cover was found at 3 percent of the area surveyed. As table 4-2 indicates, almost all species of birds were found in trees far more frequently than would be expected from their availability. We divided the species (with more than 10 observations) into three groups, depending on the proportion of total observations in which plant type was recorded when the bird was sighted in a tree (0–40 percent, 41–70 percent, 70+ percent). Roughly one-third of the 91 species were found in each group, so that 35 percent were found in trees over 70 percent of the time. This gradient of tree use is particularly steep for resident species, with 42 percent occurring in trees over 70 percent of the time. Only 19 percent of the migratory

Table 4-2

Percentage use of trees in pasture and milpa by birds in the Lacandon forest (Chajul). Number of observations; * Denotes Nearctic migrants; F is species that is commonly frugivorous.

Low (0–40 %)	Moderate (41–70 %)	High (71 + %)
Ruddy ground dove 73	Plain chachalaca 16 F	Scaled pigeon 11 F
Columbina talpacoti	*Ortalis vetula*	*Columba speciosa*
Groove-billed ani 77	Pale-vented pigeon 45 F	Aztec parakeet 43
Crotophaga sulcirostris	*Columba cayennensis*	*Aratinga astec*
Little hermit 33	White-crowned parrot 10	Long-tailed hermit 10
Phaethornis longuemareus	*Pionus senilis*	*Phaethornis superciliosus*
Rufous-tailed hummingbird 134	Mealy parrot 17	Keel-billed toucan 22 F
Amazilia tzacatl	*Amazona farinosa*	*Ramphastos sulfuratus*
Barred antshrike 27	White-bellied emerald 35	Black-cheeked woodpecker 21 F
Thamnophilus doliatus	*Amazilia candida*	*Melanerpes Pucherani*
Common tody flycatcher 16	Green-breasted mango 10	Golden-olive woodpecker 16
Todirostrum cinereum	*Anthracothorax prevostii*	*Piculus rubiginosus*
Great-crested flycatcher *10	Ruby-throated hummingbird *11	Linneated woodpecker 20
Myiarchus crinitus	*Archilochus colubris*	*Dryocopus lineatus*
White-throated flycatcher 10	White-collared manakin 12 F	Red-lored parrot 20
Empidonax albigularis	*Manacus candei*	*Amazona viridigenalis*
Least flycatcher *434	Red-capped manakin 13 F	Yellow-bellied flycatcher 41
Empidonax minimus	*Pipra mentalis*	*Empidonax flaviventris*

Spot-breasted wren 34	Greenish elaenea 13	Brown-crested flycatcher 28
Thryothorus maculipectus	*Myiopagis viridicata*	*Myiarchus tyrannulus*
House wren 54	Yellow-bellied elaenia 44 F	Social flycatcher 73 F
Troglodytes aedon	*Elaenia flavogaster*	*Myozetetes similis*
Gray catbird * 375 F	Yellow-olive flycatcher 16	Boat-billed flycatcher 20 F
Dumatella carolinensis	*Tolmomyias sulphurescens*	*Megarynchus pitangua*
White-eyed vireo *14 F	Dusky-capped flycatcher 59	Sulphur-bellied flycatcher 10 F
Vireo griseus	*Myiarchus tuberculifer*	*Myodynastes luteiventris*
Nashville warbler *32	Tropical pewee 30	Tropical kingbird 90 F
Vermivora ruficapilla	*Contopus cinereus*	*Tyrannus melancholicus*
Magnolia warbler * 237	Tennessee warbler * 30	Masked tityra 29 F
Dendroica magnolia	*Vermivora peregrina*	*Tityra semifasciata*
Ovenbird * 28	Yellow warbler * 313	Black-crowned tityra 10 F
Seiurus aurocapillus	*Dendroica petechia*	*Tityra inquisitor*
Common yellowthroat * 628	Myrtle warbler * 11 F	Brown jay 80 F
Geothlypis trichas	*Dendroica coronata*	*Psilorhinus morio*
Hooded warbler * 15	American redstart * 189	Blue-gray gnatcatcher * 93
Wilsonia citrina	*Setophaga ruticilla*	*Polioptila cacrulea*
Wilson's warbler * 301	Bananaquit 37	Tropical gnatcatcher 10
Wilsonia pusilla	*Coereba flaveola*	*Polioptila plumbea*
Yellow-breasted chat * 224 F	Red-throated ant tanager 33	Clay-colored robin 51 F
Icteria virens	*Habia fuscicauda*	*Turdus grayi*
Crimson collared tanager 26 F	Olive-backed euphonia 19 F	Lesser greenlet 75

(continues)

Table 4-2 (continued)

Low (0–40 %)	Moderate (41–70 %)	High (71 + %)
Phlogothraupis sanguinolenta	*Euphonia gouldi*	*Hylophilus decurtatus*
Scarlet-rumped tanager 185 F	Black-headed saltator 19 F	Chestnut-sided warbler * 30
Rhamphocelus passerinii	*Saltator atriceps*	*Dendroica pensylvanica*
Grayish saltator 14 F	Buff-throated saltator 110 F	Black-and-white warbler * 25
Saltator coerulescens	*Saltator maximus*	*Mniotilta varia*
Blue-black grosbeak 14 F	Melodious blackbird 64	Masked tanager 14F
Cyanocompsa cyanoides	*Dives dives*	*Tangara larvata*
Blue grosbeak * 37	Baltimore oriole * 56 F	Red-legged honeycreeper 15 F
Guiraca caerula	*Icterus galbula*	*Cyanerpes cyaneus*
Indigo bunting * 221 F		Yellow-throated euphonia 44 F
Passerina cyanea		*Euphonia hirundinaceae*
Green-backed sparrow 12		Blue-gray tanager 72 F
Arremonops chlorontous		*Thraupis episcopus*
Blue-black grassquit 416		Yellow-winged tanager 17 F
Volatina jacarina		*Thraupis abbas*
Variable seedeater 56		Summer tanager * 30
Sporophila aurita		*Piranga rubra*

White-colored seedeater 416 *Sporophila torqueloa*			Black-faced grosbeak * 10 F *Caryothraustes poliogaster*
Thick-billed seed finch 63 *Oryzoborus funereus*			Rose-breasted grosbeak * 39 F *Pheucticus ludovicianus*
Lincoln's sparrow * 28 *Melospiza lincolnii*			Chestnut-headed oropendola 28 F *Psarocolius waglen*
Yellow-billed cacique 10 *Amblycercus holosericeus*			
Orchard oriole * 248 F *Icterus spurius*			
Totals			
Species	33	25	32
Migrants	15	6	6
Residents	19	19	28
Partial frugivores	10	10	20

species occurred in the highest tree-use group and 55 percent were found in trees less than 40 percent of the time. This does not mean that migrants prefer shrubby vegetation, since the tree cover was only 4 percent; but migrants are less specialized in their use of trees. This simple exercise emphasizes what is obvious to anyone watching birds in tropical agricultural areas—trees, even a few scattered individuals, are the nucleus of activity for most avian species.

The functional composition of the tree-specialist group is quite different from those species less dependent on trees. Sixty-three percent of the specialists are commonly frugivorous—and hence could potentially disperse seeds of trees into pastures from edge vegetation. In contrast, only 29 percent of the least specialized and 40 percent of the intermediate group, commonly eat fruit. Thus isolated trees may play a key role in the establishment of woody vegetation in tropical pastures and farm fields (Guevara, Purata, and van der Maarel 1986).

The presence of trees, at a fine scale of analysis, contributes significantly to the diversity of bird species found in a pasture or field. This increase in local diversity can be demonstrated by examining the number of species found over the 36 weeks of censusing in each of 80 transect units censused in pasture and milpa (figure 4-2). Diversity increases with the proportion of the unit covered in woody vegetation. We analyzed the relationship between the number of individuals seen and the proportion of woody vegetation for the common migratory species, and all species, except the common yellowthroat (*Geothlypis trichas*), showed a positive relationship. This includes several species that at a larger scale of analysis are consistently found in open habitats rather than in forest (e.g., least flycatcher, *Empidonax minimus,* gray catbird, *Dumatella carolinensis,* yellow-breasted chat, *Icteria virens*).

Arroyo Vegetation

In eastern Chiapas, woody vegetation, which in this paper is refered to as arroyo vegetation, is most abundant along stream corridors (Salgado 1993; Warkentin et al., in press). Arroyo vegetation, often a very narrow band or line of trees along straight stretches, can widen into patches at the meeting of tributaries or ox boxes. It usually consists of second-growth species of medium to tall moist tropical forest (such as *Ficus, Trichlospermum, Cecropia, Spondias, Inga, Trema,* etc.). Rarely found in recently settled areas, arroyo vegetation still includes large relict trees from the original rainforest (also see Powell et al., 1992).

The structure of arroyo vegetation varies considerably with local management. For example, we found that communities along the La-

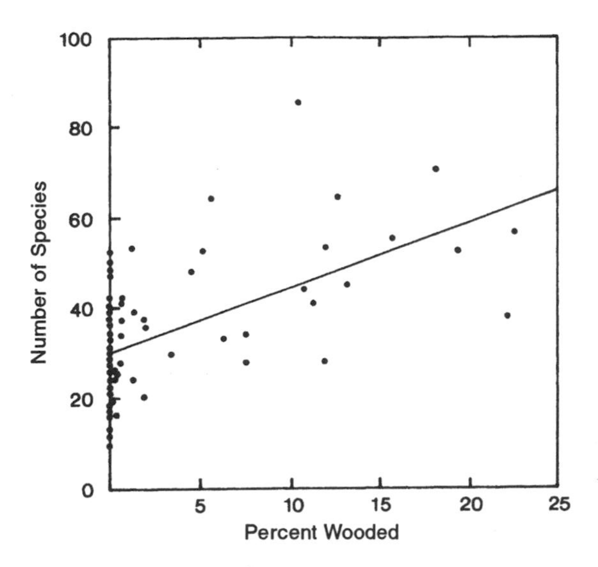

Figure 4-2

The number of species of birds recorded on individual transect units (40 by 50 meters) in milpa, pasture, or young pasture second growth in Chajul during the years 1989–1991 (cumulative over 36 censuses per transect).

cuntun River in the Marquez de Comillas had distinct "styles" of arroyo vegetation management—presumably due to the varied experiences that each community's members brought with them from their places of origin (different *ejidos* within the same area are comprised of people from different states in western Mexico, as well as from the highlands of Chiapas). There are three types of arroyo management: (1)The original forest is protected in a band along the arroyo ("forest" vegetation); (2) relict forest trees are left but the understory is removed and replaced by pasture ("shade" vegetation); and (3) second growth is allowed to grow after the original clearing ("scrub" arroyos). These different management styles are apparent in recently cleared areas but generally disappear in older landscapes as the arroyo vegetation becomes increasingly dominated by fast-growing pioneer species characteristic of tropical second growth.

Although each of these arroyo habitats has characteristic avifaunas, there is also considerable similarity. Therefore, further research on the impact of the different management styles is needed, because improving management offers a way of increasing the diversity of bird populations without taking additional lands out of production. For example, at a very fine scale (within one arroyo system), Salgado (1993) was able to relate the cumulative diversity of birds over a winter of surveys with the structural complexity and composition of the arroyo vegetation (figure 4-3).

By way of an overview, arroyo vegetation is characterized by a high density (four to five individuals per point), moderate diversity of Nearctic migrants (11 common species) and a moderate diversity of resident species. Certain forest generalist migrants are common (i.e., magnolia

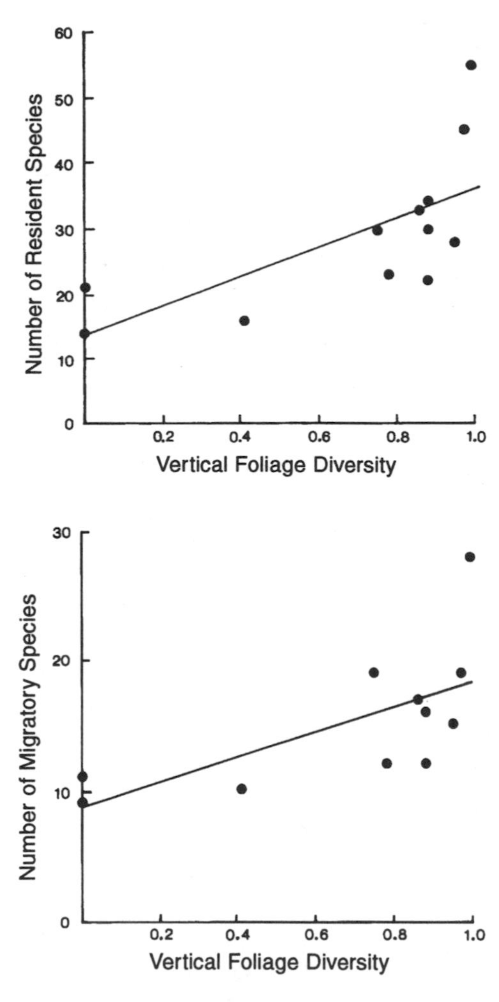

Figure 4-3

Relationship between migrant and resident diversity and foliage height diversity for transects in arroyo vegetation in cattle pastures in Chajul.
Source: from Salgado 1993.

warbler, *Dendroica magnolia,* and the american redstart, *Setophaga ruticilla*); however, with the exception of the yellow-bellied flycatcher, forest specialist migrants are rare (Warkentin et al., in press). Resident species are a mix of open country and woodland edge species, with relatively few forest specialists. However, small numbers of forest species persist in areas that have been cleared and isolated for decades (Warkentin et al., in press). The status of these "populations" needs further investigation. Still, gallery forest is most similar in species abundance and composition to other MFPs and lowland tropical second growth (Greenberg et al., ms.).

Shade Plantation

Several commodity crops in Chiapas are grown under a canopy of shade crops. The most important of these in terms of acreage is coffee, with cacao (for chocolate) a distant second. Because these crops are restricted to the understory, the canopy can vary considerably depending on the management style of the farmer. A coffee plantation can be anything from a patch of native forest that has had understory vegetation removed to a larger field with little or no shade. Most coffee production in Mexico falls between these extremes (see Flores-Fuentes 1979 for a classification scheme). In the managed forest (rustic) model, coffee production is part of a diverse agroforestry system using numerous household products. Another common form of coffee cultivation (planted) involves the management of a monoculture of a nitrogen-fixing legume (primarily *Inga* species) with useful trees, particularly fruit trees, inserted in the understory or lower canopy. In Chiapas, coffee plantations are dominated by *Inga,* with *Erythrina* used for cacao plantations found on the gulf slope.

Eastern Chiapas is outside of the main coffee-growing areas of Mexico (more important areas are the Sierra Madre of Chiapas and the mountains of Vera Cruz); therefore the more traditional approaches to coffee production on small forest plots persists more so there than in other areas. However, since the late 1970s, with the decline in maize prices, high coffee prices, and the aggressive role of INMECAFE (former Mexican National Coffee Institute), large amounts of new coffee plantation land have been established, which has required considerable investment and modernization of the cultivation system (Marquez 1988; Rice 1993).

We have studied the avifauna of rustic and planted shade plantations, as they exist in close proximity, and have found that both systems support a relatively high diversity of bird species. The cumulative total of species found on transects through both plantation types is 180. More controlled comparisons, using rarefaction techniques to compare equivalent sample sizes, showed that shade coffee supports numbers of total bird species similar to most forest habitats in eastern Chiapas, with the exception of moist tropical forest (table 4-1; Greenberg et al., in press.). Both production systems, planted and rustic, supported a similar diversity of species (110 species in approximately 100 point counts). However, there were large differences in guild (functional) composition, with rustic plantations having high densities of canopy frugivores and planted plantations supporting high densities of canopy omnivores and nectarivores.

Migratory species in rustic and planted shade plantations are both common and diverse compared to those found in other habitats in the region. We detected an average of 5 individuals within a 25-meter radius on 10-meter point counts (table 4-2), which is higher than that which was found in arroyo vegetation and second only to acacia woodlots. The number of species of migrants detected was 40, of which 14 were found at frequencies of over 0.10 per point. More important for migratory bird conservation, forest migrants commonly occur with 5 species (2 specialist) found at frequencies greater than 0.1 per point (Greenberg et al., in press).

The overall composition is most similar to the other managed forest patches of the midelevation zone (arroyo vegetation and acacia woodlots), but it is also similar to pine–oak and pine–oak liquidamber woodlands and secondary vegetation of lowland moist forest. The most distinctive feature of the coffee plantation avifauna is its high degree of seasonality (table 4-2; Greenberg et al., ms.). Coffee plantations have almost twice the total number of individuals and species during the dry season than during the late rainy season. This increase, found among migrants and residents, is restricted to omnivorous, frugivorous, and nectarivorous species. Other habitats in the region, however, show slight and usually insignificant declines in overall numbers during this period. Coffee plantations, therefore, appear to be a dry season refuge for mobile species (both Nearctic and local migrants).

Little information exists on cacao production. Other studies (Robbbins et al., 1992) have reported a higher diversity of birds in cacao plantations than coffee plantations. This difference is perhaps a result of the higher potential species pool in the lower elevation zone, where cacao is grown. We have censused small tracts of rustic cacao that has been introduced in the past decade to the Marquez de Comillas and found that for resident species this habitat was similar to the high-graded remnant forest and a depauperate version of undisturbed tropical forest. A similar result was found for rustic coffee plantations adjacent to undisturbed forest in Amazonian Ecuador (Candaday 1991). However, migrant densities were high, and the habitat supported the highest number of forest-specialist migrants of any habitat surveyed.

Acacia Woodlots

Mimosoid legumes are traditionally managed in dryer areas of Mexico, largely in midelevation valleys (Wilken 1977, 1987). In eastern Chiapas, *Acacia pennatula* is a particularly important species. Stands of acacia, known as huizichales, or, more commonly in Chiapas, ka'an chichales, are managed on grazing land at 500–1,000-meter elevations. The species occurs in disturbed portions of the live oak woodland and savanna, and is uncommon in areas dominated by mesophilous forest or pine. Acacia stands are also common in other midelevation areas. The distribution, management, and commercial value of these woodlots was thoroughly researched for the Xalapa area by Chazaro (1977) and the ecology of these woodlots has been discussed by Greenberg et al. (manuscript).

The management of acacia in the pastures of Ocosingo Valley represents one of the most advanced silvopastoral systems in eastern Chiapas. The origins of the system are unclear, but the woodlots are at least 20 to 30 years old, which predates the intensification of cattle production in the valley. The pastures of the large ranches of the valley are small enclosures of 2 to 5 hectares. Grazing is rotated between pastures, as is the growing of maize, cutting grass, and cane for cattle feed. Usually one or more of the paddocks is covered with acacia. These woodlots, which average 3.7 hectares and range from 0.5 to 20 hectares, are used primarily for fence post and other building materials. Cattle and horses are introduced into the woodlots primarily in March and April when the trees produce pods, which are a nutritious feed for the cattle. The livestock can eat the fallen pods from the woodlot floor.

Acacia woodlots support only a modest overall diversity of birds (table 4-2), similar to agricultural habitats such as milpa and scrubby pasture. But it is a particularly important habitat for overwintering Nearctic migrants, particularly small foliage-gleaning species. Larger woodlots, with well-developed understory (limited use by cattle and horses), support high densities of understory species, particularly frugivores such as yellow-breasted chat and gray catbird. We found nearly double the numbers of migratory individuals than in the next most important habitats, shade coffee and gallery woods. Migratory populations are highly diverse, with 18 species recorded at an abundance of 0.1 individuals per point or greater. In contrast, acacia woodlands support some of the lowest populations of nonmigratory species in the region. Most residents, which were associated with the shrubby border of woodlots, were birds characteristic of agricultural fields or young second growth. Acacias lose much of their foliage during the course of the winter and would seem to be more seasonal than the other habitats in the region. Although migratory bird populations declined during the winter on our repeated censuses, they did not do so at a significantly different rate from that of other habitats in the region (except coffee plantations). We examined the stability of populations through the observation of color-banded birds and found a high persistence rate (approximately 75 percent) in canopy species (those that use the acacia), which is the proportion of individuals banded in November and December present in March.

Forest Remnants

Outside of protected areas, most of the remaining forest projects are found in forest remnants in the agricultural landscape. These patches suffer from fragmentation, but more important, they are altered from the natural forest by human use. Often patches of forest are part of cooperatively managed *ejido* lands. They are most commonly used for building material and fuel wood. In lowland areas, the large trees are gradually removed and a subcanopy of fast-growing pioneer species is established under a patchy canopy. At midelevations, the understory is often burned successively to create a savanna for livestock grazing. At high elevations, slow-growing oaks are coppiced forming a chaparral of stunted trees and shrubs.

Of the managed patches, the avifauna of forest remnants most closely resembles that of the original intact forest. However, the impact of management is often difficult to access because few examples of the natural habitat remain. More research in and around protected areas can access the long-term impact of management on forest remnants.

We conducted this type of comparison in the forest remnants of the *ejidos* of Chajul and Loma Bonita in the Marquez de Comillas and the adjacent rainforest of the Montes Azules Reserve. Within 10 years of settlement, most of the large canopy trees had been removed and a dense subcanopy of *Xylopium* and other fast-growing trees dominated the forest. The understory of the forest was considerably less humid than that of the intact forest. Still, the avifauna of the high-graded forest was similar to that of the original forest—with differences found primarily in the abundance of species. Migratory species of lowland forest were found in comparable abundance between the two forest types.

However, large differences were found in the resident species. In particular, the overall abundance of resident forest birds was significantly lower with little compensatory invasion of second-growth species. Although the overall species list between remnant and undisturbed forest was similar, many forest birds were much less common in the disturbed habitat. A few selva species, including plain antvireo (*Dysithamnus mentalis)*, russet antshrike (*Thamnistes anabatinus*), and purple-crowned fairy (*Heliothryx barroti*), were not found in two years of field work in the remnant forests. Of the 151 species censuses on 174 point counts in mature forest and remnant forest, 57 were significantly more common in the former and only 1 (gray catbird) was more common in the latter (table 4-3). The results are similar to other studies of areas of selective logging in tropical forests (Johns 1992; Thiollay 1992). We found that the declining species were ecologically diverse, including understory, canopy, and forest-gap species. The loss of gap species is interesting, since it would seem that high-grading forests would mimic tree-fall gap production, a natural source of habitat for disturbance-oriented species in mature forest (Schemskie and Brokaw 1981). However, the rapid removal of most large trees may not mimic the more patchy and stochastic occurrence of natural gaps.

Since much of the habitat degradation occurred slowly over a 10-year period prior to the study, it is likely that the diversity of birds in the degraded forest was an overestimate of the long-term diversity of this habitat. Low populations of certain species may represent the "living dead"—adults unable to reproduce, a phenomenon well documented in the extinction process on Barro Colorado Island (Willis 1974).

Different Forest Patches in the Landscape Mosaic

Each forest patch type supports a somewhat different avifauna, but together they are complementary in their abilities to protect avian diversity. As figure 4-4 shows, the different types form separate clusters when overall composition is compared. Gallery forests and acacia woodlots have bird communities most similar to tropical second-growth habitats. Both systems have relatively high numbers of migrants, low numbers of resident individuals and species (table 4-2)—particularly canopy insectivores, and are valuable for providing habitat for a number of mi-

Table 4-3

The abundance (mean number of individuals per point) of species showing significant differences ($P < 0.10$) in remnant versus old-growth lowland tropical forest in the vicinity of Boca de Chajul in the Selva Lacandona in 1990–1992.

Species	Remnant forest (101 points)	Old-growth forest (73 points)	*P*	Guild
Great tinamou *Tinamus major*	0.12	0	.011	Ground omnivore
Slaty-breasted tinamou *Crypturellus boucardi*	0.21	0.01	.000	Ground omnivore
White-bellied emerald *Amazilia candida*	0.09	0.19	.087	Canopy nectarivore
Short-billed pigeon *Columba nigrirostris*	0.54	0.04	.000	Canopy frugivore
Gray-chested dove *Leptotila cassinii*	0.12	0.01	.003	Ground granivore
White-tipped dove *Leptotila verreauxi*	0.16	0.01	.000	Ground granivore
Plain chachalaca *Ortalis retula*	0.18	0	.05	Scrub frugivore
Brown-hooded parrot *Pionopsitta haematotis*	0.42	0	.000	Canopy frugivore
Red-lored parrot *Amazona autumnalis*	0.23	0.14	.014	Canopy frugivore
Blue-crowned motmot *Momotus momota*	0.20	0.07	.016	Understory omnivore
Rufous-tailed jacamar *Galbula ruficauda*	0.22	0.06	.003	Canopy insectivore
Slaty-tailed trogon *Trogon massena*	0.15	0.12	.005	Canopy frugivore
Violaceous trogon *Trogon violaceus*	0.49	0.15	.000	Canopy omnivore
Pale-billed woodpecker *Campephilus guatemalensis*	0.15	0	.000	Bark insectivore
Black-cheeked woodpecker *Melanerpes pucherani*	0.21	0.01	.000	Bark insectivore omnivore

Smoky-brown woodpecker *Venilornis fumigatus*	0.11	0	.001	Bark insectivore
Chestnut-colored woodpecker *Celeus castaneus*	0.11	0.01	.017	Bark insectivore
Barred antshrike *Thamnophilus doliatus*	0.36	0	.000	Tree-gap insectivore
Plain antvireo *Dysithamnus mentalis*	0.09	0	.01	Understory insectivore
Dot-winged antwren *Microrhopias quixensis*	0.68	0.23	.001	Canopy insectivore
Dusky antbird *Cercomacra tyrannina*	0.50	0.06	.001	Tree-gap insectivore
Black-faced antthrush *Formicarius analis*	0.69	0.12	.000	Ground insectivore
Ivory-billed woodcreeper *Xiphorhynchus flavigaster*	0.45	0.12	.000	Bark insectivore
Streak-headed woodcreeper *Lepidocolaptes affinis*	0.11	0.03	.000	Bark insectivore
Plain xenops *Xenops minutus*	0.14	0.01	.000	Canopy insectivore
Thrush-like manakin *Schiffornis turdinus*	0.30	0.03	.000	Understory frugivore
Cinnamon becard *Pachyramphus cinnamomeus*	0.06	0.01	.096	Canopy omnivore
Rufous piha *Lipaugus unirufis*	0.25	0.04	.000	Canopy frugivore
Stub-tailed spadebill *Platyrinchus mystaceus*	0.17	0.04	.016	Understory insectivore
Sulphur-rumped flycatcher *Myiobius sulphureipygius*	0.17	0	.000	Understory insectivore
Northern bentbill *Onocostoma cinereigulare*	0.60	0.40	.015	Understory insectivore
Yellow-olive flycatcher *Tolmomyias sulphurescens*	0.22	0.12	.097	Canopy omnivore

(continues)

Table 4-3 (Continued)

Species	Remnant forest (101 points)	Old-growth forest (73 points)	*P*	Guild
Ochre-bellied flycatcher *Mionectes oleagineous*	0.16	0.03	.013	Understory frugivore
Green jay *Cyanocorax yncas*	0.14	0	.008	Canopy Omnivore
White-bellied wren *Uropsila leucogastra*	0.14	0	.001	Understory insectivore
White-breasted wood wren *Henicorhina leucosticta*	0.58	0.21	.000	Tree-gap insectivore
Spot-breasted wren *Thryothorus maculipectus*	0.69	0.21	.000	Understory insectivore
Gray catbird (migrant) *Dumetella carolinensis*	0.08	0.30	.003	Canopy omnivore
Wood thrush *Hylocichla mustelina*	0.34	0.18	.049	Ground omnivore
Long-billed gnatwren *Ramphocaenus melanurus*	0.25	0.14	.086	Understory insectivore
Green shrike vireo *Smaragdonlanius pulchellus*	0.60	0.08	.000	Canopy insectivore
Lesser greenlet *Hylophilus decurtatus*	0.97	0.32	.000	Canopy insectivore
Tawny-crowned greenlet *Hylophilus ochraceiceps*	0.47	0.08	.000	Understory insectivore
Magnolia warbler (migrant) *Dendroica magnolia*	0.35	0.10	.000	Tree-gap insectivore
Golden-crowned warbler *Basileuterus culicivorus*	0.38	0.01	.000	Understory insectivore
Kentucky warbler (migrant) *Orporornis formosus*	0.13	0.06	.087	Ground insectivore

Bananaquit *Coereba flaveola*	0.22	0	.052	Canopy nectarivore
Yellow-billed cacique *Amblycercus holosericeus*	0.33	0	.000	Tree-gap insectivore
Olive-backed euphonia *Euphonia gouldi*	0.28	0.0	.001	Canopy frugivore
Yellow-winged tanager *Thraupis abbas*	0.10	0.03	.081	Canopy omnivore
Red-crowned ant tanager *Habia rubica*	0.76	0.14	.000	Understory omnivore
Blue-black grosbeak *Cyanocompsa cyanoides*	0.32	0.10	.001	Tree-gap omnivore
Black-faced grosbeak *Carothraustes poliogaster*	0.94	0.41	.016	Canopy omnivore
Orange-billed sparrow *Arremon aurantiirostris*	0.23	0.0	.010	Understory omnivore

gratory birds and resident edge species in areas that are otherwise inhospitable for anything but open-country birds. Coffee plantations support a high diversity of both migratory and resident species and are more similar to midelevation woodlands than are the other patch types. Coffee plantations are particularly well represented by omnivorous, frugivorous, and nectarivorous canopy species. Forest remnants and rustic cacao plantations (not included in figure 4-4) support an avifauna that is very similar to tropical forest, but to varying degrees depauperate. The habitats vary in their seasonal role for migratory birds. Plantations are unique in supporting a marked influx of birds during the dry season, particularly frugivorous and nectarivorous species (table 4-2). Other patch types show small, and often not significant, declines in overall bird numbers.

Important Food Plants

Individual trees may be important for the shade and cover they provide (even for species that use shrubbery). In addition, particular tree

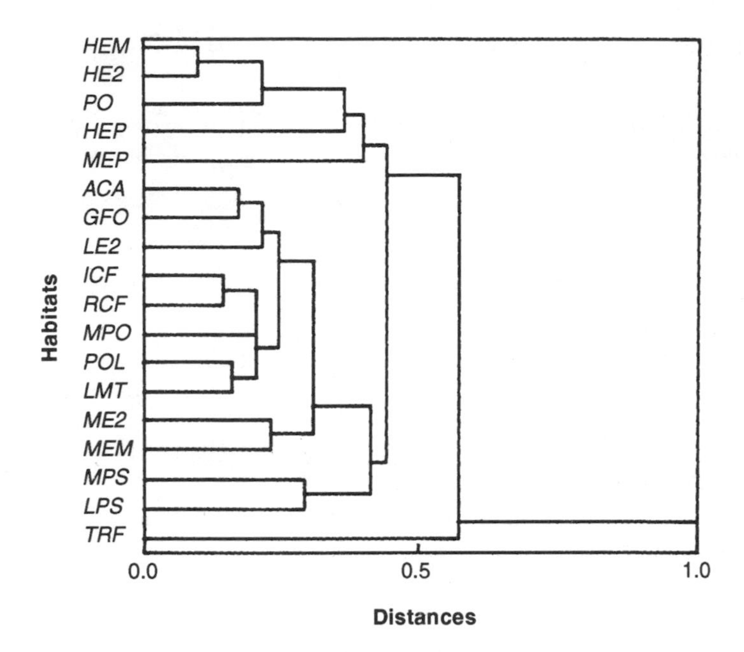

Figure 4-4

Faunal similarity of bird communities in eastern Chiapas based on Pearson correlation between species abundance. HEM: high-elevation milpa, HE2: high-elevation second growth, PO: high-elevation pine oak woodlands, HEP: high-elevation pine, MEP: midelevation pine, Aca: acacia, GFO: gallery forest, LE2: low-elevation second growth, ICF: inga coffee, RCF: rustic coffee, MPO: midelevation pine oak woodlands, LMT: lower montane forest, ME2: midelevation second growth, MEM: midelevation milpa, MPS: midelevation pasture, LPS: low-elevation pasture, TRF: low-elevation tropical forest.

species may provide a critical source of food. In our research we have identified several common trees of human-managed habitats that may play a critical role in providing food resources for either local or long-distance migrants. One striking example is *Inga,* which, when flowering, are visited by large numbers of migrant (Tennessee warblers, orioles) and resident nectarivores. This phenomenon is found to a lesser degree for *Gliricidia.* There are a number of trees that produce fruit that is heavily used by migratory and locally wandering bird populations, but none of them are as dominant a resource as the above-mentioned flowering trees. This suggests that enhancing managed forest patches by encouraging the planting of certain species is a viable strategy for improving habitat for birds. M. Foster (pers. comm.) has identified trees in the Selva Lacandona that support migratory bird populations and could be incorporated into local agroforestry practices.

Managed Forest Patches and Avian Life History

The picture portrayed of MFPs in this chapter is based largely on censuses during the temperate-zone winter months. Investigations of how the use of forest patches affects the life history of the species involved is an important area of future research. This issue has been raised often for species that are migratory from temperate-zone breeding areas. Some authors (Rappole and Morton 1985; Winker et al., 1990) have maintained that many forest-dwelling species of migrants that occur in a variety of secondary or disturbed habitats tend to show lower site tenacity, which results in lower survivorship of individuals. In particular, radio-tracked wood thrushes were shown to wander more and suffer higher predation when outside of undisturbed forests. This is an area of active research; however, I see little reason to assume that most migratory bird species show large depression in survivorship in forest patches. Several studies in shade coffee (Wunderle and Latta 1994), acacia woodlots (Greenberg et al., manuscript), and arroyo vegetation (Salgado Ortiz 1993) have shown that color-marked individuals of a variety of migratory species maintain territories or small home ranges with a high rate of overwinter persistence in MFPs. The species include forest generalists such as American redstart, magnolia warbler, black-throated blue warbler (*Dendroica caerulescens*), and black-and-white warbler (*Mniotilta varia*). In addition, analysis of repeated censuses throughout the temperate-zone winter months indicates that populations of forest migrants in MFPs show declines similar to those found in intact forest (Greenberg, in prep.). Although this does not directly assess the survivorship of individuals, a greatly reduced survival potential in MFPs should be reflected in diminished numbers for the entire pool of individuals using MFPs.

Probably a more critical issue for bird conservation is the reproductive success of resident species attempting to breed in MFPs. Perhaps the forest species detected in MFPs are wandering individuals from forested areas and their presence in surveys offers an overly optimistic estimate of their numbers when remaining forest in the Chiapas region is reduced. It is likely that the management of MFPs may eliminate or disturb the specialized conditions for nest placement for the more specialized tropical forest species. So far, little work has been completed to assess this possible effect.

Developing a Strategy for the Maintenance of Managed Forest Patches

Even with these cautionary notes it appears that at almost all scales of analysis, the maintenance of native arboreous vegetation will enhance the abundance and diversity of birds in the tropical landscape. We emphasize that in terms of managing lands that are already under agricultural use, the appropriate comparison is between a landscape devoid of the managed forest patches described earlier. A strategy for encouraging the maintenance of managed forest patches will not necessarily prevent extinction-prone species from vanishing, but it will enrich the diversity of what is now the majority of lands in northern Latin America (Gradwohl and Greenberg 1991). We believe it is most important to maintain substantial amounts of habitat for Nearctic migratory birds that crowd into the region during the temperate-zone winter. Nearctic birds generally comprise a high proportion of the birds in managed forest patches, particularly of the foliage insectivores. The presence of forest patches, with their high density and diversity of migratory birds, may compensate to a degree for overall forest loss. This may explain why observed declines in migratory birds are not more commensurate with the large amount of forest loss in the northern Neotropics (see also Lynch 1989).

Although managed forest patches may be depauperate when compared to large tracts of less-disturbed habitat, one enticing feature of developing a conservation strategy directed toward MFPs is the potential amount of land involved. Small changes in pasture management or plantation cultivation system can be broadcast over large regions in which a particular land use is found.

The impact of the management of forest patches will be primarily in the following three areas:

1. Migratory birds, even relatively specialized forest species, occur in good numbers in several of the managed forest patch types (see Greenberg et al., in press, for more thorough analysis). If populations are limited by habitat (Sherry and Holmes, in press) then the carrying

capacity of migratory birds should increase as suitable habitat is increased.

2. Nomadic or locally wandering species that depend on fruit or flower crops may find managed forest patches to be valuable stepping stones. This is particularly true when managed forest patches are found along an elevational gradient below or above protected forests (also see Guindon, chapter 9, this volume). Agroforestry systems are flexible enough to allow the incorporation of critical food plants for a large number of birds or for particular rare species, such as quetzales and bellbirds.

3. Local diversity of birds in agricultural areas increases tremendously with the addition of MFPs. Species associated with trees are disproportionately frugivorous or insectivorous and therefore provide ecological services such as reforestation and pest control.

The limits of a strategy based on managed forest patches need to be constantly kept in mind. It is likely that managed forest patches will be of limited value in protecting some of the more specialized members of the old-growth forest avifaunas. These species seem to disappear even in forest remnants (the least systematically managed of all forest patch types).

There is a need to address conservation strategies toward the diversity of MFP types for both biological and sociological reasons. Biologically, we have seen that different MFPs are distinct habitats with different avian composition. Perhaps, more important, each contains a depauperate set of food resources and a simplified structure of natural vegetation types. Because of this, each has a characteristic seasonal pattern of bird use and local movements may be required for many species, particularly Nearctic migrants, to survive through both wet and dry seasons.

From the sociological perspective, different MFPs respond to economic forces operating at different scales, from local to global. Therefore, a conservation strategy aimed at increasing the amount of native arboreous vegetation should be based on a diversity of approaches and simultaneously address the different types of MFPs. For example, gallery forest conservation may be promoted through simple extension efforts that promulgate erosion control. However, the adoption of longer forest fallow periods, the management of woodlots, and the cultivation of shade-based crops all respond to a different constellation of national and global policy and market-driven factors.

In order to increase emphasis on MFPs in future research, I present a simple lexicon for the different types of managed forest patches. These patches differ in how they are managed. A first step in developing a strategy for their conservation is to determine the value these patches might have to local landholders. The initial assessment should shape policies that develop the value of these habitats or inform outreach programs that demonstrate existing values.

Ecological Services

The incorporation of native trees into agricultural systems, whether it is for the protection of isolated trees, riparian corridors, shade canopy for plantations or woodlots, or forest patches for wood production, should provide long-term ecological services for increasing the sustainability of wood for local consumption. The most important services are soil- and water-quality protection. In addition, increased forest cover may increase local rainfall and mitigate drying. Under some circumstances, MFPs can act as fire breaks (Nepstad et al., chapter 7, this volume). Although there has been relatively little research on this topic for tropical areas, it is a common perception of landowners that streamside corridor protection provides the fire-break benefits. Shade plantations, which are often located in stream valleys, and acacia woodlots have the additional characteristic of being dominated by nitrogen-fixing trees, contributing to long-term soil fertility.

In addition, the birds themselves may provide benefits that are seldom quantified. Foremost is their potential role in the control of pest-insect populations. Although some farmers still perceive grain-eating birds as potential pests themselves, in fact, most of the species that occur in managed forest patches—even isolated trees—are insectivorous and few are specialized granivores. One of the few studies of the impact of migratory birds on arthropods in forest patch habitats found that birds can reduce the density of arthropods by as much as 75 percent (Greenberg and Salgado Ortiz 1994).

Role in the Global Economy

In contrast to the benefits mentioned above, which occur at the local level, it is becoming increasingly clear that national policy and global economic forces profoundly affect the existence and management of even the smallest managed forest patches. The influence of these broader forces should be apparent to anyone attempting habitat conservation in what is territory under continued political dispute involving indigenous groups and private landowners. Consider that in the past two decades major changes in land use in Chiapas have been influenced by the oil boom of the 1960s and 1970s (Collier et al., 1994), which led to a period of deintensification of agriculture and increased off-farm employment; successive periods of fluctuating coffee prices, leading to rapid expansion and contraction of coffee production (Rice manuscript); increase in access to cattle feedlots in U.S. border states, resulting in the expanded exportation of young cattle; decrease in maize prices; and government restrictions on harvesting trees on private lands. Each of these events has had rapid and dramatic influences on the status of MFPs in Chiapas.

Shade trees have been a critical component of the production systems for several commodities in Chiapas, the best known of which are cacao and coffee. But other crops—including cardamom, which is exported to the Middle East—are produced in shade canopy systems. In addition to the cash crop production of the understory plants in these systems is the large number of products from the canopy trees. To a large degree these products are used by the farming households or (as in the case of Inga pods) sold on local markets. However, the possibility of sending products such as macadamia nuts to a more global market remains. With the inherent tendencies of the marketplace toward global overproduction and wild price swings, the diversity of the products from the canopy trees may offset the risk of producing a single export commodity.

Managed forest patches, particularly arroyo vegetation and acacia woodlots, are integrally associated with the production of one of the most important export commodities in the region—beef. One of the fastest-growing sectors of the cattle industry in Chiapas has been, at least in the recent past, the raising of young cattle for export to feed lots in the United States.

The growth in cattle production has lead both to a greater extension in land use and a greater intensification of cattle production on land already cleared for pasture. The increased intensification generally leads to reduction of shrub and tree cover on pastures. However, the possibility of developing more ecologically sound and sustainable cattle production on intensively managed lands could include the increased incorporation of trees and managed forest patches (Venator et al., 1992; Greenberg et al., in review). These patches can provide resources for cattle production, such as feed and building material, which would reduce dependence on expensive inputs.

Conclusions

The results of our work in Chiapas underscore the common perception that the conservation of forest patches and continuous tracts of forest will have decidedly different impacts on the conservation of bird populations. Managed forest patches are particularly well suited for providing extensive habitat for long-distance migrants. Managed forest patches, individually and as a mosaic, support a high density of migrants. They can provide seasonal resources for local migrants and may be important in the buffer zone adjacent to large tracts of forest, where mobile forest-dwelling birds can find breeding habitat. Finally, they enhance the local avifaunal diversity by providing habitat for forest edge species in agricultural areas that otherwise would support only open-field species. The increase in bird diversity occurs at even the lowest levels of tree density in farm fields and pasture and reaches its apex in the diverse agroforestry systems associated with cacao and coffee production.

However, most managed forest patches support few of the most specialized forest species, particularly those that have large territories or home ranges or are the target of hunting (Inigo-Elias 1992; Robinson, chapter 6, this volume). The differences between forest patch management and forest conservation are large, and the response to these differences will have major implications for the future of biological conservation in the tropics in the coming decades. One possible response is to say that forest patches are empty forest, biologically depauperate and altered to the point that they are best considered ecological ghosts and ignored. It is sometimes said that articulating any value for biodiversity conservation to these degraded systems is dangerous, because it justifies the destruction of large tracts of forest and syphons resources from higher priority areas. On the other hand, I would argue that efforts to increase forest cover on agricultural lands are essential and complementary to efforts in protected area management in the tropics. I would also argue that the implementation of proper land stewardship, environmental protection, and soil management benefit migratory and other bird populations, making natural allies between groups working on progressive rural development and bird conservation. To the degree we can strive for similar goals for different reasons, both groups should benefit from the collaboration. In short, conservation is not necessarily a zero-sum game—we can increase habitat on used and degraded lands while continuing the fight to save the last remaining wilderness areas in the tropics.

Acknowledgments

National Geographic provided grants for exploration and research, and the National Fish and Wildlife Foundation and the Scholarly Studies Fund of the Smithsonian Institution provided financial support for the research. I benefited from discussions with John Schelhas and Robert Rice on the human dimension of managed forest patches. Much of the data were collected by Peter Bichier, Javier Salgado, Andrea Cruz, John Sterling, Hector Flores, Ian Warkentin, and Robert Reitsma.

References

Aguilar-Ortiz, F. 1982. "Estudio ecologico de las aves del cafetal." In *Estudios Ecologia en el Agroecosistema Cafetalesa,* edited by E. Avila Jimenez, pp. 103--128. Instituto Nacional de Investigaciones sobre Recursos Bioticos (INIREB), Mexico.

Bierregaard, R. O. 1986. "Changes in bird communities in virgin forest and isolated Amazonian forest fragments." *Ibis* 128: 166–167.

Breedlove, D. E. 1981. *Flora of Chiapas.* Part 1. San Francisco: California Academy of Sciences.

Calleros, G., and F. A. Brauer. 1983. *Problematica Regional de la Selva Lacandona. Direccion General de Desarrollo Forestal, Secretaria de Agricultural y Recursos Hydraulicos.* Coordinacion Ejecutiva del Programma Ecologicao de la Selva Lacandona. Palenque, Chiapas.

Candaday, C. 1991. "Effects of encroachment by industry and agriculture on Amazonian forest birds in the Cuyabeno Reserve, Ecuador." Unpublished thesis, University of Florida, Gainesville.

Chazaro, J. C. 1977. "El Huizache, *Acacia pennatula* (Schlecht. and Cham) Benth. Una invasora del centro Veracruz." *Biotica* 2: 1–18.

Collier, G. A., D. C. Mountjoy, and R. B. Nigh. 1994. "Peasant agriculture and global change." *Bioscience* 44: 398–407.

De Vos, J. 1988. *Viajes al Desierta de la Soledad: Cuando la Selva Lacandona aun Era Selva.* Secretaria de Education Publica, Mexico.

Flores-Fuentes, R. 1979. "Coffee production farming systems in Mexico." In *Workshop on Agro-Forestry Systems in Latin America,* pp. 60–71. Turrialba, Costa Rica. CATIE.

Gradwohl, J., and R. Greenberg. 1991. "Small forest reserves: Making the best of a bad situation." In *Tropical Forests and Climate,* edited by N. Myers, pp. 253–257. Dordrecht, the Netherlands: Kluwer.

Greenberg, R. 1992. "Forest migrants in non-forest habitats on the Yucatan peninsula." In *Ecology and Conservation of Neotropical Migrant Landbirds,* edited by J. M. Hagan III and D. W. Johnston, pp. 273–286. Washington, DC: Smithsonian Institution Press.

Greenberg, R., and J. Salgado Ortiz. 1994. "Interspecific defense of pasture trees by wintering Yellow Warblers." *Auk* 111: 672–682.

Greenberg, R., P. Bichier, and J. Sterling. In press. "Migratory bird populations in acacia groves in southern Mexico."

Greenberg, R., J. Salgado, I. Warkentin and P. Bichier. In press. "Migratory bird populations in managed forest patches in Chiapas, Mexico." *Conservation Biology.*

Guevara, S., S. E. Purata, and E. van der Maarel. 1986. "The role of remnant forest trees in tropical secondary succession." *Vegetatio* 66: 77–84.

Hutto, R. L. 1980. "Winter habitat distribution of migratory land birds in western Mexico with special reference to small, foliage-gleaning insectivores." In *Migrant Birds in the Neotropics: Ecology, Behavior, Distribution and Conservation,* edited by A. Keast and E. S. Morton, pp. 181–205. Washington DC: Smithsonian Institution Press.

Inigo-Elias, E. 1992. "Ecological correlates of forest fragmentation and tropical forest bird communities in the Selva Lacandona region of Chiapas, Mexico." Abstract in *Forest Remnants in the Tropical Landscape: Benefits and Policy Implications,* edited by J. Schelhas and J. Doyle, p. 95. Washington, DC: Smithsonian Migratory Bird Center.

James, F. C., and S. Rathbun. 1981. "Rarefaction, relative abundance, and diversity of avian communities." *Auk* 98: 785–800.

Johns, A. D. 1992. "Response of Amazonian rain forest birds to habitat modification." *Journal of Tropical Ecology* 7: 417–437.

Lynch, J. F. 1989. "Distribution of overwintering Nearctic migrants in the Yucatan peninsula. I. General pattern of occurrence." *Condor* 91: 515–544.

Marquez, C. 1988. "La producción agrícola de la unión de uniones ejidales y sociedades campesinas de producción de Chiapas: Problematica y perspectivas de desarrollo." Professional thesis in Agronomy, Universidad Autonoma Chapingo, Mexico.

Powell, G. V. N., J. H. Rappole, and S. A. Sader. 1992. "Neotropical migrant landbird use of lowland Atlantic habitats in Costa Rica: A test of remote sensing for identification of habitat." In *Ecology and Conservation of Neotropical Migrant Landbirds,* edited by J. M. Hagan III and D. W. Johnston, pp. 287–298, Washington, DC: Smithsonian Institution Press.

Rappole, J., and E. S. Morton. 1985. "Effects of habitat alteration on a tropical forest community." *Ornithology Monograph* 36: 1013–1021.

Rice, R. 1993. "New technology and coffee production: Examining landscape transformation and international aid in northern Latin America." Report to the Smithsonian Migratory Bird Center, Washington, DC.

Rice, R. Manuscript. "Land use patterns and agroecology of coffee and cacao in eastern Chiapas, Mexico."

Robbins, C., B. Dowell, D. Dawson, J. Colon, F. Espinoza, J. Rodriguez, R. Sutton, and T. Vargas. 1987. "Comparison of Neotropical winter bird populations in isolated patches versus extensive forest." *Acta Oecologica/Oecologica Generalis* 8: 282–292.

Robbins, C., D. Dawson, J. Colon, R. Estrada, A. Sutton, R. Sutton, and D. Weyer. 1992. "Comparison of Neotropical migrant landbird populations wintering in tropical forest, isolated forest fragments, and agricultural lands." In *Ecology and Conservation of Neotropical Migrant Landbirds,* edited by J. M. Hagan III and D. W. Johnston, pp. 207–220. Washington, DC: Smithsonian Institution Press.

Saab, V., and D. R. Petit. 1992. "Impact of pasture development on winter bird communities in Belize, Central America." *Condor* 94: 66–71.

Salgado Ortiz, J. 1993. "Utilizacion de manchones de vegetacion secundaria en areas de agostadero por una comidad de aves." Tesis profesional, University of Michoacana de San Nicolas de Hidalgo.

Sauer, J. D. 1977. *Living Fences in Costa Rican Agriculture,* publication 19. Department of Geography, University of California, Los Angeles.

Schemskie, D. W., and N. Brokaw. 1981. "Treefalls and the distribution of understory birds in a tropical forest." *Ecology* 62: 938–945.

Sherry, T. W., and R. T. Holmes. In press. "Summer versus winter limitation of Neotropical migrant populations: What are the issues and what is the evidence?" In *Ecology and Management of Neotropical Migratory Birds: A Synthesis and Review of Critical Issues,* edited by T. Martin and D. Finch. Oxford, UK: Oxford University Press.

Thiollay, J. M. 1992. "Influence of selective logging on bird species diversity in a Guianan rain forest." *Conservation Biology* 6: 47–63.

Venator, C. R., J. Glaeser, and R. Soto. 1992. "A silvopastoral strategy." In *Development or Destruction: The conversion of Tropical Forest to Pasture in Latin America,* edited by T. E. Downing, S. B. Hecht, H. A. Pearson, and C. Garcia, pp. 281–292. San Francisco: Westview Press.

Villasenor, J. F. 1993. "The importance of agricultural border strips in the conservation of North American migratory landbirds in western Mexico." M.S. Thesis, University of Montana.

Warkentin, I. G., R. Greenberg, J. Salgado Ortiz. In press. "Songbird use of gallery woodlands in recently cleared and older settled landscapes of the Selva Lacandona in Chiapas, Mexico." *Conservation Biology.*

Wilken, G. C. 1977. "Integrating forest and small-scale farm systems in Middle America." *Agro-ecosystems* 3: 291–302.

Wilken, G. C. 1987. *Good Farmers: Traditional Agricultural Resource Management in Mexico and Central America.* Berkeley: University of California Press.

Willis, E. O. 1974. "Populations and local extinctions of birds on Barro Colorado Island, Panama." *Ecology Monograph* 44: 153–169.

Winker, K., J. H. Rappole, and M. A. Ramos. 1990. "Population dynamics of the wood thrush in southern Veracruz, Mexico." *Condor* 92: 444–461.

Wunderle, J., and S. C. Latta. 1994. "Population biology and turnover of Nearctic migrants wintering in small coffee plantations in the Dominican Republic." *Journal für Ornithologie* 135: 477.

Chapter 5

Arthropod Diversity in Forest Patches and Agroecosystems of Tropical Landscapes

Alison G. Power

Introduction

In most of the tropics, managed ecosystems, especially agroecosystems, dominate the landscape, and remnant primary forest exists in patches embedded in that agricultural landscape. As forested landscapes have given way to agricultural landscapes throughout much of the tropics, conservation efforts have mostly focused on protecting large tracts of natural primary forest. Yet much of the remaining forest exists in small, unprotected patches that are utilized, and sometimes managed, by humans. In fact, a mere 5 percent of the terrestrial environment is unmanaged and uninhabited (Western and Pearl 1989), and only 3.4 percent is currently fully protected in national parks (WRI/UNEP/UNDP 1994). Thus most species that survive in forest remnants must interact with agricultural systems. To understand how the organisms within these patches function, it is essential to understand their interactions with agricultural ecosystems. Furthermore, although the biodiversity of any particular managed ecosystem may be low in comparison to unmanaged ecosystems, a large proportion of the total species of a region is likely to be located in such systems (Pimentel et al., 1992). This suggests that more attention should be paid to understanding patterns of biodiversity in managed systems and to understanding how these patterns shape the ecological processes occurring within preserved areas and forest patches.

It is common for conservation biologists to use the metaphor, if not the theory, of island biogeography (MacArthur and Wilson 1967) as a model for understanding the impacts of forest isolation or fragmentation on populations (Burgess and Sharp 1981; Wilcove et al., 1986; Shafer 1990). Because of the higher likelihood of extinction in small patches, forest patches are dependent on larger, undisturbed forest tracts for colonizing species. Yet the island model implies that populations interact only with the mainland or other islands. Unlike true island floras and faunas, which do not interact in any important way with the sea around them, the flora and fauna of forest patches undoubtedly do interact with the "sea" of agricultural ecosystems surrounding them. The extent of this interaction varies tremendously with the species and the type of agroecosystem.

Interactions between forest patches and agricultural landscapes occur in both directions. Carroll (1990) pointed out that agriculture impacts natural areas in three ways: as a political and economic instrument, by means of commodity prices or subsidies; as a production technology using pesticides, fertilizer, and soil disturbance; and as a biological process via fragmentation, species invasions, etc. Clearly, these impacts are not unidirectional, and forest remnants can also influence processes that occur within agricultural systems.

In this chapter I focus on the diversity of arthropods, primarily insects, in managed ecosystems and the movements of these organisms between managed ecosystems and forest patches; other chapters in this volume address other taxa such as birds. After a brief overview of arthropod diversity in tropical forests and forest fragments, I address the following questions: To what extent do agricultural habitats play an important role in supporting arthropod biodiversity? Do agroecosystems differ in their ability to support a diverse arthropod fauna? Do forest patches provide any important "ecosystem services" to surrounding agroecosystems with respect to arthropod pest control?

Arthropod Diversity in Tropical Forests and Forest Patches

While the controversy continues over the magnitude of total insect species richness, particularly in the tropics (Erwin 1991; Gaston 1991), we do know a few things about tropical arthropod biodiversity. We know, for example, that the diversity of insects is remarkably high in those tropical forests where attempts have been made to document it (e.g., Erwin 1983; Knight and Holloway 1990; Wilson 1987a). We also know that in the Amazonian rainforest, arthropods can constitute 93 percent of the total animal biomass (Wilson 1987b).

But what we don't know about arthropod biodiversity in tropical landscapes is perhaps even more striking than what we do know. Although studies of temperate zone arthropods suggest that habitat fragmentation can have significant impacts on population abundance and distribution (Hopkins and Webb 1984; Turin and den Boer 1988; Desender and Turin 1989), we know little about the effects of forest fragmentation on tropical forest arthropods. Fragmentation studies of tropical insects are rare, and most have been carried out in the Brazilian Amazon as part of the Minimum Critical Size of Ecosystems project, which has been examining species composition and ecosystem processes in forest fragments ranging from 1 to 1,000 hectares (Lovejoy et al., 1986). These studies have shown that the species richness of forest understory butterflies declined dramatically with area in the forest fragments, but overall butterfly species richness increased in the isolated fragments due to invasion by butterflies typical of secondary growth (Brown 1991; Lovejoy et al., 1986). Species richness and abundance of dung and carrion beetles were significantly lower in the forest fragments than in contiguous forest, but significantly higher in fragments than in clear-cut areas (Klein 1989). The abundance of many species of euglossine bees declined in forest fragments, and overall visitation rates of bees to baits were lower in the fragments than in contiguous forest (Powell and Powell 1987).

Daily and Erlich (1995) examined butterfly diversity in fragments of midelevation moist forest in Costa Rica and reported significant differences in butterfly diversity in patches of different sizes. Butterflies were trapped in seven small forest fragments (3–30 hectares) and one larger fragment (227 hectares), with traps placed approximately 23 meters from the forest edge in all patches to partially control for edge effects. Although overall butterfly species diversity was only weakly correlated with the size of forest fragments, the distribution of butterfly subfamilies was significantly different between the small fragments and the larger patch. In particular, species in the Charaxinae and Nymphalinae, considered forest interior species, were almost entirely limited to the large forest patch and a smaller patch adjacent to it. In contrast, the remaining patches were dominated by satyrines characteristic of open areas or forest edges, particularly *Cissia satyrina.* Moreover, these results undoubtedly underestimated the differences in butterfly fauna between small and large patches. Due to trap placement near the edge, the entire area of small patches was sampled whereas the interior of the large patch was not.

Because of its impact on insect diversity, species composition, and abundance, fragmentation may have significant impacts on ecological processes mediated by insects within forests. Butterflies and euglossine bees are important pollinators of forest plants, and changes in species composition and abundance may result in declines in plant species richness within forest fragments (Lovejoy et al., 1986). Studies in the temperate zone indicate the likelihood that pollinator declines in habitat fragments will result in such an outcome (Jennersten 1988). Insects also play important roles in decomposition and nutrient cycling. Klein (1989) showed that Amazonian forest fragments had fewer and smaller dung and carrion beetles and that these changes resulted in lower rates of decomposition of dung.

Although information on arthropod movement between forest patches is scarce, it appears that forest patches may either facilitate or interfere with insect movement. Forest butterflies use corridors to move between forest patches in temperate agricultural landscapes (Dover 1990; Dover et al., 1992), but it is not clear whether this movement is essential for population persistence (Fry 1994) or the extent to which it is important for tropical forest butterflies. The movement of dung and carrion beetles of Amazonian rainforest is highly restricted by clear-cuts. Klein (1989) found no movement of forest beetles into or across clear-cuts, even when he placed baited pitfall traps within 15 meters of contiguous forest.

Forest patches and corridors can present barriers to movement of insects adapted to grasslands and agricultural habitats. Wooded hedgerows can reduce the dispersal and increase the isolation of populations of meadow butterflies (Fry and Main 1993). Studies in the temperate zone have shown that forests can also be relatively impermeable to some species of carabid beetles that are important predators of pests in agricultural systems (Jepson 1994). In agricultural fields these natural enemies are frequently subject to local extinctions caused by activities such as plowing, and they depend on dispersal for recolonization (den Boer 1990). Forests may act as barriers to movement of these predators and prevent recolonization of agricultural habitats after disturbance.

Do Managed Ecosystems Support Biodiversity?

The general paucity of information about insect diversity in the tropics is even more apparent for tropical managed ecosystems. There have been few comparisons of overall species richness of arthropods in adjacent natural and agricultural systems, and the degree of species overlap between adjacent habitats is largely unknown. We assume that different types of agricultural ecosystems are likely to differ in their ability to support arthropod biodiversity, but few studies have compared arthropod diversity among agroecosystems. Before examining the available data, it is useful to consider the range of agroecosystems that are common in tropical landscapes.

Types of Agroecosystems

Agricultural systems in the tropics range from simplified monocultures of either annual (e.g., rice, maize, cotton) or perennial (e.g., rubber, oil palm, sugarcane) crops to complex mixtures of plant species of varying life form and life history, such as those found in home gardens. They also range from extensive, mobile systems such as pastoral livestock farming or swidden systems (shifting cultivation), to intensive, permanent crop and/or animal farming. A particular crop such as common beans may be grown in a variety of systems, including monocultures, simple polycultures with two or three crop species, or complex home garden systems with 40 to 100 species. A perennial tree crop such as coffee may be grown under a variety of conditions, from traditional polycultures of coffee and shade trees to monocultures of coffee under intensive management with high inputs of fertilizers and herbicides. Attributes of these systems that may significantly impact insect biodiversity include plant taxonomic diversity, plant structural diversity, abundance and diversity of litter, and diversity of microclimates.

Arthropod Diversity and Abundance in Managed Ecosystems

For most insect groups, species richness appears to be higher in undisturbed habitats than in agricultural systems, but total abundance is often lower. Arthropod sampling programs on the island of Sulawesi in Indonesia found higher moth diversity in undisturbed forest than in pure coconut plantations or cultivated areas of rice, maize, and soybean (Holloway et al., 1990). Diversity in mixed coconut–clove agroforestry systems was intermediate. Moreover, the representation of different moth families depended on habitat: the Geometridae, Notodontidae, and Lymantriidae, whose larvae are largely arboreal feeders, declined substantially in agricultural areas, whereas the Noctuidae and Sphingidae with herb-feeding larvae increased (Holloway et al., 1990; Holloway and Stork 1991).

A similar pattern of moth diversity in forest and agricultural systems was seen on Seram (Holloway 1991). The modern agricultural systems showed reduced diversity and shifts in family proportions that resembled those seen in Sulawesi. However, the traditional swidden systems exhibited a more complex pattern. In swidden systems, temporary forest clearings are planted for a few years with annual or short-lived crops and then allowed to remain fallow for a period longer than the cultivation cycle (National Research Council 1993). In these systems, moth diversity was low in the recently cleared and cultivated areas, higher in the young fallows, and nearly equivalent to undisturbed forest in the older fallows of regenerating forest. Interestingly, the moths encountered in the swidden systems were not those characteristic of agricultural areas, but rather early gap phase species common in natural forest. The representation of moth families was similar to that found in undisturbed forest in all phases of the swidden cycle.

Total insect abundance was greater in perennial agricultural systems than in primary forest, in both lowland and higher elevation comparisons on Sulawesi (Holloway and Stork 1991; Stork and Brendell 1990). At low elevations, a garden plot containing coffee, kapok, cloves, coconut, mango, and lemon trees had higher insect abundance than that in humid forest. At higher elevations (1,100–1,150 meters), a coffee plantation with *Acacia* shade trees had higher insect abundance than undisturbed forest. However, different insect taxa exhibited different patterns of abundance. For instance, ants were consistently more abundant in the agricultural systems than in forest, but flies were much less abundant in the managed systems.

In Costa Rica, Janzen (1973a,b) carried out sweep samples near San Vito in primary forest and in adjacent bean fields and corn fields six months after harvest. In a single sample of 800 sweeps, Janzen found fewer species of adult insects but more individuals in the corn than in primary forest (409 species and 2,424 individuals in corn versus 969 species and 1,976 individuals in forest). Moreover, the proportion of juvenile insects was higher in corn (14 percent) than in primary forest (6–9 percent), suggesting the potential for rapid population increase in the agricultural system. Similarly, in Nigeria, Badejo (1990) found a greater proportion of juvenile mites, as well as greater overall mite abundance, in cassava fields than in adjacent 20-year-old secondary forest.

In Janzen's Costa Rican study, a single sample of 800 sweeps from beans or corn contained fewer beetle species (59–79 and 63 species, respectively) than a sample from primary forest (242–312 species), but the number of individuals in the forest was equal to or greater than the number of individuals in the agricultural fields (Janzen 1973a,b). When two samples of 800 sweeps each were combined, the cumulative number of beetle species rose to 99 in beans and 455 in primary forest. Clearly, the initial sample underestimated the total number of beetle species in all systems. In addition, there was very little overlap between beetles collected in corn fields and the natural forest; only 2 of the 63 species in corn were also found in the forest samples. In contrast, the two primary forest samples of 242 and 312 species had 99 species in common.

Differences between the numbers of bugs (Hemiptera/Heteroptera) from Costa Rican agroecosystems and forest were somewhat less dramatic: one sample of 800 sweeps included 14–18 species from beans, 42 species from corn, and 21–26 species from primary forest (Janzen 1973a,b). The combined forest samples had 41 species, and again, species overlap between forest and agroecosystems was low, with only 5 species found in both corn and forest. The number of individual bugs was greatest in corn and least in forest. In this taxon, then, species richness was as high or higher in corn as in the forest, but species richness was much lower in the bean fields. An obvious conclusion is that the diversity of insect species found in agricultural systems depends on the crop species and the resources that they present to insects. The enormous quantity of pollen produced by corn probably provides an important resource for many insects.

The diversity and abundance of soil arthropods also vary between agricultural systems and undisturbed habitats. Dangerfield (1990) found reduced species richness, abundance, and biomass of soil macrofauna in maize fields compared to natural miombo forests in Zimbabwe. Earthworms, woodlice, and millipedes were rare or absent in maize fields, and the abundance of aboveground predators such as spiders and ants was also quite low. Of the fauna sampled by Dangerfield, only beetle larvae appeared to be unaffected by these habitat changes.

Ant Diversity in Forest Patches and Agroecosystems

Ant diversity and community structure have been particularly well studied in tropical habitats, including both agricultural and forest ecosystems. Ants are both numerically and ecologically dominant in tropical forest systems, where the nonoverlapping territories of dominant ants form a three-dimensional mosaic that significantly impacts the distribution and abundance of other arthropods (Majer 1993). Gilbert (1980) has argued that the "ant mosaic" is one of four key biotic organizational features that influence the diversity and species composition of Neotropical forest biota, including plants and vertebrates as well as invertebrates. Thus, the ant community has the potential to play a highly significant role in controlling herbivorous insects in both forest patches and agroecosystems in tropical landscapes.

Ants tend to dominate the abundance and species richness of the arthropod fauna of tropical forests. In Brazil, ants constitute 26–47 percent of the arboreal arthropod biomass in rainforest (Adis et al., 1984; Fittkau and Klinge 1973) and 10–33 percent in cacao plantations (Majer et al., 1994). They comprise 51 percent of total arboreal insect numbers in Brazilian rainforest (Adis et al., 1984) and 18 percent in Bornean rainforest (Stork 1987). In Ghanaian cacao plantations, ants constitute up to 70 percent of arthropod biomass (Majer 1976) and make up 89 percent of the total insect numbers (Leston 1973). Species diversity is also high, with arboreal samples including 67 and 88 species in cacao plantations in Ghana and Papua New Guinea, respectively (Room 1971, 1975b), and 135 species in Peruvian rain forest (Wilson 1987a). Estimates of species richness increase substantially when ground-foraging ants are included in the samples, rising to 128 species in Ghanaian cacao plantations (Room 1971) and over 500 species in a 10-square-kilometer area of Peruvian rainforest (Verhaagh 1990).

Like other insects groups, ants often, but not always, exhibit reduced species richness in agricultural systems. Studies of ground-foraging ants in seven habitats in Papua New Guinea showed that ant species richness was highest in primary forest and in 15- to 25-year-old rubber plantations with a diverse understory of saplings and herbs (Room 1975a). Ant species richness was somewhat lower in cacao and coffee plantations, which contained a variable diversity of shade trees and understory herbs, and lower still in grasslands with low plant species diversity. Lowest ant species richness occurred in oil palm plantations, which were effectively monocultures since understory herbs were weeded frequently, and in urban mowed lawns dominated by a single grass species. The species composition of the forest ant fauna was most similar to that of rubber and coffee plantations, whereas cacao and oil palm had only moderate species overlap with forest. There was relatively little species overlap between forest and grassland and only one ant species was found in both forest and urban lawns.

A comparison of ground-foraging ants that attended tuna baits in humid forest and perennial agricultural systems in Costa Rica suggested that even relatively diverse agroforestry systems, such as the traditional cacao system, have lower ant diversity than nearby undisturbed forest (Roth et al., 1994). Ant diversity was significantly lower in monocultural banana plantations and in active cacao plantations than in either undisturbed humid forest or cacao plantations that had been abandoned for 25 years. Among the active cacao plantations, the plantation that included the most diverse shade tree assemblage also had the highest ant diversity. It is possible that using ground-foraging ants as an indicator of overall ant diversity underestimates differences between habitats, because the ants that visit tuna baits are likely to be generalists, which may be more prevalent in disturbed habitats (Roth et al., 1994).

As the Costa Rican cacao example illustrates, ant diversity may vary substantially among agroecosystems with the same primary crop. Perfecto and Snelling (1995) compared the species richness and diversity of ants in five types of coffee production systems in Costa Rica, ranging from traditional, low-input coffee plantations with a high diversity of shade trees and some annual crops, to intensive, high-input monocultural coffee systems containing no shade trees or annual crops. In the first year of the study, 23 ant species were collected foraging on the ground and 28 species were collected foraging on coffee bushes; species overlap was high, with 14 species recorded from both ground and bush. Over two years, a total of 33 species of ground-foraging ants were collected, although only 16 were collected in both years. This suggests either that ant communities in these systems were not stable from year to year or that the sample size was inadequate for estimating diversity.

Species richness and diversity of ground-foraging ants were positively correlated with the amount of shade in the five types of coffee systems, as well as with the vegetational and structural diversity of the habitat. The amount of shade in the different systems was in turn strongly correlated with the gradient of vegetational and structural diversity, ranging from an average of 99 percent shade in traditional coffee systems to an average of 32 percent shade in intensively managed coffee monocultures. An average of nine to ten ant species were collected from diverse traditional systems, whereas two to four were collected in coffee monocultures. In comparisons of ant communities within a particular type of coffee system, the species composition of ground-foraging ants was more similar among coffee monocultures than among vegetationally diverse systems.

Highly shaded systems are likely to have more moderate temperature and moisture regimes than systems with little shade, and this may influence ant diversity. Furthermore, systems with many shade trees also provide a more diverse resource base for insect prey, including a diverse layer of decaying wood and leaf litter, as well as a variety of plants. Perfecto and Snelling (1995) found higher densities of beetles, cockroaches, and collembolans in diverse, traditional coffee systems than in coffee monocultures, although orthopterans were more abundant in monocultures. In preliminary samples of the arboreal fauna using insecticidal fogging, Perfecto (pers. comm.) found twice as many beetle species in traditional coffee systems as in modern coffee monocultures, and species richness of both beetles and ants in the shade trees was comparable to that found in tropical rainforest trees.

In these Costa Rican coffee systems, species richness and diversity of ants foraging on the coffee bushes themselves were correlated with neither the amount of shade nor the vegetational diversity of the habitat (Perfecto and Snelling 1995). Species numbers ranged from 8 to 12 along the gradient from coffee monoculture to diverse traditional system and appeared to be largely unaffected by intensive coffee management, despite changes in insect prey resources. Both microclimate and habitat are probably more constant for bush foragers than for ground foragers, for whom soil temperatures and the structure of the litter layer are known to be important (Andersen 1990; Torres 1984).

In contrast to the pattern of ant species richness in natural and agricultural habitats in Costa Rica and Papua New Guinea, ant species richness in Puerto Rico was greater in grassland (37 species) and agricultural land (38 species) than in a subtropical wet forest or traditional coffee plantations (20 species each) (Torres 1984). Like most islands, Puerto Rico is relatively depauperate, and overall ant species richness is low. Species overlap was high between grassland and agroecosystems (34 species in common) and virtually complete between forest and coffee systems, but only 10 species were found in all habitats. Forest ants had narrow temperature tolerances, and Torres (1984) suggested that lower diversity of climatic microhabitats in the forest and the cacao systems led to lower species richness. In this study, disturbance of agricultural lands, such as annual plowing, had little long-term impact on ant species richness; ant communities recovered quickly and community organization was unaffected (Torres 1984).

Thus ant species richness varies with plant species diversity and structural diversity in the range of agroecosystems in which it has been examined. Diversity of food resources, microclimatic diversity, amount of shade, and the structure of the litter layer may all be important determinants of the observed patterns in ant species richness (Andersen 1990; Torres 1984). However, the ecological factors that influence the diversity of arboreal and ground-foraging ants probably differ. Ground-foraging ants may be more sensitive to factors that affect microclimate at the soil surface, such as the amount of shade and changes in the litter layer.

Do Managed Ecosystems Support Forest Arthropods?

Whether or not agroecosystems are useful in supporting fauna from forest patches depends on the degree to which forest species actually utilize such nonforested areas. One way to address this question is to examine the degree of species overlap between forest patches and various agricultural systems. Species overlap between tropical forests and annual agroecosystems appears to be relatively low. Janzen (1973a,b) found very little species overlap between recently abandoned corn fields and the natural forest for beetles and bugs in Costa Rica. Similarly, there was low ant species overlap between forest and yam fields in Puerto Rico (Torres 1984). In Mexico 12 out of the 25 ant species collected in forest surrounding an annual polyculture were also detected within the polyculture during the first year after planting (Risch and Carroll 1982a).

Species overlap between forest and perennial agroecosystems such as tree crops and agroforestry systems tends to be higher than for annual agroecosystems, but overlap still varies among systems. Torres (1984) reported complete overlap of the ant species in humid tropical forest and coffee plantations in Puerto Rico. In Costa Rica, ants of undisturbed forest had high species overlap with ants in abandoned cacao plantations, but relatively little overlap with ants in either of the active agricultural systems, banana or cacao (Roth et al., 1994). Room (1975a) reported intermediate species overlap among ants in forest and cacao, but high overlap between forest, rubber, and coffee systems. Differences in the degree of species overlap between forest and various agroforestry systems may reflect differences in the vegetational diversity within those systems. Roth et al. (1994) reported that ant species overlap between active cacao and forest was greatest for the cacao plantation with the highest vegetational diversity.

One type of agricultural system that appears to have relatively high species overlap with intact forest is the swidden system. Holloway and Stork (1991) reported high species overlap in the moths encountered in forest and swidden systems in Indonesia, and low species overlap between swidden and modern agricultural systems. Species found in the early stages of the cultivation cycle were common in natural forest gaps, and throughout the swidden cycle, moth families in the swidden mirrored those in undisturbed forest. Holloway and Stork (1991) suggest that traditional swidden agriculture or agroforestry systems may be the managed ecosystems most compatible with biodiversity conservation.

The size and intensity of forest clearing undoubtedly influence the potential diversity within the swidden. In a study of the impact of forest clearing on arthropods in the southern Appalachian Mountains, Shure and Phillips (1991) found that forest arthropods moved readily into small clearings (0.016 hectares) but not into larger clearings. The impact of traditional swidden systems was probably limited since clearings were small (less than 1 hectare) and distributed at low densities throughout a vast matrix of undisturbed forest. Moreover, many useful trees were left in the system and the soil was seldom turned over at planting, so the intensity of disturbance was relatively low. These characteristics of swidden systems are consistent with the notion that they may effectively conserve the diversity of forest arthropods.

Do Forest Patches Support Natural Enemies of Agricultural Pests?

Several authors have suggested that natural enemies of agricultural pests may depend on resources in natural habitats adjacent to agricultural systems (e.g., Glass and Thurston 1978; Matteson et al., 1984). Such a role for natural areas has been well established in some temperate systems (Emden 1965; Wainhouse and Coaker 1981; Altieri and Schmidt 1986; Boatman 1994), but good data from tropical systems are rare. There is, however, considerable anecdotal evidence that natural areas adjacent to cultivated fields can support natural enemies of crop pests in the tropics.

During the rapid increase in cotton production in Latin America in the 1950s, it was noted that uncultivated areas could provide food and shelter for the predators and parasitoids that control pests in Peruvian cotton (Beingolea 1959). Smith and Reynolds (1972) attributed the relative lack of pest problems in Costa Rican cotton production during the 1950s and 1960s to the fact that Costa Rican cotton fields were small and often surrounded by natural areas or native pasture. In contrast, in other Central American countries such as Guatemala, where pest problems were severe, cotton was planted in huge, contiguous monocultures.

The predators that move readily into agricultural systems are likely to be habitat generalists like many of the ground-foraging ants. In a study of the movement of insect predators between agricultural fields (wheat and corn) and "seminatural" habitats (wetlands, meadows, and pastures) in Switzerland, Duelli (1990) found that many of the species encountered in agricultural fields were habitat generalists that moved freely among all habitats, whereas a greater proportion of species encountered in natural habitats were restricted to those habitats.

Forest patches, including patches of secondary forest, may be useful as reservoirs of natural enemies of tropical tree crops as well. In Malaysia, many pests of cacao appear to have moved to cacao from wild hosts in secondary forest (Conway 1972). While this could be viewed as a negative consequence of having secondary forest in the landscape, the secondary forest also serves as a reservoir of the natural enemies of these pests. Conway (1972) argued that the benefits of having patches of natural vegetation outweigh the potential negative impacts, and thus intact blocks of forest should be integrated into the agricultural landscape in Malaysia.

A comparison of the ant fauna in two annual agroecosystems in Mexico may have provided the best evaluation to date of the potential role of forest in enhancing biological control in agricultural systems in the tropics (Risch and Carroll 1982a). Ground-foraging ants were sampled in two corn–bean–squash polycultures, one surrounded by 40-year-old forest (the forest milpa) and one surrounded by 1-year-old secondary growth (the field milpa). During the year following planting, the number of ant species in the forest milpa rose from 3 to 14 species, whereas the number of ant species in the field milpa fell from 3 to 1 species. Sampling of the habitats surrounding the two milpas detected 28 species in the forest and only 6 species in the 1-year-old secondary growth.

However, the diversity of predators does not necessarily correlate with their efficiency as biological control agents. In the Mexican agroecosystems, ant foraging efficiency, measured by the occupancy and removal rates of baits, was higher in the field milpa than in the forest milpa (Risch and Carroll 1982a). Unlike the diverse ant community of the forest milpa, the ant community of the field milpa was dominated by a single, abundant, highly aggressive ant species, *Solenopsis geminata*. This ant prefers open habitats and colonizes new agricultural fields rapidly but does not persist in habitats where succession is allowed to proceed. In recently plowed agricultural fields in Nicaragua, *S. geminata* dominated the ant fauna between 20 and 60 days after plowing but then declined in abundance as revegetation occurred (Perfecto 1991). In the forest milpa in Mexico, *S. geminata* dominated the ant community soon after planting but gradually declined as the crops matured during the year. Thus the forest milpa supported higher biodiversity than the field milpa, but the potential for effective biological control appeared to be higher in the agroecosystem surrounded by fields rather than by forest. It is important to keep in mind that this study included only one system of each type, with no replication. Studies are needed to evaluate the generality of this outcome across a range of annual agroecosystems and types of surrounding vegetation.

In coffee systems, ant-foraging efficiency is higher in monocultures than in traditional diverse systems. Nestel and Dickschen (1990) studied the foraging efficiency of ground-foraging ants in coffee agroecosystems with and without shade trees in Mexico. Ants discovered tuna baits more quickly in coffee systems with no shade trees compared to those with either a single species of shade tree or multiple species. In the unshaded coffee systems, the ant fauna was again dominated by *S. geminata*, whereas other ant species were also present in the shaded system.

Thus undisturbed forest can serve as an important refuge for natural enemies of agricultural pests in tropical areas, and the diversity of natural enemies in agroecosystems is likely to be higher when fields are interspersed with natural vegetation. Yet the effectiveness of the natural enemy community depends both on the ability to exploit a range of resources and on foraging efficiency in agricultural systems. Where an agroecosystem provides a suitable habitat for a single abundant, aggressive ant species, such as *S. geminata*, that can respond rapidly to pest availability, the diversity of natural enemies supported by forest patches may not be essential for effective pest control. Risch and Carroll (1982b) found significant increases in the diversity and abundance of both herbivores and nonant predators when *S. geminata* was excluded from annual agroecosystems, but the beneficial effects of the ant in reducing insect pests outweighed the negative impact of reducing other predators in the system.

Conclusions

Despite the scarcity of good data, it is clear that the changing tropical landscape is having dramatic effects on arthropod diversity and abundance. Deforestation and forest fragmentation will undoubtedly lead to the loss of forest arthropods, as habitats are lost or invaded by species of secondary growth. Yet some managed ecosystems appear to have the potential to support high insect diversity, including a high diversity of forest specialists. Although studies are few and usually very limited in scope, they indicate the potential for managing existing agroecosystems to enhance their potential for conservation of arthropod diversity.

Agricultural systems in the tropics differ significantly in the extent to which they support arthropod biodiversity. The diversity of insects in annual agroecosystems is typically much lower than in forest or in perennial agroecosystems. Species overlap between these systems and forest is low, and the insects that colonize these systems are usually open-habitat specialists or generalists. In contrast, insect diversity can be quite high in some perennial agroecosystems, especially those with relatively high vegetational diversity. Traditional cacao and coffee systems appear to function reasonably well as surrogates of forest, so long as they retain a diversity of shade trees and understory herbs. Even these systems, however, are not likely to provide appropriate habitat and resources for all forest arthropods. Traditional swidden and shifting cultivation systems may support much of the biodiversity of the original forest, but this is unlikely to be true if fallows are shortened because of increasing land scarcity.

Anecdotal information suggests that forest remnants within the tropical landscape may serve as reservoirs for natural enemies of agricultural pests. Good experimental tests of this hypothesis are rare, however, and firm conclusions would be premature. It appears that natural enemy diversity is higher in areas with greater vegetational diversity and adjacent to forest patches, presumably because of the diversity of prey resources offered by these habitats. Nonetheless, the effectiveness of these diverse natural enemies in terms of limiting pest insect populations is not necessarily higher than that of the less diverse, open-habitat natural enemy fauna that invades simple monocultures. Further research and more extensive experimental data are needed to resolve this issue.

References

Adis, J., Y. D. Lubin, and G. C. Montgomery. 1984. "Arthropods from the canopy of inundated terra firme forests near Manaus, Brazil, with critical considerations on the pyrethrum fogging technique." *Studies on Neotropical Fauna and Environment* 19: 223–236.

Altieri, M. A., and L. L. Schmidt. 1986. "The dynamics of colonizing arthropod communities at the interface of abandoned, organic and commercial apple orchards and adjacent woodland habitats." *Agriculture, Ecosystems and Environment* 16: 29–43.

Andersen, A. N. 1990. "The use of ant communities to evaluate change in Australian terrestrial ecosystems: A review and a recipe." *Proceedings of the Ecological Society of Australia* 16: 347–357.

Badejo, M. A. 1990. "Seasonal abundance of soil mites (Acarina) in two contrasting environments." *Biotropica* 22: 382–390.

Beingolea, O. 1959. "Notas sobre la bionomica de aranas e insectos beneficos que occuren en el cultivo de algodon." *Revista Peruana de Entomología Agricola* 2:36–44.

Boatman, N. 1994. "Field margins: Integrating agriculture and conservation." British Crop Protection Council, Farnham, Surrey.

Brown, K. S. 1991. "Conservation of Neotropical environments: Insects as indicators." In *The Conservation of Insects and Their Habitats,* edited by N. M. Collins and J. A. Thomas, pp. 349–404. London: Academic Press.

Burgess, R. C., and D. M. Sharp. 1981. *Forest Island Dynamics in Man-dominated Landscapes.* New York: Springer-Verlag.

Carroll, C. R. 1990. "The interface between natural areas and agroecosystems." In *Agroecology,* edited by C. R. Carroll, J. H. Vandermeer, and P. Rosset, pp. 365–383. New York: McGraw-Hill.

Conway, G. R. 1972. "Ecological aspects of pest control in Malaysia." In *The Careless Technology: Ecology and International Development,* edited by M. T. Farvar and J. P. Milton, pp. 467–488. Garden City, New York: Natural History Press.

Daily, G. C., and P. R. Erlich. 1995. "Preservation of biodiversity in small rainforest patches: Rapid evaluations using butterfly trapping." *Biodiversity and Conservation* 4: 35–55.

Dangerfield, J. M. 1990. "Abundance, biomass and diversity of soil macrofauna in savanna woodland and associated managed habitats." *Pedobiologia* 34: 141–150.

den Boer, P. J. 1990. "The survival value of dispersal in terrestrial arthropods." *Biological Conservation* 54: 175–192.

Desender, K., and H. Turin. 1989. "Loss of habitats and changes in the composition of the ground and tiger beetle fauna in four west European countries since 1950 (Coleoptera: Carabidae, Cicindelidae)." *Conservation Biology* 48: 277–294.

Dover, J. W. 1990. "Butterflies and wildlife corridors." *The Game Conservancy Review of 1989* 21: 62–64.

Dover, J. W., S. A. Clarke, and L. Rew. 1992. "Habitats and movement patterns of satyrid butterflies (Lepidoptera: Satyridae) on arable farmland." *Entomologist's Gazètte* 43: 29–44.

Duelli, P. 1990. "Population movements of arthropods between natural and cultivated areas." *Biological Conservation* 54: 193–207.

Emden, H. F. V. 1965. "The role of uncultivated land in the biology of crop pests and beneficial insects." *Scientific Horticulture* 17: 121–126.

Erwin, T. L. 1983. "Beetles and other arthropods of the tropical forest canopies at Manaus, Brasil, sampled with insecticidal fogging techniques." In *Tropical Rain Forests: Ecology and Management,* edited by S. L. Sutton, T. C. Whitmore, and A. C. Chadwick, pp. 59–75. Oxford, UK: Blackwell.

Erwin, T. L. 1991. "How many species are there? Revisited." *Conservation Biology* 5: 330–333.

Fittkau, E. J., and H. Klinge. 1973. "On the biomass and trophic structure of the Central Amazonian rain forest ecosystem." *Biotropica* 5: 2–14.

Fry, G. L. A. 1994. "The role of field margins in the landscape." In *Field Margins: Integrating Agriculture and Conservation,* edited by N. Boatman, pp. 31–40. Farnham, Surrey, UK: British Crop Protection Council.

Fry, G. L., and A. R. Main. 1993. "Restoring seemingly natural communities on agricultural land." In *Reconstruction of Fragmented Ecosystems,* edited by D. A. Saunders, R. J. Hobbs, and P. R. Ehrlich, pp. 225–241. Chipping Norton, UK: Surrey Beatty & Sons.

Gaston, K. J. 1991. "The magnitude of global insect species richness." *Conservation Biology* 5: 283–296.

Gilbert, L. E. 1980. "Food web organization and conservation of Neotropical diversity." In *Conservation Biology: An Evolutionary-Ecological Perspective,* edited by M. E. Soulé and B. A. Wilcox, pp. 11–33. Sunderland, MA: Sinauer.

Glass, E. H., and H. D. Thurston. 1978. "Traditional and modern crop protection in perspective." *BioScience* 28: 109–115.

Holloway, J. D. 1991. "Aspects of the biogeography and ecology of the Seram moth fauna." In *The Natural History of Seram,* edited by I. Edwards and J. Proctor, pp. 37–62. Andover, MD: Intercept.

Holloway, J. D., and N. E. Stork. 1991. "The dimensions of biodiversity: The use of invertebrates as indicators of human impact." In *The Biodiversity of Microorganisms and Invertebrates: Its Role in Sustainable Agriculture,* edited by D. L. Hawksworth, pp. 37–62. London: CAB International.

Holloway, J. D., G. S. Robinson, and K. R. Tuck. 1990. "Zonation in the Lepidoptera of northern Sulawesi." In *Insects and the Rain Forests of South East Asia (Wallacea),* edited by W. J. Knight and J. D. Holloway, pp. 153–166. London: The Royal Entomological Society of London.

Hopkins, P. J., and N. R. Webb. 1984. "The composition of the beetle and spider fauna on fragmented heathlands." *Journal of Applied Ecology* 21: 935–946.

Janzen, D. H. 1973a. "Sweep samples of tropical foliage insects: Description of study sites, with data on species abundances and size distributions." *Ecology* 54: 659–686.

Janzen, D. H. 1973b. "Sweep samples of tropical foliage insects: Effects of seasons, vegetation types, elevation, time of day, and insularity." *Ecology* 54: 687–708.

Jennersten, O. 1988. "Pollination in *Dianthus deltoides* (Caryophyllaceae): Effects of habitat fragmentation on visitation and seed set." *Conservation Biology* 2: 359–367.

Jepson, P. C. 1994. "Field margins as habitats, refuges and barriers of variable permeability to Carabidae." In *Field Margins: Integrating Agriculture and Conservation,* edited by N. Boatman, pp. 67–76. Farnham, Surrey, UK: British Crop Protection Council.

Klein, B. C. 1989. "Effects of forest fragmentation on dung and carrion beetle communities in central Amazonia." *Ecology* 70: 1715–1725.

Knight, W. J., and J. D. Holloway. 1990. *Insects and the Rain Forests of South East Asia (Wallacea).* London: The Royal Entomological Society of London.

Lestòn, D. 1973. "The ant mosaic—tropical tree crops and the limiting of pests and diseases." *Pest Abstracts and News Summaries* 19: 311–341.

Lovejoy, T. E., R. O. Bierregaard Jr., A. B. Rylands, J. R. Malcolm, C. E. Quintela, L. H. Harper, K.S. Brown Jr., A. H. Powell, G. V. N. Powell, H. O. R. Schubart and M. B. Hays. 1986. "Edge and other effects of isolation on Amazon forest fragments." In *Conservation Biology: The Science of Scarcity and Diversity,* edited by M. E. Soulé, pp. 257–285. Sunderland, MA: Sinauer.

MacArthur, R. H., and E. O. Wilson. 1967. *The Theory of Island Biogeography.* Princeton, NJ: Princeton University Press.

Majer, J. D. 1976. "The ant mosaic in Ghana cocoa farms: Further structural considerations." *Journal of Applied Ecology* 13: 145–156.

Majer, J. D. 1993. "Comparison of the arboreal ant mosaic in Ghana, Brazil, Papua New Guinea and Australia: Its structure and influence on arthropod diversity." In *Hymenoptera and Biodiversity,* edited by J. Lasalle and I. D. Gauld, pp. 136–141. Wallingford, England, UK: CAB. International.

Majer, J. D., J. H. C. Delabie, and M. R. B. Smith. 1994. "Arboreal ant community patterns in Brazilian cocoa farms." *Biotropica* 26: 73–83.

Matteson, P. C., M. A. Altieri, and W. C. Gagne. 1984. "Modification of small farmer practices for better pest management." *Annual Review of Entomology* 29: 383–402.

National Research Council. 1993. *Sustainable Agriculture and the Environment in the Humid Tropics.* Washington, DC: National Academy Press.

Nestel, D., and F. Dickschen. 1990. "The foraging kinetics of ground ant communities in different Mexican coffee agroecosystems." *Oecologia* 84: 58–63.

Perfecto, I. 1991. "Dynamics of *Solenopis geminata* in a tropical fallow field after ploughing." *Oikos* 62: 139–144.

Perfecto, I., and R. Snelling. In press. "Biodiversity and tropical ecosystem transformation: Ant diversity in the coffee agroecosystem in Costa Rica." *Ecological Applications.*

Pimentel, D., U. Stachow, D. A. Takacs, H. W. Brubaker, A. R. Dumas, J. J. Meaney, J. A. S. O'Neil, D. E. Onsi, and D. B. Corzilius. 1992. "Conserving biological diversity in agricultural/forestry systems." *BioScience* 42: 354–362.

Powell, A. H., and G. V. N. Powell. 1987. "Population dynamics of male euglossine bees in Amazonian forest fragments." *Biotropica* 19: 176–179.

Risch, S. J., and C. R. Carroll. 1982a. "The ecological role of ants in two Mexican agroecosystems." *Oecologia* 55: 114–119.

Risch, S. J., and C. R. Carroll. 1982b. "Effect of a keystone predaceous ant, *Solenopsis geminata,* on arthropods in a tropical agroecosystem." *Ecology* 63: 1979–1983.

Room, P. M. 1971. "The relative distribution of ant species in Ghana's cocoa farms." *Journal of Animal Ecology* 40: 735–751.

Room, P. M. 1975a. "Diversity and organization of the ground foraging ant faunas of forest, grassland and tree crops in Papua New Guinea." *Australian Journal of Zoology* 23: 71–89.

Room, P. M. 1975b. "Relative distributions of ant species in cocoa plantations in Papua New Guinea." *Journal of Applied Ecology* 12: 47–61.

Roth, D. S., I. Perfecto, and B. Rathcke. 1994. "The effects of management systems on ground-foraging ant diversity in Costa Rica." *Ecological Applications* 4: 423–436.

Shafer, C. L. 1990. *Nature Reserves: Island Theory and Conservation Practice.* Washington, DC: The Smithsonian Institution Press.

Shure, D. J., and D. J. Phillips. 1991. "Patch size of forest openings and arthropod populations." *Oecologia* 86: 325–334.

Smith, R. F., and H. T. Reynolds. 1972. "Effects of manipulation of cotton agroecosystems on insect pest populations." In *The Careless Technology: Ecology and International Development,* edited by M. T. Farvar and J. P. Milton, pp. 373–406. Garden City, NY: Natural History Press.

Stork, N. E. 1987. "Guild structure of arthropods from Bornean rain forest trees." *Ecological Entomology* 12: 69–80.

Stork, N. E., and M. J. D. Brendell. 1990. "Variation in the insect fauna of Sulawesi trees with season, altitude and forest type." In *Insects and the Rain Forests of South East Asia (Wallacea),* edited by W. J. Knight and J. D. Holloway, pp. 173–190. London: The Royal Entomological Society of London.

Torres, J. A. 1984. "Diversity and distribution of ant communities in Puerto Rico." *Biotropica* 16: 296–303.

Turin, H., and P. J. den Boer. 1988. "Changes in the distribution of carabid beetles in the Netherlands since 1880. II. Isolation of habitats and long-term time trends in the occurrence of carabid species with different powers of dispersal (Coleoptera, Carabidae)." *Biological Conservation* 44: 179–200.

Verhaagh, M. 1990. "The Formicideae of the rain forest in Panguana, Peru: The most diverse local ant fauna ever recorded." In *Social Insects in the Environment,* edited by G. K. Veeresh, B. Mallik, and C. A. Viraktamath, pp. 217–218. New Delhi: Oxford and IBH Publishing Co.

Wainhouse, D., and T. H. Coaker. 1981. "The distribution of carrot fly (*Psila rosae*) in relation to the fauna of field boundaries." In *Pests, Pathogens and Vegetation: The Role of Weeds and Wild Plants in the Ecology of Crop Pests and Diseases,* edited by J. M. Thresh, pp. 263–272. Boston: Pitman.

Western, D., and M. C. Pearl. 1989. *Conservation for the Twenty-First Century.* New York: Oxford University Press.

Wilcove, D. S., C. H. McLellan, and A. P. Dobson. 1986. "Habitat fragmentation in the temperate zone." In *Conservation Biology: The Science of Scarcity and Diversity,* edited by M. E. Soulé, pp. 237–256. Sunderland, MA: Sinauer.

Wilson, E. O. 1987a. "The arboreal ant fauna of Peruvian Amazon forests." *Biotropica* 19: 245–251.

Wilson, E. O. 1987b. "The little things that run the world (the importance and conservation of invertebrates)." *Conservation Biology* 1: 344–346.

WRI/UNEP/UNDP. 1994. *World Resources 1994–95.* New York: Oxford University Press.

Chapter 6

Hunting Wildlife in Forest Patches: An Ephemeral Resource

John G. Robinson

The human impact on tropical forests has been recognized for many years (Myers 1980). Conversion of forests into agricultural and pastoral landscapes is driven by the imperatives of national development, the subsistence needs of landless peoples, and the economic objectives of national and international corporations. The frequent result is forest patches distributed across a human-dominated landscape.

These forest patches might have value for conserving biological diversity, and ecologists have been able to apply existing theory to predict the amount of biodiversity remaining in these forest fragments. Making an analogy between these forest patches and true islands (e.g., Diamond 1976; Terborgh 1976), and drawing on the theory of island biogeography (MacArthur and Wilson 1967), a number of authors suggested an approach to understand patterns of species diversity based on the extinction and colonization curves affecting biological communities. Rates of species extinction on real islands could be estimated (Diamond 1972) and extrapolated to habitat islands (Wilcox 1980). Experimental tests of species loss through time (e.g., Lovejoy et al., 1984) generally confirmed the overall prediction of theory. Patterns of species loss from habitat islands revealed that certain types of species were more susceptible to extinction than others (Terborgh 1974), and it was recognized that the secondary consequences following the primary loss of key species had a significant effect on community structure and species diversity (Putz et al., 1990; Terborgh 1992).

This understanding was incomplete however: Human impact on the tropical forest is not limited to the initial fragmentation. Once humans have created their agricultural landscapes, they do not turn their backs on the remaining forest patches. Rural peoples continue to use the forest for food, fuel, medicine, material for housing construction, and other subsistence needs as well as for items important in local and national trade. As a result, the biological processes predicted by ecological theory do not have the opportunity to proceed unimpeded.

Understanding the impact of forest fragmentation therefore requires an understanding of how people continue to alter the biological community in forest patches—and people are active in most forest patches, including those that are protected. This is important for the following two reasons: (1) The influence of human activities on that community is much more immediate than that of the biological processes stimulated by fragmentation, and (2) human influence on the biological community is different from the effects of fragmentation alone.

Of all human activities, hunting is arguably the most important determinant of the diversity and abundance of larger-bodied wildlife species in tropical forest patches. Hunters take mammal, bird, and reptile species, generally selecting those individuals with body mass over a kilogram. Hunting depresses species densities, often driving local extinction. While game species account for only a small proportion of the mammalian and avian species in most forests, because they are larger bodied they account for a high proportion of the biomass. Thus the loss of game species through hunting has profound secondary consequences on biological diversity and community structure.

The loss of these species also affects the potential of forest patches to serve as reservoirs of game for human consumption. This is not an idle concern because forest patches today allow many rural people access to animal protein. If hunting is not sustainable, game populations in most patches are not viable over the long term, and thus their potential to act as resources is ephemeral.

Impacts of Fragmentation

Even in the absence of hunting, fragmentation has a significant impact on game species. The short-term impact varies with the living space that animals of a species require. Game species, which tend to be larger bodied, can therefore be limited by inadequate living space within smaller forest patches. Very small patches—even in the total absence of hunting—tend to lose most game species almost immediately. For instance, the 1- and 10-hectare plots in the Minimum Critical Size of Ecosystems (MCSE) project in central Amazonia are too small to contain any of the avian game species (species from the families Tinamidae, Cracidae, Phasianidae, Psophiidae, and Ramphastidae) (Bierregaard and Lovejoy 1989). Similarly, isolation of a 1-square-kilometer plot resulted in the immediate loss of the spider monkey (*Ateles*) and the saki (*Chiropotes*) (Rylands and Keuroghlian 1988). Many important game species have very large home ranges. White-lipped peccary herds in South America, for instance, range over 20–100 square kilometers depending on group size (Fragoso 1994), while bearded pigs in Borneo range over hundreds of square kilometers (Pfeffer and Caldecott 1986). At 860 hectares, the forest at Fazenda Montes Claros, in the Atlantic forest of Brazil, is apparently too small to support ocelot (Fonseca and Robinson 1990). While Barro Colorado Island does have a small ocelot population, this 15 square kilometer island is apparently too small to support jaguar and puma (Glanz 1982). A single female tiger requires a range of 20 square kilometers and a male at least twice that (Sunquist 1981); a jaguar requires a range of perhaps 50 square kilometers (Schaller and Crawshaw 1980). In addition, many species use a range of different habitats over the course of a year, and if a forest fragment does not contain the appropriate selection of habitats, it would not be able to support the species. For instance, Branch (1981) describes how many primate species in Amazonia seasonally range between *terra firma* and *varzea* habitats, and require both. Proboscis monkeys in Borneo move between, and require both, riverine forest and mangrove (Bennett and Sebastian 1988).

The long-term impact of fragmentation varies with the population density of the species. Large-bodied species tend to occur at lower densities (Robinson and Redford 1986b), and thus their population sizes in forest fragments tend to be smaller than those of smaller-bodied species. As indicated above, many game species are large bodied, and thus isolated populations of these species are susceptible to inbreeding and the loss of genetic heterogeneity, demographic instabilities, and unpredictable catastrophes. Even independent of hunting, populations of many game species isolated in forest fragments might be too small to survive for long periods of time.

Using Neotropical mammalian examples, figure 6-1 illustrates the area needs of a number of species. The figure plots the area required to

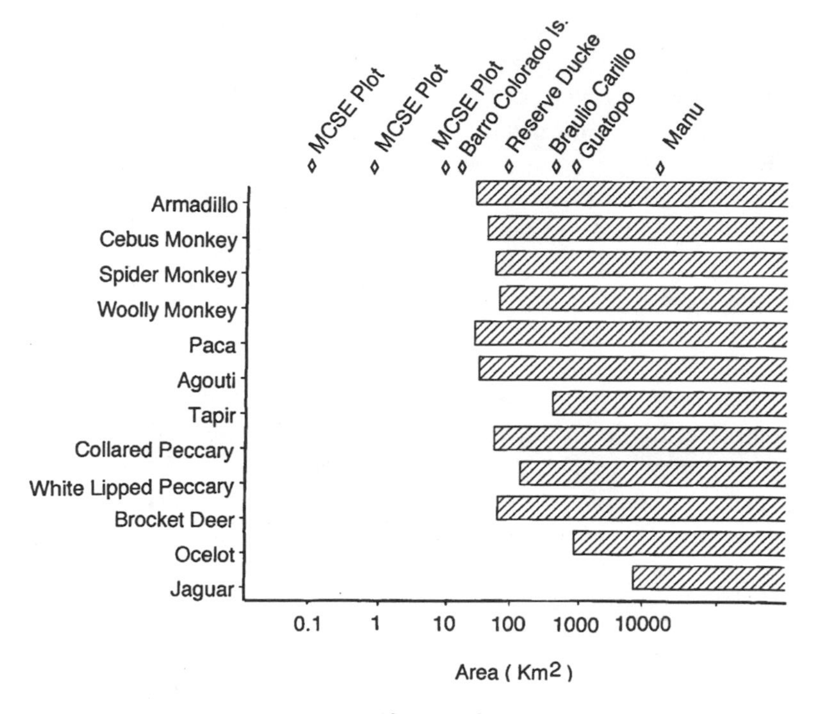

Figure 6-1

Minimum areas required to support 500 individuals of selected neotropical game species. Areas plotted on a $\log_{10}$ scale, with areas of selected forest patches identified: Minimum Critical Size of Ecosystems (MCSE) plots (10, 100, 1,000 hectares); Barro Colorado Island, Panama; Reserve Ducke, Amazonas, Brazil; Braulio Carillo National Park, Costa Rica; Guatopo National Park, Venezuela; and Manu National Park, Peru.

support 500 individuals of each species at the average population density for that species (Robinson and Redford 1986b). The number 500, which is more or less arbitrary, derives from the arguments that populations at an effective size of 50 or fewer are subject to inbreeding depression, and at an effective size of 500 or fewer will tend to lose genetic heterozygosity (Soulé and Wilcox 1980). In addition, the sizes of a range of some selected protected areas in the Neotropics are plotted. All of the game species selected here require forest fragments larger than the 15-square-kilometer island of Barro Colorado in Panama. With the exception of tapirs and the two felids, all of the species should be able to maintain populations of 500 individuals in fragments of 100 square kilometers (similar in size to the Reserve Ducke in Brazil). Only the jaguar requires an area as large as the Manu National Park in Peru (over 10,000 square kilometers).

Therefore, even in the absence of other influences on the wildlife, fragmentation by itself can severely limit the species composition of the forest and thus the potential of the forest to conserve wild game.

Wild Game Exploitation from Tropical Forests

Forest fragmentation allows greater exploitation of forest wildlife. The immigration of people into forested areas, and the conversion of forest into agricultural land, gives people increased access to forest wildlife. This process derives from, or fuels, the development of a transportation infrastructure, which allows easier transport of forest products to markets. The high rate of tropical forest conversion over the last 20 years is most probably associated with increased exploitation and consumption of forest wildlife. Although this inference has not been demonstrated conclusively on a global basis, it is nonetheless supported by the large number of studies of wildlife harvesting from tropical forest which indicate that present-day harvest of many species is not sustainable (e.g., Caldecott 1986; Bodmer et al., 1990; Robinson and Redford 1991, 1994; Fa et al., 1995).

Tropical forest wildlife are hunted by rural peoples for food, hides, fur, feathers, and as a source of medicinal products (Redford and Robinson 1991), and the extent of commercialization of this wildlife varies greatly. Many of these products never enter the market economy but instead are consumed within the rural communities. Many rural communities exist in relatively continuous forest, with little easy access to markets. Documenting the impact of this subsistence hunting is therefore difficult; but extrapolating from available case studies, Redford and Robinson (1991) estimate that some 19 million mammals, birds, and reptiles are consumed in rural communities throughout the Brazilian Amazon every year. Even when people have easy access to meat from domestic animals, wild meat can still be a significant part of the diet (Ayres 1991).

Wildlife is also hunted by rural peoples for sale in local urban markets. People can supplement their cash income, if adequate transportation is available, by selling wild game. From the perspective of the rural peoples, it is still appropriate to consider this activity more subsistence than commercial, because the income generated is used for the personal needs of the hunters rather than for the accumulation of capital. For instance, bush meat and fish provided the primary source of income to 73 percent of rural hunters sampled in Lahm's (1993) study in northeastern Gabon. The Mbuti pygmy of central Zaire traditionally exchange duiker meat and other forest products for local agriculturalists' iron implements, tobacco, and cultivated food (Hart 1978). In Gadsby's (1990) sample in southeast Nigeria, 90 percent of interviewees sold their hunted game. The ribereño communities living around the Tamshiyacu-Tahuayo reserve in northern Peru consume most of the primates, rodents, birds, and smaller-bodied game species themselves, but they transport much of the hunted ungulates (peccary, deer, and tapir) by river to the commercial markets of nearby Iquitos (Bodmer et al., 1988). The annual sale of wild game from the Iquitos markets was estimated by Gardner (1982) from data provided by Castro et al. (1976) to exceed 200 metric tons. More recent estimates of sales yield comparable figures (Bodmer et al., 1990).

In many areas, this local market hunting of forest wildlife supplies more commercial operations and attracts "middlemen," who purchase game from rural peoples. Where bush meat is an important source of protein for urban dwellers, a commercial network frequently develops to expedite the movement of forest products from the surrounding countryside. In Sarawak, for instance, a major commercial trade supplies the town of Sibu. In one sample year—1984—commercial traders brought in some 10,000 bearded pig carcasses and 1,500 deer (Caldecott 1986). Since the 1970s the Mbuti pygmy in central Zaire have regularly exchanged a high proportion of their wildlife harvest with commercial meat dealers (Hart 1978). In southeast Nigeria, most meat is sold to traders who resell the carcasses in urban centers (Gadsby 1990). Frequently there is increased capitalization associated with this exchange. Gadsby reports a common practice is for professional bush meat traders to supply ammunition and shotguns to the best hunters in local communities, with the agreement that all game is sold to the contractor. In other areas, the hunt is carried out by professionals. In equatorial Guinea, professional hunters bring game meat directly into markets (Fa et al., 1995; Juste et al., in press). Over a 212-day period in 1990 and 1991, Fa and his colleagues recorded 10,812 carcasses of 13 species brought into the markets of Rio Muni. Commercial hunters supplement the meat supply to the Iquitos markets (Bodmer et al., 1990).

Hunting of wildlife for food is therefore very important for the economies of many rural communities living in and around forested areas. The examples noted above illustrate the gradation from largely subsistence to largely commercial. The commercialization of game depends on an efficient transportation system to get harvests to market, which frequently requires the opening up of forest. Because commercialization in general increases harvest rates (Geist 1988), hunting will have a large impact on wildlife densities.

Impacts of Hunting

People living in tropical forests worldwide generally prefer to hunt, for meat and leather, the larger-bodied species. Among mammals these are normally the ungulates. In the Neotropics, peccary, deer, and tapir are preferred game (e.g., Bodmer 1994; Wetterberg et al., 1976). Where these animals are abundant, they constitute the most important part of the harvest by weight (e.g., Smith 1976; Vickers 1984). In Africa, forest duikers and pigs are preferred prey and contribute a significant fraction of animal biomass harvested (e.g., Colyn et al., 1987; Infield 1988; Fa et al., 1995). In Asia, it is the deer and, in Borneo, the bearded pig (e.g., Caldecott 1986). Primates and rodents are important game species worldwide, and armadillos are frequently hunted in Latin America. Avian species preferred by hunters tend to be large and gallinaceous, although many birds are also taken for their bright feathers and bills, which are used in traditional decorations and ceremonies (e.g., Bennett and Dahaban 1995; Kwapena 1984). In contrast to Latin America and Asia, birds in African forests are seldom hunted (J. Hart, pers. comm.). Redford (1992, 1993) has made the point that a large proportion of game species, both mammalian and avian, in tropical forests are frugivorous. These are often described as "fat" or "tasty" by hunters (Freese et al., 1982; Silva and Strahl 1991; Mittermeier 1991). Among reptiles, turtles and crocodilians are the most important game species (Redford and Robinson 1991).

Rural peoples also hunt wild species to supply the fur trade and the markets for folk and traditional medicines. Both of these demands put hunting pressure on carnivores such as spotted cats and otters, which are important in the fur trade (Smith 1976; Fitzgerald 1986), and bears and cats are important in the medicinal trade (Fitzgerald 1986; Gaski and Johnson 1994).

Hunting generally reduces the densities of game species (Freese et al., 1982; Johns 1986; Peres 1990; Glanz 1991; Payne 1992; Bennett and Dahaban, 1995), and the extent of the reduction depends on the resilience of the species to hunting and to the intensity of hunting. Species with lower intrinsic rates of natural increase, such as primates and carnivores (see Robinson and Redford 1986a), are the least resilient to hunting. Large-bodied species in general have lower rates of increase; but pigs and peccaries have a higher rate than would be expected from their body mass, and their populations are therefore more resilient to hunting.

Intensity of hunting varies regionally with the cultural preferences for different species and with local variations in human dependence on wild game. Redford (1992), summarizing existing density data from hunted and unhunted sites in the Neotropics, found that even light hunting can decrease most densities of preferred species by 70–80 percent. Heavier hunting can literally decimate a population and empty the forest of large mammals and birds. When preferred game species are available, the densities of small-bodied species, which are frequently not hunted as heavily, sometimes increase, presumably because their larger-bodied competitors have been eliminated (Johns 1986; Mitchell and Raez 1991). If and when preferred game species are extirpated, hunters will turn to these less-preferred prey (e.g., Vickers 1991).

Game species account for a significant proportion of mammalian and avian biomass in those tropical forests in which hunting is absent. Redford (1992), citing the example of Manu National Park, calculates that over 50 percent of the avian biomass and over 75 percent of the mammalian biomass is composed of species preferred by hunters in other locales. Therefore, when these game species are removed, and their ecological contribution lost, the change in ecosystem processes is expected to be significant. Redford coined the phrase "the empty forest" to describe the impact of hunting on the biological community.

Impacts of Both Fragmentation and Hunting

In the absence of hunting, except in the smallest patches, fragmentation probably has little direct short-term effect on the population densities of most game species (figure 6-2). Some species, such as those restricted to forest interiors, might show density declines, while others might benefit from the creation of edges. Fragmentation does however isolate populations, which in the long term makes them more vulnerable to local extinction through small population processes (loss of genetic variance, demographic instability, local catastrophes). Hunting, however, is ubiquitous in fragmented forests. Only in a minority of strictly protected ares, where regulations are effectively enforced, is hunting absent. Hunting has an immediate short-term effect on population densities of game species, with the extent of population decline depending on both

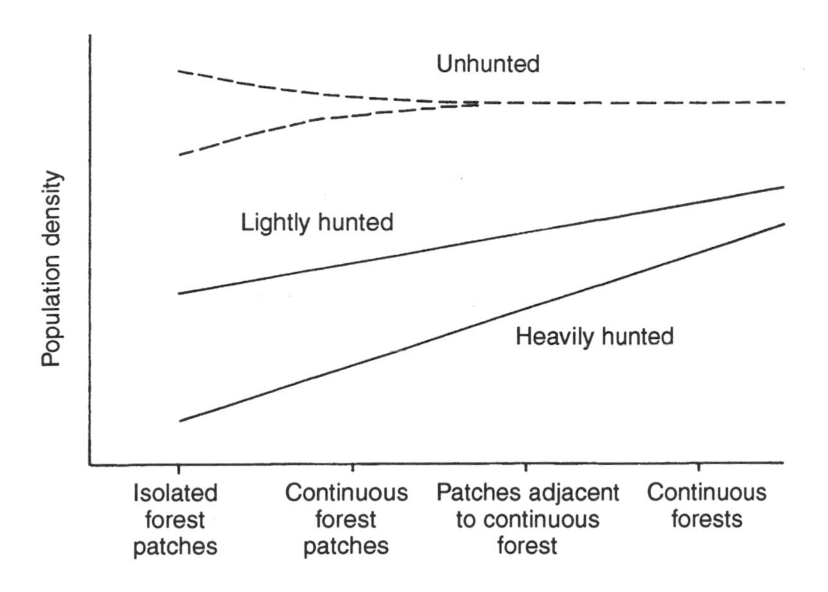

Figure 6-2
Schema of impact of hunting on the population density of a game species under different conditions of forest fragmentation. In the absence of hunting, population density, depending on the species, might increase or decline with forest fragmentation. With hunting, the extent of the decline in population density depends on both the intensity of hunting and the species resilience to hunting. Hunting intensity tends to be higher and to have greater impact in forest patches.

the intensity of hunting and the resilience of the species population to hunting (figure 6-2). The impact of hunting is greater in forest patches because fragmentation prevents hunted populations from being replenished through immigration and limits their movements across the landscape. Hunting also exacerbates the long-term impact of fragmentation by reducing population density, and thus population size in forest patches.

These patterns are evident when one considers mammalian communities in Neotropical forests subject to varying fragmentation. In forests that have been fragmented to the stage where only isolated patches remain, hunting drives many species into local extinction. The forests of central Panama, for example, have largely been converted to agricultural and urban uses, leaving only patches of forest. Forests like those on

Barro Colorado Island (BCI) are strictly protected, while mainland forest patches have been hunted to varying degrees. Comparing mammalian densities (Glanz 1991) patches on mainland Panama with those on BCI reveals that the remaining mammalian community depends on hunting intensity. Species preferred by hunters, such as the ungulates, are almost always eliminated from forest patches, even when hunting is relatively light. Species like primates, large-bodied rodents, and xenarthrans, which were not preferred in this area, only persist when hunting is light.

Where forests still retain some connectivity, even preferred species can persist when hunting is light. At most of the sites in extensive primate surveys in the Peruvian Amazon (Freese et al., 1982), the forests had been opened up by colonists to form a cultivated mosaic, but the forest patches were not isolated from one another. In this region, large-bodied primates, such as woolly, spider, and howler monkeys, which are preferred prey, are still found in forest patches, though at reduced densities where hunting was light. Where the forest patches are adjacent, and connected to large tracts of unfragmented forest, species that are hunted occur at very reduced densities, but the connectivity to a large faunal reservoir appears to reduce the probability of local extinctions. Johns (1986) examined the mammal communities at Ponta da Castanha, a terra firma area on the edge of Lake Tefé in central Amazonia. The forest had been fragmented by a mixture of logging and the establishment of Brazil nut plantations and agricultural fields. Tapir, peccary, and deer were rare because of heavy hunting prior to the study; however, agoutis and pacas were able to persist, despite heavy hunting. Large-bodied primates were also rare, because of occasional hunting. But populations of these game species were still present.

Where forests are still continuous, hunted species are better able to withstand hunting. Density estimates provided by Richard Bodmer and his associates for Tahuayo-Blanco in the Peruvian Amazon (Bodmer 1994) from a regularly hunted area revealed that wildlife densities, while reduced, are still significant. However, even in continuous forest, one will see local extinctions if sites have been hunted for long periods of time and/or if the hunting pressure was intense. For example, Peres (1990) reports the absence of woolly and spider monkeys, species preferred by hunters, at a number of sites in continuous Amazonian terra firma forest where hunting had been prolonged and intense.

The generalization that the extent of fragmentation predicts the persistence of game species in forests depends mostly on the correlation between fragmentation and hunting intensity. In general, the intensity and impact of hunting is greater in fragmented forests where more humans are present. The impact of that hunting is also greater. The local extinction of game species in more fragmented forests is not the consequence of the biological processes associated with small population sizes, which tend to have an impact on populations only over the long term. Instead population declines and local extinctions are mostly the consequence of hunting, which affects populations over the short term.

Indirect Effects on Community Structure

Loss of certain components from a biological community reverberates through the community. Forest fragmentation affects different species differently (Terborgh 1992). In the absence of hunting, fragmentation has more of an effect on species that occur at low population densities. Predators, in particular, tend to occur at lower densities than their prey, and as such tend to be more vulnerable to fragmentation. Loss of predators has striking consequences on the rest of the community structure (Terborgh 1992). The loss of large felids allows densities of prey species to increase. The mechanism for this increase is a little unclear: it might be simply predator release (Terborgh 1992) or it might also involve resource competition among prey species (Fonseca and Robinson 1990). Whichever, the greater densities of these species, frequently medium-sized frugivorous and granivorous mammals, has effects on forest seed dispersal and predation. One generalization is that predation on large seeds increases with a concomitant drop in regeneration of these tree species (Putz et al., 1990).

In forest patches with hunting, the patterns of species loss are not the same as those predicted from consideration of biological processes alone (figure 6-3) for the following three reasons: (1) hunters target certain classes of species—ungulates, primates, large gallinaceous birds; (2) those species which have low intrinsic rates of natural increase are susceptible to overexploitation—again primates, carnivores and large gallinaceous birds; and (3) many of these species are frugivorous (Redford 1992) and herbivorous (Dirzo and Miranda 1990, 1991) mammals and birds. In the absence of hunting, many of these species would maintain relatively dense populations in forest patches and thus would not be directly susceptible to fragmentation. When there is hunting, these game species are frequently the first to disappear.

While game species only account for a minority of species, they make up a majority of the frugivorous biomass in many tropical forests (Red-

Diet \ Body Size	Large-bodied (over 10 kg)	Medium-sized (1–10 kg)	Small-bodied (under 1 kg)
Carnivore	Jaguar, Puma	Large raptors	
Myrmecophage	Giant Anteaters	Tamandua, Armadillos	
Insectivore–Omnivore		Armadillos	Callithrichids, Opossums
Frugivore–Omnivore		Cebus monkeys, didelphis, coatis, cracids, toucan	Pigeons
Frugivore–Granivore		Pithecline monkeys, paca, agouti, macaws	Acouchi, other rodents, parrots
Frugivore–Herbivore	Tapir, Peccaries, Brocket deer	Howler, spider, and woolly monkeys	
Herbivore–Browser	Capybara, White-tailed deer	Sloths	

Figure 6-3

Comparison of the impacts of fragmentation and hunting on mammalian and avian game species found in Neotropical forests. Forest fragmentation negatively affects species occurring at low-population densities (shaded area), while hunting especially targets large and medium-sized frugivores and herbivores (enclosed in thick-lined box).

ford 1992) and thus have a significant impact on ecosystem processes. Emmons (1989) and Redford (1992) have pointed out that a majority of tree species in most tropical forests have seeds dispersed by these same frugivorous mammals and birds. Loss of these seed dispersers will be expected to significantly alter tree composition in forest fragments (Howe 1984). Predators that prey on these species are also expected to be secondarily influenced as their prey base is lost. While this loss can be exacerbated when the predators themselves are hunted, as they sometimes are for the commercial markets in skins, bones, and other body parts (Mills and Jackson 1994; Smith 1976), loss of their prey base frequently appears to be more immediate. An instructive example is provided by Karanth (1991) and Karanth and Sunquist (1992), who argued that despite active hunting for tiger to supply the traditional Chinese medicine markets, the factor most responsible for the decline of tiger populations in India is the loss of the prey populations—axis deer, gaur and sambur—through human hunting.

Species loss in most forest patches in the modern landscape will be determined by short-term processes. In the very smallest patches, some species will immediately drop out because of the lack of sufficient living space. In larger forest patches, hunting, another short-term process, determines which species will be most quickly lost, which in turn will have secondary effects on the whole biological community. Few forest patches will have the luxury to "relax" down to new community equilibria through the long-term biological processes determined by forest patch size and isolation.

Potential of Forest Fragments as Game Reservoirs

The importance of forests to provide animal protein to rural peoples around the world cannot be overstated. Yet what is the potential of forest patches to act as reservoirs for game? Small fragments (less than 10 square kilometers) probably offer little potential. Such forests are too small to support viable populations of most important game species even in the absence of hunting. Hunting will tend to decrease population densities significantly, which will exacerbate the situation. Only less-preferred game species with high densities and rates of population increase (e.g., *Didelphis* opossums, Fonseca and Robinson 1990) would be expected to survive.

In intermediate-sized forest patches (10–100 square kilometers), many game species would survive in the absence of hunting. When there is hunting, survival of local populations will depend on the hunting intensity and species-specific intrinsic rates of natural increase. When hunting is light, species with high densities and rates of population increase, such as some of the larger rodents (e.g., agoutis in the Neotropics), armadillos, and perhaps some ungulates (e.g, blue duiker in Africa), would be able to sustain the pressure. When hunting is heavy, most game species preferred by hunters would most probably be eliminated. Less-favored species, especially those whose populations are unaffected by or benefit from human presence, can exist at densities high enough to sustain viable populations. An example would be "grasscutters" (*Thryonmys swinerianus*), a rodent heavily exploited in Ghana

(Falconer 1992). Even in forest patches between 100 and 1,000 square kilometers in size, preferred game populations are susceptible to local extinction when hunting is heavy.

Some important game species have the potential of surviving over the long term only in fragments larger than 1,000 square kilometers, even in the absence of hunting. In the Neotropical data set illustrated in figure 6-1, game species such as tapir, ocelot, and jaguar need forest patches at least this large to sustain viable populations. When there is hunting, populations will be more vulnerable to local extirpation. However, in forest patches larger than 1,000 square kilometers, access by hunters to all areas is difficult, which regulates the overall intensity of hunting.

Conclusions

The wildlife communities resident in forest patches depend to a large extent on the intensity of human hunting, which is ubiquitous in tropical forests. Hunting intensity increases with forest fragmentation, as hunters usually have greater access to the forest and to transportation systems, particularly roads, that allow them to move carcasses to markets. Access to transportation systems goes hand in hand with increased participation in market economies, and one of the consequences of this is the acquisition of new hunting technologies. Shotguns, cartridges, and batteries for night hunting allow hunters to take a greater diversity of game. The populations of the species hunters prefer decline and are more likely to go locally extinct. Because game species account for a high proportion of mammalian and avian biomass in tropical forests, these changes have secondary impacts on many other species. Except in the case of very large forest patches, hunting will significantly alter the composition of the biological community.

Forest patches do not have a great potential to act as reservoirs for game species, which appears to contradict the observation that at the present time forests are supplying vast quantities of wild meat to rural communities. However, over the last 30 years, tropical forests around the world have been fragmented at unprecedented rates, thereby giving people unprecedented access to those forests. This access has allowed high levels of wild game exploitation, which, with the exception of locales where hunting is light, is not sustainable for most species. (e.g., Caldecott 1986; Fa et al., 1995; King 1994; Robinson and Redford 1994). Rural peoples are exploiting an ephemeral resource, made available to them because of forest fragmentation, which will ultimately become exhausted. As this happens, the impact on rural economies and rural protein consumption will likely be profound.

Acknowledgments

The ideas presented here derive from discussions with many people, including Kent Redford, Liz Bennett, Ullas Karanth, Richard Bodmer, Gustavo Fonseca, and Carlos Peres. I especially thank Liz Bennett, John Hart, and Linda Cox for critical comments on this manuscript.

References

Ayres, J. M. 1991. "On the track of the road: Changes in subsistence hunting in a Brazilian Amazonian village." In *Neotropical Wildlife Use and Conservation,* edited by J. G. Robinson and K. H. Redford, pp. 82–91. Chicago: University of Chicago Press.

Bennett, E. L, and Z. Dahaban. 1995. "Responses of wildlife to different types of disturbance in Sarawak, and implications for forest management." In *Ecology, Conservation, and Management of Southeast Asian Rainforest,* edited by R. B. Premack and T. Lovejoy. New Haven, CT: Yale University Press.

Bennett, E. L., and A. C. Sebastian. 1988. "Social organization and ecology of proboscis moneys (*Nasalis larvatus*) in mixed coastal forest in Sarawak." *International Journal of Primatology* 9: 233–255.

Bierregaard, R. O., and T. E. Lovejoy. 1989. "Effects of forest fragmentation on Amazonian understory bird communities." *Acta Amazonica* 19: 215–241.

Bodmer, R. E. 1994. "Managing wildlife with local communities: The case of the Reserva Communal Tamshiyacu-Tahuayo." In *Natural Connections: Perspectives in Community-Based Conservation,* edited by D. Western, M. Wright, and S. Strum, pp. 113–134. Washington DC: Island Press.

Bodmer, R. E., T. G. Fang, and L. Moya I. 1988. "Ungulate management and conservation in the Peruvian Amazon." *Biological Conservation* 45: 303–310.

Bodmer, R. E., N. Y. Bendayàn, A., L. Moya I., and T. G. Fang. 1990. "Manejo de ungulados en la Amazonia Peruana: Analisis de su caza y comercializacion." *Boletin de Lima* 70: 49–56.

Branch, L. C. 1981. "Seasonal and habitat differences in the abundance of primates in the Amazon (Tapajos) National Park, Brazil." *Primates* 24: 424–431.

Caldecott, J. 1986. *Hunting and Wildlife Management in Sarawak.* WWF Malaysia report. Kuching, Sarawak, Malaysia.

Castro, N., J. Revilla, and M. Neville. 1976. "Carne de monte como una fuente de proteinas en Iquitos, con referencia especial a monos." *Revista Forestal del Peru* 6(1–2): 19–32.

Colyn, M. M., A. Dudu, and M. Mankoto ma Mbaelele. 1987. "Donnees sur l'exploition du 'petie et moyen gibier' des forets ombrophiles du Zaire." In *International Symposium and Conference on Wildlife Management in Sub-Saharan Africa,* pp. 110–145. Paris: UNESCO.

Diamond, J. M. 1972. "Biogeographic kinetics: Estimation of relaxation times for avifaunas of southwest Pacific islands." *Proceedings of the National Academy of Science USA* 69: 2742–2745.

Diamond, J. M. 1976. "Island biogeography and conservation: Strategy and limitations." *Science* 193: 1027–1029.

Dirzo, R., and A. Miranda. 1990. "Contemporary Neotropical defaunation and forest structure, function and diversity—a sequel to John Terborgh." *Conservation Biology* 4: 444–447.

Dirzo, R., and A. Miranda. 1991. "Altered patterns of herbivory and diversity in the forest understory: A case study of the possible consequences of contemporary defaunation." In *Plant–Animal Interactions: Evolutionary Ecology in Tropical and Temperate Regions,* edited by P. W. Price, T. M. Lewishon, G. W. Fernandes and W. W. Benson, pp. 273–287. New York: Wiley.

Emmons, L. H. 1989. "Tropical rain forests: Why they have so many species and how we may lose this biodiversity without cutting a single tree." *Orion* 8(3): 8–14.

Fa, J. E., J. Juste, J. Perez del Val, and J. Castroviejo. 1995. "Impact of market hunting on mammal species in Equatorial Guinea." *Conservation Biology* 9: 1107–1115.

Falconer, J. 1992. *Non-timber Forest Products in Southern Ghana.* ODA Forestry No. 2. 23pp. London: Overseas Development Administration.

Fitzgerald, S. 1986. *International Wildlife Trade: Whose Business Is It?* Washington, DC: World Wildlife Fund.

Fonseca, G. A. B., and J. G. Robinson. 1990. "Forest size and structure: Competitive and predatory effects on small mammal community." *Biological Conservation* 53: 265–294.

Fragoso, J. M. V. 1994. *Large Mammals and the Community Dynamics of an Amazonian Rain Forest.* Ph.D. dissertation, University of Florida, Gainesville.

Freese, C. H., P. G. Heltne, N. Castro, and G. Whitesides. 1982. "Patterns and determinants of monkey densities in Peru and Bolivia, with notes on distributions." *International Journal of Primatology* 3: 53–90.

Gadsby, E. L. 1990. "The status and distribution of the drill, *Mandrillus leucophaeus,* in Nigeria: A report focusing on hunters and hunting and their threat to remaining populations of drills and other forest primates in southeast Nigeria." Report to Wildlife Conservation International, New York Zoological Society.

Gardner, A. L. 1982. "Wildlife management and tropical biology." Paper presented at the 33rd annual AIBS meeting, Penn State University, Pennsylvania.

Gaski, A. L., and K. A. Johnson. 1994. "Prescription for extinction: Endangered species and patented oriental medicines in trade." TRAFFIC report, Washington, DC.

Geist, V. 1988. "How markets in wildlife meat and parts, and the sale of hunting privileges, jeopardize wildlife conservation." *Conservation Biology* 2: 15–26.

Glanz, W. E. 1982. "The terrestrial mammal fauna of Barro Colorado Island: Censuses and long-term changes." In *The Ecology of a Tropical Forest,* edited by E. G. Leigh Jr., A. S. Rand, and D. M. Windsor, pp. 455–468. Washington, DC: Smithsonian Institution Press.

Glanz, W. E. 1991. "Mammalian densities at protected versus hunted sites in central Panama." In *Neotropical Wildlife Use and Conservation,* edited by J. G. Robinson, K. H. Redford, pp. 163–173. Chicago: University of Chicago Press.

Hart, J. A. 1978. "From subsistence to market: A case study of the Mbuti net hunters." *Human Ecology* 6: 325–353.

Howe, H. F. 1984. "Implications of seed dispersal by animals for tropical reserve management." *Biological Conservation* 30: 261–281.

Infield, M. 1988. "Hunting, trapping and fishing in villages and on the periphery of the Korup National Park." Washington, DC: Report to World Wide Fund for Nature.

Johns, A. D. 1986. "Effects of habitat disturbance on rainforest wildlife in Brazilian Amazonia." Washington, DC: Report to World Wildlife Fund.

Juste, J., J. E. Fa, J. Perez-del Val, and J. Castroviejo. In press. "Market dynamics of bushmeat species in Equatorial Guinea." *Journal of Applied Ecology.*

Karanth, K. U. 1991. "Ecology and management of the tiger in tropical Asia." In *Wildlife Conservation: Present Trends and Perspectives for the 21st Century,* pp. 156–159. Proceedings of the International Symposium on Wildlife Conservation, The V International Congress of Ecology, Japan.

Karanth, K. U., and M. E. Sunquist. 1992. "Population structure, density and biomass of large herbivores in the tropical forests of Nagarahole, India." *Journal of Tropical Ecology* 8: 21–35.

King, S. 1994. "Utilization of wildlife in Bakossiland, West Cameroon, with particular reference to primates." *TRAFFIC Bulletin* 14: 63–73.

Kwapena, N. 1984. "Traditional conservation and utilization of wildlife in Papua New Guinea." *The Environmentalist* 4(7): 22–26.

Lahm, S. A. 1993. "Ecology and economics of human/wildlife interaction in northeastern Gabon." Ph.D. dissertation, New York University.

Lovejoy, T. E., J. M. Rankin, R. O. Bierregard Jr., K. S. Brown Jr., L. H. Emmons, and M. van der Voort. 1984. "Ecosystem decay of Amazon forest remnants." In *Extinctions,* edited by M. H. Nitecki, pp. 295–325. Chicago: University of Chicago Press.

MacArthur, R. H., and E. O. Wilson. 1967. *The Theory of Island Biogeography.* Princeton, NJ: Princeton University Press.

Mills, J. A., and P. Jackson. 1994. *Killed for a Cure: A Review of the Worldwide Trade in Tiger Bone.* Cambridge, UK: TRAFFIC International.

Mitchell, C. L., and E. Raez L. 1991. "The impact of human hunting on primate and game bird populations in the Manu biosphere reserve in southeastern Peru." Report to Wildlife Conservation International, New York Zoological Society.

Mittermeier, R. 1991. "Hunting and its effect on wild primate populations in Suriname." In *Neotropical Wildlife Use and Conservation,* edited by J. G. Robinson and K. H. Redford, pp. 93–107. Chicago: University of Chicago Press.

Myers, N. 1980. *Conversion of Tropical Moist Forests.* Washington, DC: National Academy of Sciences.

Payne, J. C. 1992. "A field study of techniques for estimating densities of duikers in Korup National Park, Cameroon." M.S. thesis, University of Florida, Gainesville.

Peres, C. A. 1990. "Effects of hunting on western Amazonian primate communities." *Biological Conservation* 54: 47–59.

Pfeffer, P., and J. O. Caldecott. 1986. "The bearded pig (*Sus barbatus*) in East Kalimantan and Sarawak." *Journal of the Malaysian Branch of the Royal Asiatic Society* 59: 81–100.

Putz, F. E., E. G. Leigh Jr., and S. J. Wright. 1990. "The arboreal vegetation on 70-year old islands in the Panama Canal." *Garden* 14: 18–23.

Redford, K. H. 1992. "The empty forest." *BioScience* 42: 412–422.

Redford, K. H. 1993. "Hunting in Neotropical forests: A subsidy from nature." In *Tropical Forests, People and Food,* edited by C. M. Hladik, A. Hladik, O. F. Linares, H. Pagezy, A. Semple, and M. Hadley, pp. 227–246. Paris: UNESCO.

Redford, K. H., and J. G. Robinson. 1991. "Subsistence and commercial uses of wildlife in Latin America." In *Neotropical Wildlife Use and Conservation,* edited by J. G. Robinson and K. H. Redford, pp. 6–23. Chicago: University of Chicago Press.

Robinson, J. G., and K. H. Redford. 1986a. "Intrinsic rate of natural increase in Neotropical forest mammals: Relationship to phylogeny and diet." *Oecologia* 65: 516–520.

Robinson, J. G., and K. H. Redford. 1986b. "Body size, diet, and population density of Neotropical forest mammals." *American Naturalist* 128: 665–680.

Robinson, J. G., and K. H. Redford. 1991. "Sustainable harvest of Neotropical forest animals. In *Neotropical Wildlife Use and Conservation,* edited by J. G. Robinson and K. H. Redford, pp. 415–429. Chicago: University of Chicago Press.

Robinson, J. G., and K. H. Redford. 1994. "Measuring the sustainability of hunting in tropical forests." *Oryx* 28: 249–256.

Rylands, A. B., and A. Keuroghlian. 1988. "Primate populations in continuous forest and forest fragments in central Amazonia." *Acta Amazonica* 18: 291–307.

Schaller, G. B., and P. G. Crawshaw Jr. 1980. "Movement patterns of jaguar." *Biotropica* 12: 161–168.

Silva, J. L., and S. D. Strahl. 1991. "Human impact on populations of chachalacas, guans, and currasows (Galliformes: Cracidae) in Venezuela." In *Neotropical Wildlife Use and Conservation,* edited by J. G. Robinson and K. H. Redford, pp. 37–52. Chicago: University of Chicago Press.

Smith, N. J. H. 1976. "Spotted cats and the Amazon skin trade." *Oryx* 13: 362–371.

Soulé, M. E., and B. A. Wilcox. 1980. *Conservation Biology. An Evolutionary–Ecological Perspective.* Sunderland, MA: Sinauer.

Sunquist, M. E. 1981. "The social organization of tigers (*Panthera tigris*) in Royal Chitawan National Park, Nepal." *Smithsonian Contributions to Zoology* 336: 1–98.

Terborgh, J. 1974. "Preservation of natural diversity: The problem of extinction prone species." *BioScience* 24: 715–722.

Terborgh, J. 1976. "Island biogeography and conservation: Strategy and limitations." *Science* 193: 1029–1030.

Terborgh, J. 1992. "Maintenance of diversity in tropical forests." *Biotropica* 24: 283–292.

Wetterberg, G. B., M. Ferreira, W. L. S. Brito, V. G. Araújo. 1976. *Fauna Amazónica preferida como alimento,* p. 17. Technical series #4, FAO Brazil.

Wilcox, B. A. 1980. "Insular ecology and conservation." In *Conservation Biology: An Evolutionary–Ecological Perspective,* edited by M. E. Soulé and B. A. Wilcox, pp. 95–118. Sunderland, MA: Sinauer.

Vickers, W. T. 1984. "The faunal components of lowland South American hunting kills." *Interciencia* 9: 366–376.

Vickers, W. T. 1991. "Hunting yields and game composition over ten years in an Amazonian village." In *Neotropical Wildlife Use and Conservation,* edited by J. G. Robinson and K. H. Redford, pp. 53–81. Chicago: University of Chicago Press.

Part II

Regional Landscapes

Chapter 7

The Ecological Importance of Forest Remnants in an Eastern Amazonian Frontier Landscape

Daniel C. Nepstad, Paulo Roberto Moutinho, Christopher Uhl, Ima Célia Vieira, and José Maria Cardosa da Silva

Introduction

Tropical forests are globally significant regulators of water, energy, carbon and nutrient fluxes, and the habitat of half of the world's plant and animal species. Following a pattern that is very similar to the historical land-use patterns of the temperate zone (Cronon 1983), tropical forest landscapes are being converted to mosaics of cattle pastures, agricultural fields, secondary forests, logged forests, and primary forest remnants as people turn to these ecosystems for sustenance and wealth. In this chapter, we evaluate the ecological importance of forest remnants within the context of a frontier landscape in eastern Amazonia, where one-third of the original forest cover has been cleared or selectively logged. We analyze the role of forest remnants in maintaining the hydrologic cycle, in restricting the spread of fires, and as sanctuaries for plant and animal species that might recolonize abandoned lands. This analysis does not discuss the health of forest remnants per se, which has been addressed elsewhere (Bierregaard et al., 1992; and chapter 10 of this volume). Rather, it examines the role of remnants in maintaining the biological integrity of a regional ecosystem.

Forest Remnants in the Context of Eastern Amazonia

Amazonia is unique among the world's major tropical forest formations in that more than 85 percent of the pre-Columbian, closed-canopy forest is still intact (Fearnside 1993; Skole and Tucker 1993). The most prevalent habitat patches in most of the region are agricultural clearings that stretch along roads and rivers in forested landscapes. It is along the eastern and southern margins of Brazilian Amazonia where the more typical tropical forest landscape is found, with patches of primary forest surrounded by fields and secondary forest; even here, however, most landholdings have at least 50 percent forest cover, as is required by Brazilian law.

One rationale for examining the ecological role of forest remnants in Amazonia is found in the areal coverage of forested landscapes that are fragmented or undergoing fragmentation. The 300,000 square kilometers of Amazonian forest that have been cleared are equivalent to approximately six nations the size of Costa Rica. Each year, an area of Amazon forest that is one-fifth to one-half of a Costa Rica in size is cleared (Fearnside 1993; Skole and Tucker 1993). This clearing has created a large area of forest fragments. In 1988, there was an estimated 16,000 square kilometers of forest remnants (less than 100-square-kilometer blocks) in Amazonia that were isolated from larger blocks of forest by pastures, agricultural clearings, and roads (Skole and Tucker 1993). In the 11,000-square-kilometer Bragantine Region east of Belém, deforestation has reduced primary forest cover to approximately 2 percent of the land area (Vieira et al., in press). In sum, the potential ecological role of forest remnants is increasing in importance in Amazonia and is already important in places such as the Bragantine region.

As we discuss Amazonian forest remnants, it is also important to move beyond the dichotomous "forest versus nonforest" view of tropical landscapes. Tropical forest remnants are not islands in ecological deserts but rather patches in a matrix of ecosystems that vary greatly in their degree of ecological impoverishment. The dichotomous view of tropical landscapes is reinforced by most satellite-based analyses of tropical deforestation, in which it is difficult to distinguish between different forms of cleared land and in which the forest scars resulting from logging and fire disappear from scenes recorded by the LANDSAT Thematic Mapper satellite within one to five years (Lefebvre and Stone 1994). As we shall show, the functional integrity and species composition of different types of anthropogenic ecosystems are highly variable and must be understood to evaluate the ecological importance of forest remnants.

The studies that are the basis of this analysis were conducted at the Fazenda Vitoria (Victoria Ranch), 7 kilometers from the town of Paragominas and 300 kilometers southeast of Belém (figure 7-1). An-

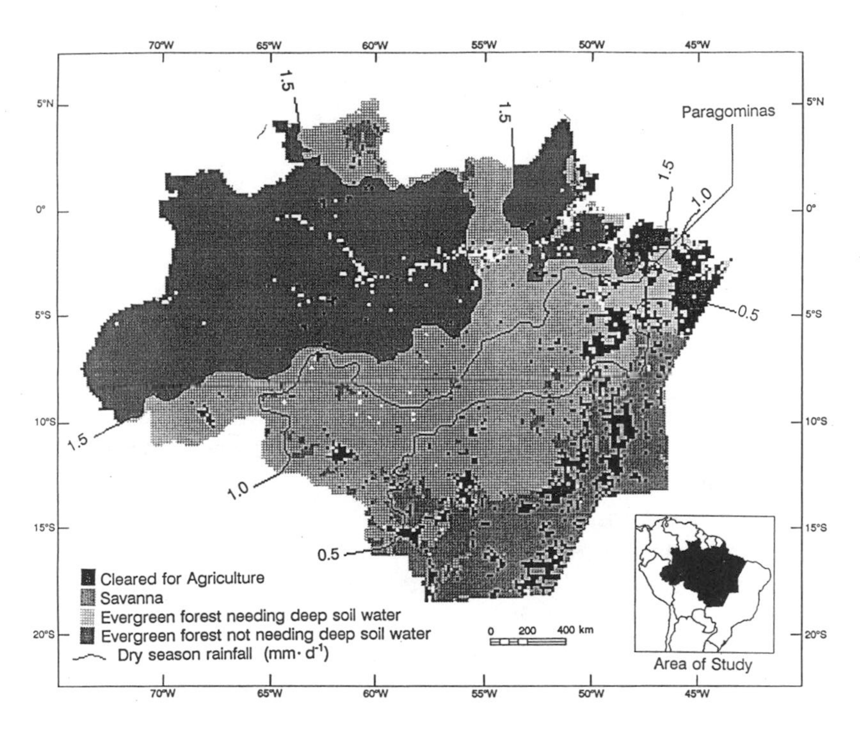

Figure 7-1

Map of Brazilian Amazonia with isopleths of average annual rainfall. The black areas have been cleared for agriculture and cattle pastures (Stone et al. 1994). Evergreen forests that depend on deep-rooting are particularly susceptible to fire should the predicted reduction in rainfall become a reality or should the recent high frequency of ENSO-related drought events continue. In their current state, these forests act as fire breaks through the landscape. The main site of the studies included in this chapter is near Paragominas. *Source:* Based on Nepstad et al. 1995.

nual rainfall is 1,800 millimeters, with less than 20 percent falling during the six-month dry season. The deeply weathered clay soil at the site is common in Amazonia. After 30 years of occupation, this relatively old Amazonian frontier is a mosaic of logged forests, active and abandoned pastures, regrowing secondary forests, and remnants of intact primary forest. In recent years, bulldozers, fertilizer, and new forage species have been employed by some ranchers to reform degraded pastures. These rejuvenated cattle pastures, which are profitable and appear to be more long lasting than the earlier types (Mattos and Uhl 1994), run the risk of reducing the recuperative capacity of the forest where they are formed.

The ecosystems that were studied include a mature, closed-canopy forest remnant of approximately 140 hectares, a 50-hectare secondary forest that extended in a 300- to 500-meter swath along the forest boundary, and a 50-hectare abandoned pasture adjacent to the secondary forest. The secondary forest was on a former cattle pasture site that was grazed until 1976, when the growth of woody vegetation lead to pasture abandonment. The abandoned pasture was formed in 1969 through forest clearing. It had been grazed heavily and burned three times prior to abandonment in 1984 (figure 7-2).

Functional Importance

Regional Climate

The most serious threat to the integrity of forest ecosystems in Amazonia is regional climate change. Amazonia is characterized by what is perhaps the most continental climate on the planet, with a large percentage of rainfall originating as water vapor released into the atmosphere by forests located "upwind," that is, to the east (Salati et al., 1979; and Victoria et al., 1991). Recent basinwide modeling studies that examined the regional climatic effect of complete forest conversion to pasture predict a 20 percent reduction of rainfall and an increase in air temperature resulting from forest conversion (Lean and Warrilow 1989; Shukla et al., 1990). While the scenario of a complete conversion to pasture is certainly extreme, the trend is of vital interest, for even slight reductions in rainfall—or, more likely, the increased frequency of severe drought years—could render primary forests in eastern and southern Amazonia vulnerable to fire (Nepstad et al., 1994; Nepstad et al., 1995; Nobre et al., 1991; Meggers 1994; Uhl and Kauffman 1990), as is discussed below.

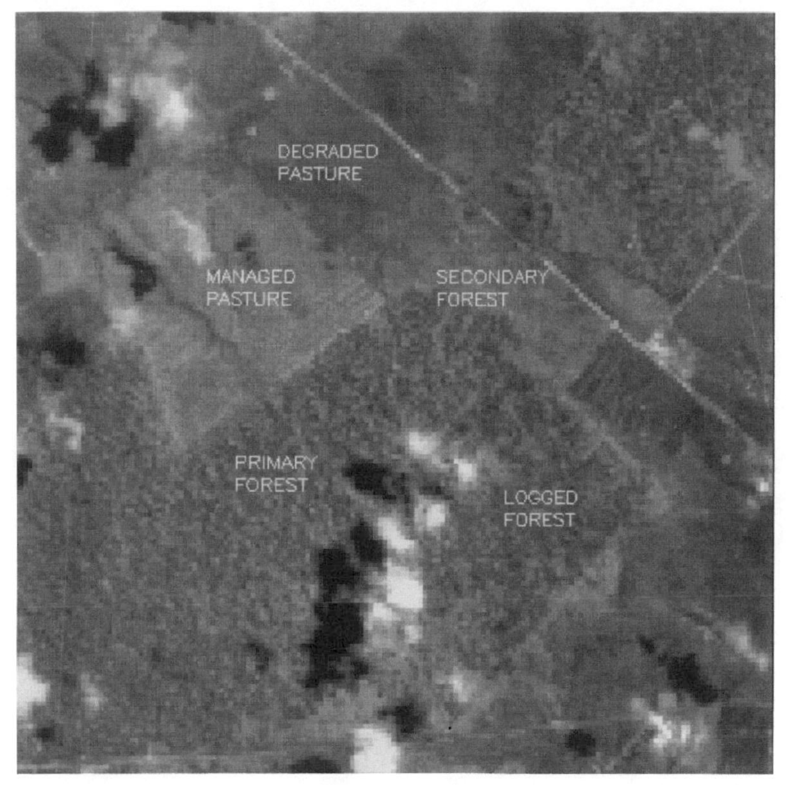

Figure 7-2

Satellite image (LANDSAT Thematic Mapper, September 1988) of the Fazenda Vitoria study site near Paragominas. The managed, active pastures have been reformed through bulldozing. The secondary forest swathe extends between the primary forest and pasturelands. The primary forest has been interrupted at its southeastern corner by logged forest. The town of Paragominas is visible in the lower right. White and accompanying black patches are clouds and cloud shadows, respectively.

Within the context of the regional climate system, the importance of forest remnants is determined by (1) the hydrologic differences between primary forests and the ecosystems that are replacing them and (2) the areal extent of each new ecosystem. In this sense, forest remnants are most important when they cover large portions of the landscape and when the hydrologic functions that they maintain are sharply different from those performed by new ecosystems. The amount of water vapor leaving an ecosystem through evapotranspiration illustrates this point. Pastures release 20–40 percent less vapor to the atmosphere during the dry season than do primary forests (Nepstad et al., 1994; Wright et al., 1992), which becomes important for regional climate only if these pastures cover substantial portions of a watershed or landscape. Moreover, when pastures are abandoned, the rate of evapotranspiration returns to within 90 percent of the primary forest rate within 15 years of regrowth (Nepstad et al., 1995). Hence, for evapotranspiration and other area-dependent ecosystem functions, such as runoff (and streamflow), carbon storage, and trace gas emissions, the ecological importance of forest remnants is determined by their areal extent and the relative functional behavior of the new ecosystems. The role of forest fragments in maintaining the regional climate of Amazonia is therefore small because they cover only 5 percent as much land surface as is covered by secondary forests and agricultural lands (Skole and Tucker, 1993).

Fire Regime

Perhaps the most important function performed by primary forest remnants in eastern Amazonia is their role in limiting the spread of fires. These forest remnants extend like giant fire breaks through the frontier landscape, preventing the spread of fires used to clear agricultural plots and reduce woody growth in pastures, as well as accidental or recreational fires. The very low flammability of primary forests relative to the pastures and scrub growth that replace them is caused by the low mass and high moisture content of the forest litter layer. The forest maintains a dense leaf canopy 20–50 meters above the ground throughout the year (figure 7-3), preventing all but 2 or 3 percent of the sun's energy from reaching the fine fuels on the forest floor. The leaves, twigs, and small branches that comprise this fine fuel remain moist, never drying substantially because of the low temperature (under 28°C) and high relative humidity (greater than 70 percent) of the understory air (Uhl and Kauffman 1990). Significant drying of the fine fuel layer only takes place during exceptionally dry periods, when plant-available soil moisture is depleted to depths of greater than 8 meters, provoking leaf-shedding by some trees (Nepstad et al. 1994, Nepstad et al. 1995). As leaves are shed by drought-stressed trees, more sunlight reaches the forest floor, elevating the air temperature, lowering the relative humidity, and drying the fine fuel layer.

The new ecosystems of eastern Amazonia are more flammable than the primary forests they replace because their fine fuel layer near the ground is either larger, or drier, or both. Pasture is highly flammable be-

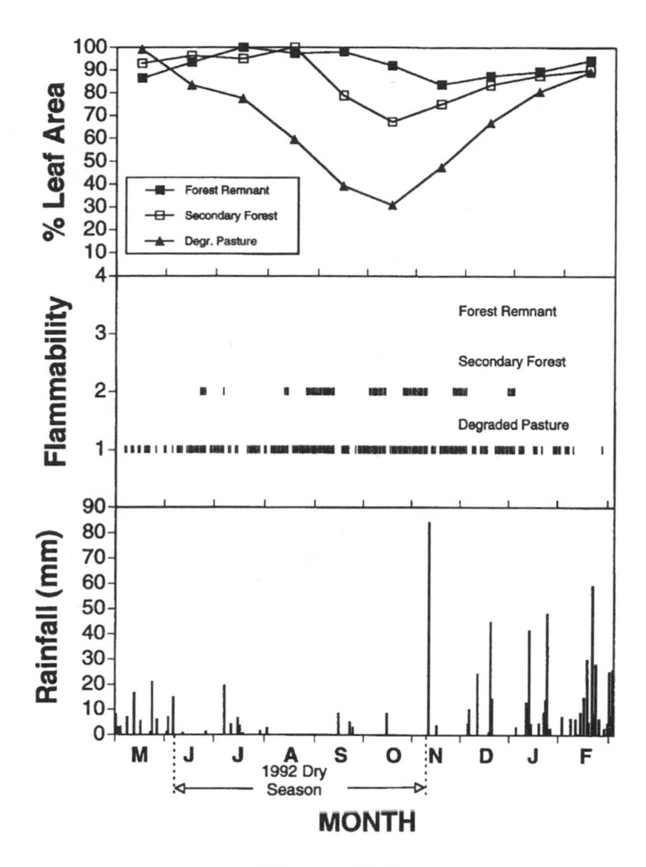

Figure 7-3
Seasonal trend of leaf area (percentage of maximum) and flammability during the 1992 ENSO event at the Fazenda Vitoria. The primary-forest remnant showed little leaf shedding relative to secondary forest and abandoned pasture, despite the very low rainfall during that year (40 percent below average). This resulted in a very small number of days during which it was flammable, while abandoned pasture and secondary forests were flammable for several weeks each.

cause the drying energy of the sun shines directly on the abundant fuel layer—the grasses and shrubs that comprise the pasture. When pastures are abandoned and trees establish and overtop the grasses and shrubs, ecosystem flammability is gradually reduced as the tree layer provides shade to the fuel layer, slowing its drying, and as grasses give way to the taller trees.

Taking into account both the effects of short-term rainfall regime on the moisture content of fine fuels (Uhl and Kauffman 1990) and the effects of long-term rainfall regime on deep-soil moisture supplies, and hence on leaf retention by the primary-forest canopy (Nepstad et al., 1995), we have estimated the periods during which the three ecosystems at the Fazenda Vitoria were flammable during a very dry year (1992, figure 7-3). The flammability of the pasture and secondary-forest ecosystems is determined by the short-term rainfall history (i.e., the number of consecutive days with little or no rainfall). Pastures can be ignited within a day or two of a dry-season rainfall event, while secondary forests can be burned within 10 days of dry-season rain, independent of the total amount of rain that fell during the preceding rainy season (Uhl and Kauffman 1990). Primary forests become flammable only under conditions such as those of late 1992, when the low rainfall of the preceding rainy season (40 percent below average) was not sufficient to replenish deep soil moisture supplies; some tree species became drought stressed during the subsequent dry season and began to shed their leaves. The net effect of short-term and long-term rainfall history in 1992 is that the pasture was flammable for more than 150 days and the secondary forest for less than 50 days. Based on observations of fire in primary forests of neighboring ranches, we estimate that the primary forest of Fazenda Vitoria was flammable for only a few days prior to the end of the 1992 dry season in November (figure 7-3).

The effectiveness of forest remnants in preventing the spread of fires depends on their dimensions. A single block of forest in the middle of a large expanse of pasture and secondary forest will do little to prevent fire propagation. If the same area of forest were stretched across the clearing, however, then fires started on one side of the remnant could not spread to the pastures and scrub forest on the other side. There also appears to be a minimum width of primary-forest strips that are able to resist the spread of fire, however. Near Paragominas, ground fires can penetrate the edges of primary forests (Uhl and Buschbacher 1985). The ability of forest remnants to resist fire entry is also reduced through selective logging, which opens the leaf canopy and introduces a large amount of fine fuel to the forest floor (Uhl and Kauffman 1990).

The evergreen forests that are resistant to fire because of their utilization of deep-soil moisture may be common, representing approximately 50 percent of closed-canopy forests in Brazilian Amazonia (figure 7-1). If rainfall declines in the region, as predicted, the late-1992 flammability of the Paragominas forest (figure 7-3) could become a more frequent, and widespread, phenomenon. Fires will spread into primary forests with greater frequency, killing trees and animals. Since burned forests release less water into the atmosphere than primary forests, an increased frequency of forest fires could trigger a positive feedback; the reduction in forest evapotranspiration following fire could lead to lower rainfall, which, in turn, leads to even more fire (Nepstad et al., 1995). Evidence from charcoal dating, sediment studies, linguistic diversification, and ceramics indicates that episodes of large-scale, drought-related fire appear to have befallen Amazonia at 400- to 700-year intervals during the last 2,000 years (Meggers 1994). This potential positive feedback between the predicted reduction in rainfall associated with deforestation and increased fire frequency in the region could supersede even the most valiant efforts to establish ecological reserves and defend remnants of primary forest in Amazonia

Trees, Birds, Bats, and Ants

Human occupation of Amazonian landscapes causes the proliferation of some plants and animals and the restriction or elimination of others. It is in the context of humans' dual effects on a regional biota that we now evaluate the importance of forest remnants as insularized habitats of plants and animals that might recolonize abandoned lands. In this analysis, we examine the tree flora, and the bird, bat, and ant faunas of adjacent primary forest, secondary forest, and degraded pasture at the Fazenda Vitoria. We discuss the characteristics of the species that depend on forest remnants, of those that thrive in nonforest ecosystems, and the interactions between these two groups.

Tree inventories of the three ecosystems at the Fazenda Vitoria were conducted in 1989 to 1992, with identifications made by Nelson Rosa (Museu Paraense Emilio Goeldi). The primary-forest inventory included all trees larger than 20 centimeters diameter at breast height (dbh) in a 50 by 1000-meter (5-hectare) transect, trees between 11 and 20 centimeters dbh in ten 25 by 25-meter (0.62-hectare) subplots, and trees between 1 and 10 centimeters dbh in seven 10 by 25-meter (0.17 hectare) subplots. In the secondary forest, there were no trees larger than 20 centimeters dbh. All trees larger than 5 centimeters dbh were inventoried in a 50 by 250-meter (1.25-hectare) transect, and trees between 1 and 5 centimeters dbh were inventoried in five 25 by 25-meter (0.31-hectare) plots. An additional three to four days were spent in each ecosystem collecting those species not encountered in the sample plots. In the degraded pasture, located next to the secondary forest, all trees were identified that emerged above the grass/shrub vegetation.

Of the 268 species of trees with a diameter at breast height equal to or larger than one centimeter found in the primary forest, only 69 (26 percent) occurred in the adjacent, 14-year-old secondary forest and 13 (5 percent) in the degraded pasture. Another 75 and 6 nonforest tree species were found in the secondary forest and degraded pasture, respectively. Hence, roughly one-fourth of the primary-forest tree flora had appeared in the secondary forest of the Vitoria following 14 years of regrowth.

Many of the tree species encountered in the secondary forest and degraded pasture had persisted on those sites through vegetative sprouting. Based on field observations and excavations of trees in secondary forests (Kauffman 1989; Nepstad et al., in press; N. Rosa, unpublished; Uhl et al., 1988), approximately 60 percent of the forest species found in the secondary forest were capable of sprouting from roots and stumps and may have persisted with little or no seedling establishment. These tree species may survive in the frontier landscape independently of primary-forest remnants. The remaining 28 forest tree species found in the secondary forest, which have little or no ability to sprout following cutting and burning, depend on new seedling establishment to recolonize abandoned fields. The number of such remnant-dependent tree species that occur in the secondary forest presumably increases with time as additional species colonize this ecosystem from the neighboring primary forest.

In the Fazenda Vitoria degraded pasture, the establishment of forest tree seedlings depends on successful dispersal of seeds from the forest remnant into the pasture, because cycles of grazing, weeding, and burning have depleted the pasture soil of tree seeds (Nepstad et al., 1991, in press). Birds and bats are the most important vectors by which the seeds of forest trees can arrive in the degraded pasture because only 11 percent of the forest tree flora is dispersed by wind and many of these anemocoric species are unlikely to travel more than 100 meters. Moreover, there is little movement of ground-dwelling forest mammals into the degraded pasture. We therefore studied the types of tree seeds arriving in the degraded pasture from the primary-forest remnant by analyzing the fecal contents of birds and bats captured in the primary forest, the secondary forest, and at 100 and 400 meters into the degraded pasture during both the wet and dry seasons. The rain of seeds into 1.5 by 1.5-meter cloth seed traps was also measured beneath fruit-bearing treelets (*Solanum crinitum*) and shrubs (*Cordia multispicata*) in the degraded pasture and above the grass/shrub matrix, approximately 100 m from the secondary-forest edge (Nepstad et al., in press; Vieira et al.,

1994). In addition to providing information about the mechanisms by which forest tree species arrive in degraded pastures, this study also provided information on the types of forest bird and bat species that venture into surrounding secondary forests and pastures.

Of the 22 species of trees found in the seed traps and in the feces of birds and bats captured in the abandoned pasture, only 3 were native to the primary forest. More important, the largest seeds that we found dispersed into the degraded pasture weighed only 16 milligrams (*Rollinia exsucca*)—about the size of a rice grain. While our sampling of birds, bats, and seed rain was not conducted throughout the year, nor was it conducted with sufficient intensity to detect all tree species that are deposited in the degraded pasture, it appears that the number of forest tree species that arrive in the degraded pasture is quite small. The primary-forest remnant contains many tree species that could potentially colonize the degraded pasture but have no mechanism by which to disperse their seeds into the pasture. This is not surprising given the relatively small numbers of forest bird species that were found in the degraded pasture. Of 248 bird species encountered in visual and audio censuses of the forest, only 18 occurred in the degraded pasture. Of these forest species, 119 were observed in the secondary forest (figure 7-4) (J. M. C da Silva, unpublished data; Silva et al., in press).

The forest remnant at the Fazenda Vitoria played an important role in maintaining populations of forest bird species, but its importance for maintaining bat populations was proportionally less. Half of the 14 bat species captured with mist nets in the forest were also encountered in the degraded pasture, and 8 of the 14 forest species were encountered in the secondary forest (figure 7-4). Forest bats may be better represented in the pasture than forest birds because they fly at night, when microclimatic conditions are very similar in the two habitats. The bat species that occurred in pasture and secondary-forest may depend on primary-forest for roosting and feeding.

The frugivorous bird species that occur in secondary forest and degraded pasture in the Paragominas region are small bodied and have generalist feeding behavior. Eighty-three percent of the degraded pasture bird species weighed less than 40 grams (Silva et al., in press), thus explaining the small sizes of tree seeds that were dispersed into the degraded pasture. Two of these species, *Ramphocelus carbo* and *Tachyphonus rufus,* were the most important vectors for the movement of tree seeds into the degraded pasture because of their habits of flying between the secondary forest and fruiting shrubs in the degraded pasture (Silva et al., in press). The bird species that depend on forest remnants

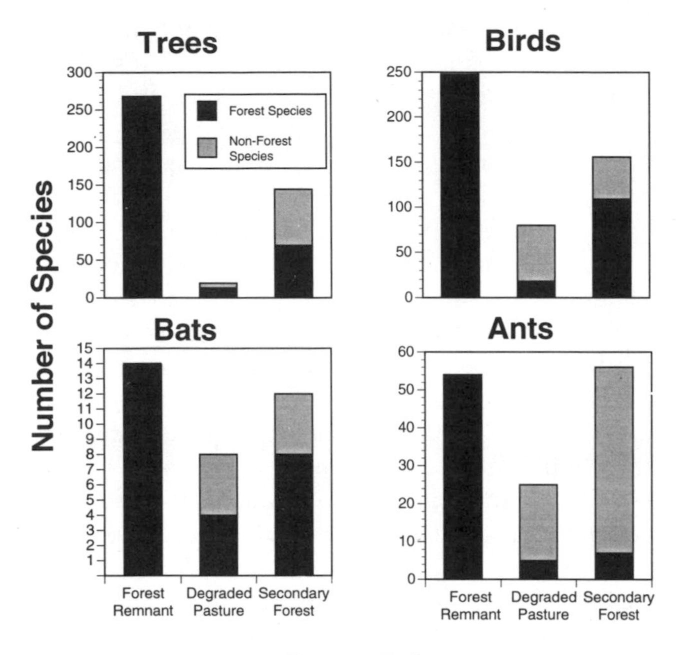

Figure 7-4

Number of forest and nonforest tree, bird, bat, and ant species in the primary forest-remnant, the degraded pasture, and the secondary forest on abandoned pasture (figure 7-2) at the Fazenda Vitoria, Paragominas.

for their survival in the frontier landscape are more specialized in their feeding habits and span a broad range of body sizes.

The seed-dispersing, forest bat species that ventured beyond the primary-forest remnant also have generalist feeding behavior and small body size. *Carollia perspicillata* was the most frequently captured species in both the forest (60 percent of all captures) and the nonforest ecosystems (42 percent of all captures in the pasture and secondary forest). This 20-gram bat is the dispersal agent for at least 25 plant species in Trinidad alone (Goodwin and Greenhall 1961) and disperses seeds of such tree genera as *Cecropia, Solanum, Cassia, Vismia, Banara,* and *Rollinia* at the Fazenda Vitoria. It roosts in bridges and other manmade structures and will fly more than two kilometers between roosting and feeding areas (Heithaus and Fleming 1978). Forest bat species that were not captured outside of the primary-forest remnant include species of the genera *Lonchophylla, Phylloderma,* and *Tonatia.*

Ant faunas, like bird and bat faunas, can be used to evaluate the role of forest remnants in the maintenance of species populations in frontier landscapes and to understand the patterns of plant succession on abandoned lands. We studied the ant fauna at the Fazenda Vitoria by collecting ants in pitfall traps in each of the three ecosystems and in both wet and dry seasons. In contrast with the bird and bat fauna, only a small percentage of the 54 forest ant species were found in the secondary forest (13 percent) and abandoned pasture (9 percent). This is not to say that ant species diversity was not high in the anthropogenic ecosystems. A total of 20 nonforest ant species were found in the degraded pasture and 49 were found in the secondary forest (figure 7-4). These results clearly underestimate the number of forest ant species that depend on forest remnants in frontier landscapes since the pitfall trap method that was employed in this study does not capture many of the arboreal ant species, such as several genera in the subfamilies Dolichoderinae, Formicinae, Pseudomyrmicinae, and Ponerinae.

Comparison of numbers of ant species among ecosystems does not provide information on population densities or ecological roles. The ant community of the degraded pasture was dominated by three ant species, none of which was found in the forest remnant. More than 90 percent of the 1,692 ants collected in the pasture pitfall traps belonged to the following species of the subfamily Myrmicinae: *Wasmannia auropunctata* (68 percent), *Pheidole* sp. (18 percent), and *Solenopsis* sp. (7 percent). Ant population shifts leading to the dominance of these three species have implications for forest succession in the abandoned pasture, for these species are generalist seed eaters that move large numbers of small tree seeds to underground nests, potentially limiting the rate of forest recovery (Moutinho et al. 1993; Nepstad et al., in press). These ant species, which are highly efficient at recruiting around food resources that appear on the ground and at defending these resources from competing species, may therefore prevent the colonization of other forest ant species that are less effective in gaining control of such resources.

The ant communities of the degraded pasture and secondary forest are also functionally different from that of the forest remnant because of the high concentration of the cutter ant, *Atta sexdens*—a species that was not found in the primary forest. *A. sexdens* is capable of moving seeds as large as corn grains and of cutting open larger seeds; it also defoliates and clips tree seedlings, preferring them over grass and shrub seedlings (Nepstad et al., in press). Against these activities that reduce tree seedling establishment in abandoned pastures, this voracious species also constructs nests that extend more than 5 meters deep into the soil. These nests provide microsites rich in organic matter refuse, in which roots proliferate and may thereby increase the rate of tree growth. Nest density of cutter ants in the secondary forest is more than 20 times greater than nest density of cutter ants (*Atta cephalotes* and *Acromyrmex* sp.) in the primary forest (Moutinho 1995).

In the case of trees, birds, bats, and ants, primary-forest remnants play an important role in maintaining residual populations of species that are unsuccessful in colonizing and persisting in degraded pastures and secondary forests. These species include (1) nonsprouting trees that are not effectively disseminated by wind or by those generalist birds and bats that carry seeds into surrounding ecosystems, (2) specialized birds and bats that depend on disturbance-sensitive forest trees for their food or shelter, and (3) arboreal and leaf litter ant species (table 7-1). Forest remnants may be necessary for the maintenance of 90 percent of the native tree flora, half of the forest bird and bat faunas, and 85 percent of the forest ant fauna in eastern Amazonian landscapes

Table 7-1

General ecological characteristics of the tree, bird, bat, and ant species that are (a) dependent on forest remnants and (b) capable of thriving in abandoned pastures or secondary forests.[a]

	Ecological characteristics	
Organism type	Species that depend on forest remnants	Species that do not depend on forest remnants
Trees	Large seeds no sprouting ability	Sprout after fire and cutting; or small seeds dispersed to pastures and secondary forest
Birds	Specialized diets, variable body size	General diets, small body size
Bats	Specialized diets (??)	General diets (??)
Ants	Arboreal ants, some litter ants	Generalist ants, aggressive seed harvesters, cutter ants

[a]Based on studies at the Fazenda Vitoria, Paragominas, Brazil.

of active and abandoned pasturelands. As secondary forests mature, they may gradually accumulate additional species of forest trees, birds, bats, and ants, reducing the importance of primary-forest remnants. However, in the Paragominas region, secondary forests older than 14 years of age are difficult to find.

Conclusions

Even when primary-forest remnants in frontier Amazonia are surrounded by human populations that hunt large mammals and birds, they provide important habitat for hundreds of plant and animal species not present in degraded pastures and that are slow to colonize secondary forests. The Fazenda Vitoria remnant, which is situated adjacent to a sawmill with 50-laborer households and which is a mere seven kilometers from the Paragominas urban center, still supports populations of the region's five primate species (S. Ferrari and P. Moutinho, unpublished data), hundreds of tree species, and a rich fauna of birds, bats, and ants. It is unclear if these species populations are reproductively viable, but in the short term they provide a source of organisms to many species that can colonize secondary forests in the event of land abandonment.

Forest remnants are best perceived as one component of a larger strategy to conserve tropical forests. They reduce the risk that fire will sweep across frontier landscapes. Moreover, they provide habitat for many plant and animal species that can recolonize secondary forests growing on land abandoned from agriculture. Large ecological reserves are an indispensable component of any strategy to conserve Amazonian forests, but they are by no means the only component, and efforts to implement parks in Amazonia have met with very limited success (Peres 1994). In Amazonia, the biggest strides in forest conservation will be made by steering rural development toward land uses that do not eliminate the primary-forest remnants that, by law but not necessarily in practice, must occupy at least half of every rural landholding.

Acknowledgments

This research was supported by grants form the U.S. Agency for Internaitonal Development, U.S. National Science Foundation, The MacArthur Foundation, and The Brazilian *Conselho Nacional de Desenvolvimento cientifico e Tecnologico.* We thank K. Schwalbe for preparing the graphics.

References

Bierregaard, R. O., T. E. Lovejoy, V. Kapos, A. A. dos Santos, and R. W. Hutchings. 1992. "The biological dynamics of tropical rainforest fragments: A prospective comparison of fragments and continuous forest." *BioScience* 42(11): 859–866.

Cronon, W. 1983. *Changes in the Land: Indians, Colonists, and the Ecology of New England.* New York: Hill & Wang.

Fearnside, P. M. 1993. "Deforestation in Brazilian Amazon: The effect of population and land tenure." *Ambio* 22(8): 537–545.

Goodwin, G. G., and A. M. Greenhall. 1961. "A review of the bats of Trinidad and Tobago." *Bulletin of the American Museum of Natural History* 122: 191–301.

Heithaus, E. R., and T. H. Fleming. 1978. "Foraging movements of a frugivorous bat, *Carollia perspicillata* (Phyllostomatidae)." *Ecology Monograph* 48: 127–143.

Kauffman, J. B. 1989. "Survival by sprouting following fire in tropical forests of the eastern Amazon." *Biotropica* 23(3): 219–224.

Lean, J., and D. A. Warrilow. 1989. "Simulation of the regional climatic impact of Amazon deforestation." *Nature* 342: 411–413.

Lefebvre, P. A., and T. A. Stone. 1994. "Monitoring selective logging in eastern Brazilian Amazonia using multi-temporal LANDSAT thematic mapper imagery." Proceedings of the International Symposium of Resource and Environmental Monitoring, ISBRS Commission 7, Rio de Janeiro, Brazil.

Mattos, M. M., and C. Uhl. 1994. "Economic and ecological perspectives on ranching in the eastern Amazon." *World Development* 22(2): 145–158.

Meggers, B. J. 1994. "Archeological evidence for the impact of mega-Niño events of Amazonia during the past two millennia." *Climate Change* 28: 321–338.

Moutinho, P. R. S., D. C. Nepstad, K. Araujo, and C. Uhl. 1993. "Formigas e floresta [Ants and forests]." *Ciencia Hoje* 88: 59–60.

Moutinho, P. R. S., D. C. Nepstad, and E. A. Davidson. 1995. "Acabar com a saúva, mas nem tanto [Eliminate cutter ants, but not that much]." *Ciencia Hoje* 18: 10–11.

Nepstad, D. C., C. Uhl, and E. A. S. Serrão. 1991. "Recuperation of a degraded Amazonian landscape: Forest recovery and agricultural restoration." *Ambio* 20: 248–255.

Nepstad, D. C., C. R. de Carvalho, E. A. Davidson, P. Jipp, P. A. Lefebvre, G. H. Negreiros, E. D. da Silva, T. A. Stone, S. Trumbore, and S. Vieira. 1994. "The role of deep roots in the hydrological and carbon cycles of Amazonian forests and pastures." *Nature* 372: 666–669.

Nepstad, D. C., P. Jipp, P. R. S. Moutinho, G. H. Negreiros, and S. Vieira. 1995. "Forest recovery following pasture abandonment in Amazonia: Canopy seasonality, fire resistance and ants." In *Evaluating and Monitoring the Health of Large-Scale Ecosystems,* edited by David Rapport, pp. 333–349. NATO ASI Series. New York: Springer-Verlag.

Nepstad, D. C., C. Uhl, C. Pereira, J. M. C. Silva. In press. "A comparative study of tree seedling establishment in abandoned pasture and mature forest of eastern Amazonia. *Oikos.*

Nobre, C. A., P. J. Sellers, and J. Shukla. 1991. "Amazonian deforestation and regional climate change." *Journal of Climate* 4: 957–988.

Peres, C. A. 1994. "Indigenous reserves and nature conservation in Amazonian forests." *Conservation Biology* 8: 586–588.

Salati, E., A. Dall'Olio, J. Gat, and E. Natsui. 1979. "Recycling of water in the Amazon basin: An isotope study." *Water Resources Research* 15: 1250–1258.

Shukla, J., C. A. Nobre, and P. Sellers. 1990. "Amazon deforestation and climate change." *Science* 247: 1322–1325.

Silva, J. M. C., C. Uhl, and G. Murray. In press. "Plant succession, landscape management, and the ecology of frugivorous birds in abandoned Amazonian pastures." *Conservation Biology.*

Skole, D., and C. Tucker. 1993. "Tropical deforestation and habitat fragmentation in the Amazon satellite data from 1978 to 1988." *Science* 260: 1905–1910.

Stone, T. A., P. Schlesinger, R. A. Houghton, and G. M. Woodwell. 1994. "A satellite-based map of South American vegetation." *Photogrammetric Engineering and Remote Sensing.* 60(5): 541–551.

Uhl, C., and R. Buschbacher. 1985. "A disturbing synergism between cattle ranching burning practices and selective tree harvesting in the eastern Amazon." *Biotropica* 17: 265–68.

Uhl, C., and J. B. Kauffman. 1990. "Deforestation effects on fire susceptibility and the potential response of tree species to fire in the rain forests of the eastern Amazon." *Ecology* 71: 437–449.

Uhl, C., R. Buschbacher, and E. A. S. Serrão. 1988. "Abandoned pastures in eastern Amazônia, I: Patterns of plant succession." *Journal of Ecology* 76: 663–68

Victoria, R. L., Luiz A. Martinelli, J. Mortatti, and J. E. Richey. 1991. "Mechanisms of water recycling in the Amazon basin: Isotopic insights." *Ambio* 20(8): 384–387.

Vieira, I. C. G., C. Uhl, and D. C. Nepstad. 1994. "The role of the shrub *Cordia multispicata* cham. as a 'succession facilitator' in an abandoned pasture, Paragominas, Amazônia." *Vegetatio* 115: 91–99.

Vieira, I. C. G., D. Nepstad, R. Salomão, N. Rosa. In press. "A floresta Amazônica após um seculo de agricultura: O caso da Zona Bragantina [The Amazon forest after a century of agriculture: The case of the Bragantine Zone.] *Ciencia Hoje.*

Wright, I., J. Gash, H. da Rocha, W. Shuttleworth, C. A. Nobre, G. Maitelli, C. Zamparoni, and P. Carvalho. 1992. "Dry season micrometeorology of central Amazonian ranchland." *Quarterly Journal of the Royal Meteorological Society* 118: 1083–1099.s

Chapter 8

Biology and Conservation of Forest Fragments in the Brazilian Atlantic Moist Forest

Virgílio M. Viana and André A. J. Tabanez

Introduction

Tropical forest fragmentation, a worldwide phenomenon that is generating increasing interest, is important for biodiversity conservation and sustainable development (Gradwohl and Greenberg 1991; Harris 1984; Janzen 1988; Shafer 1990; Wilcox and Murphy 1985).

One of the most threatened tropical forest ecosystems in the world is the Atlantic moist forest of Brazil (Mata Atlântica), which extends along the Brazilian coast, where the largest part of that country's population and economic activities are located. At the beginning of this century, the forest covered about 1.1 million square kilometers, or 12 percent of the Brazilian territory (Blockhus et al., 1992; Fonseca 1985; SOS Mata Atlântica and INPE 1993). The Mata Atlântica, having one of the highest levels of biological diversity in the world (Mori et al., 1981), is representative of humid tropical forests and their associated ecosystems.

Most remaining forest cover in the Mata Atlântica is on the hillsides along the coast, with very little remaining in the plateau region, where agricultural expansion has resulted in the loss of more than 95 percent of the forest (DEPRN 1991; SOS Mata Atlântica and INPE 1993). Plateau forests (*mata de planalto*) are some of the most fragmented and threatened ecosystems of the Atlantic forest domain. Despite their importance and uniqueness, these small, isolated forest fragments have received little attention in the literature (Bertoni et al., 1988; Leitão Filho 1987; Viana et al., 1992). Most vegetation studies have focused on floristic and phytosociological descriptions of old-growth forest patches and very little on the dynamics associated with forest fragmentation. As a consequence, there is little crucial biological information to guide forest conservation programs in the region. The plateau forests provide an excellent opportunity to develop a theoretical understanding of the biological consequences of fragmentation since they have a long and rather well-documented history of isolation (Viana 1990).

This chapter provides an overview of fragmentation in the Atlantic Moist forests of Brazil and presents a case study. We conclude with a discussion of the implication of current research to the conservation of forest fragments in the region.

Fragmentation of the Atlantic Moist Forest of Brazil

Contemporary deforestation of the Atlantic moist forests began with the settlement of Portuguese colonists along the coast in the sixteenth century and progressed as the agricultural frontier moved toward the interior. In the state of São Paulo, the greatest losses in forest cover were in the beginning of the twentieth century (table 8-1), during a period of rapid expansion of agricultural frontier that was mainly driven by coffee plantations. Current deforestation levels are high, ranging from 69 percent in the state of Santa Catarina to 99 percent in the state of Alagoas (CIMA 1991).

The challenge of conserving the Atlantic moist forest is not accurately represented in average deforestation levels for individual states because deforestation varies across the landscape within states. In the state of São Paulo, for example, deforestation is highest on the plateau, where topography, road access, and soil fertility are particularly favorable for agricultural expansion (SOS Mata Atlântica and INPE 1993; Viana, in press). Less than 2 percent of the forest cover remains there.

Table 8-1

Historical process of deforestation in the state of São Paulo.

Year	Forest Cover (%)	Loss/year (%)
1500 (original)	81.2	–
1854	79.7	0.001
1886	70.5	0.29
1907	58.0	0.60
1920	44.8	1.02
1935	26.2	1.16
1952	18.2	0.47
1962	13.7	0.45
1973	8.3	0.49
1985	7.4	0.07
1990	7.2	0.04

Source: SOS Mata Atlântica & INPE 1993.

Therefore, the representativeness of forest remnants in the plateau region is poorer than in the coastal areas. In addition, areas with the highest deforestation rates are the most fragmented. During the years 1985–90, 10.8 percent of the remaining forest cover was lost, leaving only 284,654 hectares of forest (SOS Mata Atlântica and INPE 1993). In the Piracicaba region, the remaining forest cover of old-growth (0.81 percent of the area) and secondary forests (1.23 percent of the area) is distributed among 29 and 73 fragments, respectively. Small forest fragments (less than 50 hectares) comprise 89.9 percent of remaining old-growth patches and 87.1 percent of secondary-forest patches (DEPRN 1991). Thus the forest landscape is highly fragmented (figure 8-1).

Forest ecosystems in the plateau region are the most threatened of the Atlantic moist forest. However, most protected areas of the state of São Paulo are located along the coast, not on the plateau (DEPRN 1991). Furthermore, a high proportion of the forest reserves in the plateau region are small and isolated fragments in a predominantly agricultural landscape (table 8-2). Therefore, the consequences of fragmentation, such as edge effects, are important for protected areas in the plateau forests. The same pattern applies to the majority of the Mata Atlântica—most remaining forest cover and protected areas are located in areas poorly suited for agriculture.

Most of the remaining forest cover (74 percent) in the plateau region of the state of São Paulo is privately owned. Protected reserves account

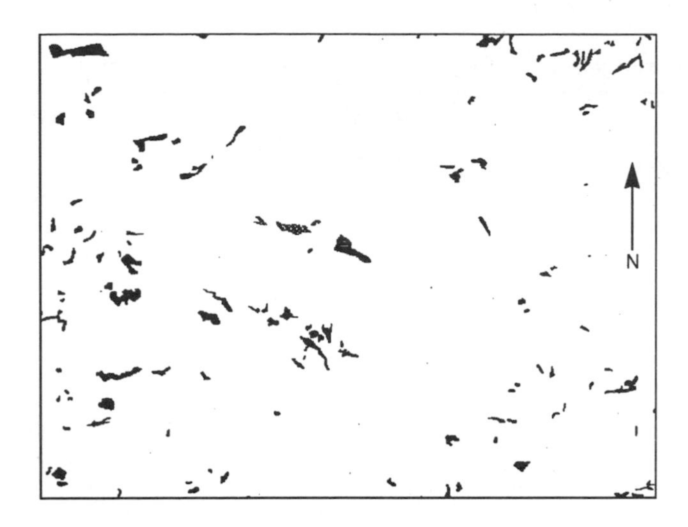

Figure 8-1
Distribution of forest fragments in the region of Piracicaba, São Paulo.

for less than 1 percent of the area and 26 percent of the remaining forest cover. Because of this, innovative approaches must be used to develop a conservation strategy for this region (Viana, in press). These approaches should make use of stakeholder analysis and a farming systems framework to understand the complex decision-making processes that lead landowners to clear forested areas and protect or restore forest fragments. These privately owned remnants often provide critical habitats for endangered species not protected in existing reserves. Moreover, fragments of plateau forest in intensively cultivated agricultural landscapes are not simply remnants of presettlement ecosystems. They are unique biological entities shaped by a history of increasing isolation, selective logging, and fire and landscape effects of agricultural, industrial, and urban activities.

Forest fragments maintain key ecological functions that are important for sustainable development strategies for the Atlantic moist forest region. Among these, two should be highlighted: conservation of biodiversity and protection of watersheds. Biodiversity conservation encompasses not only protection of endangered species, but also conservation of germ plasm for forest restoration. Forest fragments provide propagules of plant and animal species that can be reintroduced to areas in which they are now extinct and which private and public institutions are striving to restore. Moreover, forest fragments serve as "islands of

Table 8-2
Size and number of protected areas of plateau forests and associated ecosystems in the state of São Paulo.

Size (hectares)	Number
< 100	5
100–500	10
500–1,000	5
1,000–5,000	7
5,000–10,000	0
> 10,000	2
Total	29

Source: DEPRN 1991.

biodiversity" in agricultural landscapes and as a source of colonizers for many plants and animals rarely introduced through restoration schemes. It is critical that we improve our understanding of the biology of endangered species and how they respond to habitat fragmentation (Brown and Brown 1992; Bernardes et al., 1990). In addition, forest fragments, especially those on hillsides and along rivers and streams, are an important component of integrated watershed protection strategies; water availability is a key limitation to economic development in the state of São Paulo now and for the decades to come.

Deforestation is still going on at an alarming rate, given the small fraction of forest remaining (SOS Mata Atlântica and INPE 1993). Land-use policies and environmental education programs must be improved to reverse the current situation. There is some cause for hope. Both governmental and nongovernmental organizations have improved their monitoring structure and increased their environmental awareness. However, because the Atlantic moist forest is so fragmented, especially in the plateau region, particular attention from the scientific and environmental communities is urgently required.

How well forest fragments provide ecological functions for the Atlantic moist forest region will depend on their long-term sustainability. Indeed, if most fragments are not self-sustainable, restoration and conservation measures become a high priority. Therefore, a better understanding of the factors leading to forest degradation, coupled with new strategies for forest restoration is of a central importance to the development of a conservation plan for the region (Viana et al., 1992).

Case Study: The Santa Rita Forest

The Santa Rita forest is a privately owned forest fragment that has been studied since 1990 by the University of São Paulo's Laboratory of Tropical Silviculture in the "Escola Superior de Agricultura Luiz de Queiroz" (ESALQ) as part of their Biology and Management of Forest Fragments project. Our hypothesis, supported by this study, is that forest fragments are often not self-sustainable and require restoration practices to maintain their functions over the long run. This hypothesis is based on the understanding that edge effects (Lovejoy et al., 1986; Kapos 1989; Lawrence 1991), vine colonization following logging and fire disturbances (Baur 1964; Putz 1992), and reduced population sizes (Soulé 1987; Viana et al., 1992) all contribute to continuing degradation of forest fragments. We test the predictions that, if forest fragments are degrading, we should find (1) poor or nonexistent regeneration and critically small populations of several species, (2) poor forest structure dominated by low diversity eco-units (*sensu* Oldeman 1983), and (3) an arrested forest succession process.

Methods

The study site is located in the central part of the state of São Paulo. The site, characterized as "plateau forest," is on the western part of the Atlantic moist forest. The forest is semideciduous, with high plant-species diversity and emergent trees up to 55 meters in height. The soil, a mixture of ultisols, cambisols, and alfisols, has moderate to high fertility and good drainage. The climate is seasonal, characterized by a pronounced dry season (Cwa - Köppen: average rainfall of 1,250 millimeters per year), with six dry months, coinciding with the winter season, that receive less than 100 millimeters of rain per month.

The Santa Rita fragment is a 9.5-hectare forest, well protected, with no record of logging. The former owner practiced intensive agriculture on the remainder of the farm but chose not to plant sugarcane around the fragment to avoid damage from escaped fires. The area was recently sold and the new owner has planted sugarcane to the edge of the forest. A fire destroyed a portion of the fragment in mid-1994; illegal hunting, although present, is limited.

The forest was sampled in 1991 in three 10-meter-wide transects that cut across the fragment (figure 8-2). A total of 7,671 square meters (or about 8 percent) of the forest area was surveyed. All trees above 5 centimeters diameter at breast height (dbh) were measured, tagged, and identified. Voucher specimens were deposited at the herbarium at ESALQ. Soil seed bank samples of 0.04 square meters were collected in the edge, intermediate, and center of the fragment, five replicates in each area. Samples were taken to a nursery, where seeds were germinated under full sun. In ten random samples taken from all three transects, leaf area cover was obtained using a 0.5 by 0.5 meter grid at 2 meters height.

Results and Discussion

The Santa Rita fragment has a relatively high diversity (101 tree species, Shannon-Weiner = 3.84[1] with a highly variable structure (figure 8-3a,b). The forest is dominated by *Astronium graveolens* Jacp. (Anacardiaceae) and *Securinega guaraiuva* Kuhlm (Euphorbiaceae). In comparison with other fragments being studied in the region, this is a relatively diverse and well-protected forest (Tabanez et al., submitted).

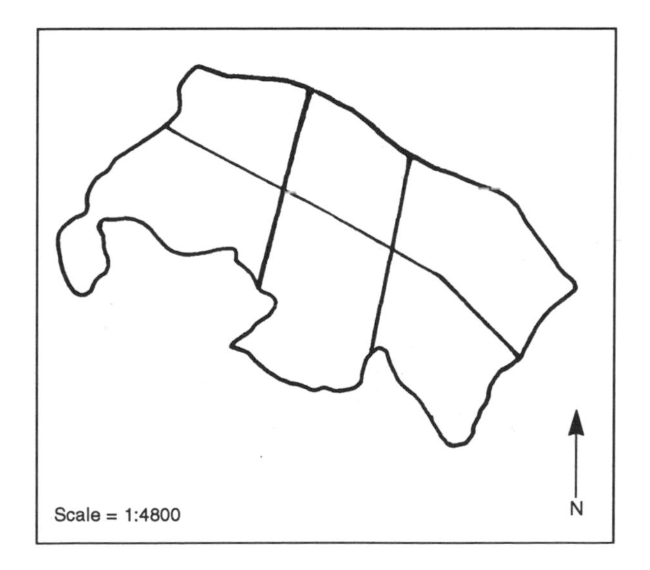

Figure 8-2
Santa Rita forest fragment with the location of transects.

---

[1]The Shannon-Weiner quantity is based on the equation $H = \sum_{i=1}^{s} (P_i)(\log_2 P_i)$; where s = the number of species; P_i = proportion of the total sample belonging to the ith species.

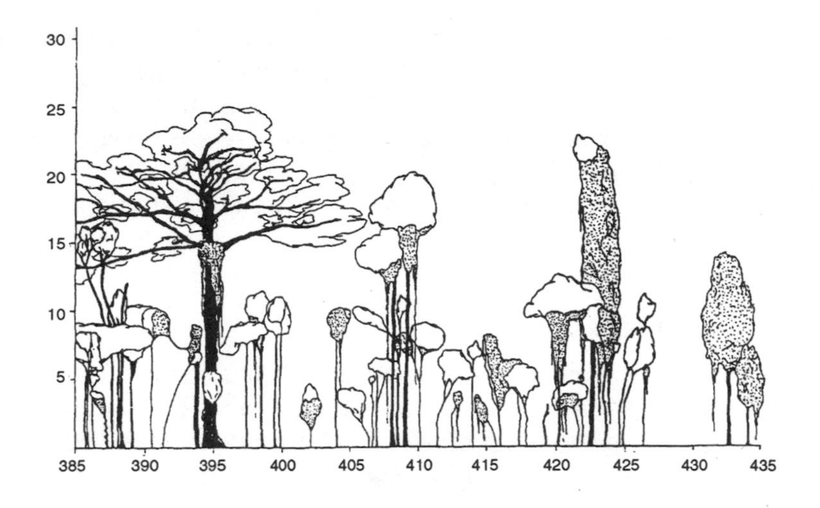

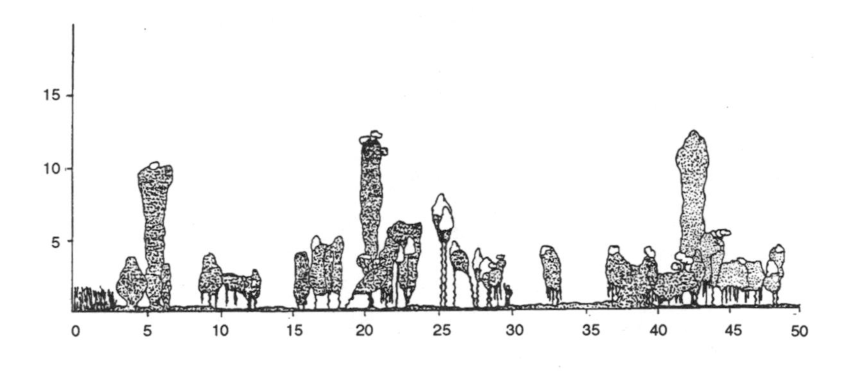

Figure 8-3

Forest profile for the Santa Rita forest: (top) indicating an area in the center of the fragment, and (bottom) indicating an area in the edge of the fragment.

The Santa Rita fragment, much like other plateau forests, is a mosaic of patches, or "eco-units" (*sensu* Oldeman 1983). Eco-units are forest patches within forest fragments that have similar structure and successional stage. Four eco-units were described from the fragment: low forest ("capoeira baixa"), high forest ("capoeira alta"), old-growth forest ("mata madura"), and bamboo-dominated forest. These eco-units vary in their extent, structure, and diversity (table 8-3). The low forest eco-units are 2–5 meters high, with low diversity and basal area, and dominated by vines and climbers such as *Celtis iguanae* (Ulmaceae). The high forest eco-units are 5–15 meters high, with higher diversity and basal area than the low forest eco-units. Approximately 60 percent of the area sampled was in secondary forest; the old-growth forest and bamboo-dominated eco-units are minor components of this fragment. The low proportion of the fragment in old-growth forest eco-units indicates that the fragmentation process has resulted in an ecosystem structure that is probably poorer than that of precolonial forests or that of larger and less-disturbed forest fragments (which are extremely rare today in the plateau region). These results show that (1) this fragment is not simply a remnant patch of the mature forest ecosystem but has been highly modified during the fragmentation process and (2) forest succession does not seem to be effective in recovering ecosystem structure, as indicated by a low proportion of old-growth eco-units. On the contrary, low diversity eco-units may be increasing in size at the expense of high diversity eco-units. Changes in the relative proportion of different eco-units will be examined in the future with the analysis of data from permanent plots.

A key element of an effective successional process leading to forest recovery is the establishment of a tree canopy layer, usually dominated by large-gap pioneer species (Denslow 1987; Martinez-Ramos 1985). In old-growth and self-sustaining forests, pioneer species are expected to dominate large gaps, which are particularly favorable environments, given their high light availability. In forest fragments dominated by high light capoeira eco-units, pioneer species are expected to have high den-

Table 8-3

Characteristics of eco-units in the Santa Rita Fragment in terms of percentage of the total area sampled, tree species diversity (Shannon-Weiner) and basal area (m^2/ha)

Eco-unit	Area sampled (%)	Diversity	Basal area	Trees/ha
Low secondary	26.08	1.17	9.05	445.00
High secondary	60.44	2.02	12.61	985.76
Old growth	6.19	1.89	18.49	926.32
Bamboo	7.29	2.01	19.33	1,001.79
Total/Average	100.00	3.84	12.53	842.24

sity. However, we found only 13 individuals of pioneer species, such as *Bauhinia forficata* Link. (Cesalpinaceae), *Vernonia diffusa* Less (Compositae), *Trema micrantha* (l.) Blum (Ulmaceae), and *Solanum granulosum-leprosum* Dunal (Solanaceae), that are favored by large-gap environments. This is equivalent to 16.95 trees per hectare, or less than 2 percent of the total density (930.78 trees per hectare) for all species. They had importance indexes of 5.4, 1, 0.9, and 0.4, and rankings of 17, 33, 47, and 90, respectively. Estimated population sizes for these species ranged from 12 to 36 individuals in the fragment.

We examined two possible explanations for this low importance of pioneers: (1) lack of seed availability in the soil seed bank and (2) competition with other large-gap guilds such as climbers and vines. We found that the soil seed bank demonstrated high seed density of pioneers (table 8-4), suggesting that lack of seed availability is not likely to be a strong barrier to the regeneration success of large-gap pioneers.

The relative success of other large-gap guilds, such as climbers and vines, was analyzed using leaf area indices along transects. In the low forest eco-unit, 65.3 percent of the total leaf area was attributed to vines and climbers. In contrast, 39.9 percent of the total leaf area in high for-

Table 8-4

Structure of soil seed banks (number of germinated seedlings/m^2) for different periods of sampling (time 1 and 2) and distances from forest edge (D1, D2, D3).

	Time 1			Time 2			
Species	D1	D2	D3	D1	D2	D3	Total
Pioneer	29	57	102	127	30	81	426
Solanum	8	19	41	11	21	40	140
Trema	12	7	4	32	1	3	59
Cecropia	6	21	20	9	3	9	68
Abutilum	0	0	32	0	0	6	38
Other pioneer	0	0	0	75	6	26	107
Herbaceous	175	14	12	54	39	32	326
Vines	0	0	0	19	2	5	26
Nonpioneers	0	0	0	0	0	12	12
Unknowns	105	48	27	12	3	3	198
Contaminants	99	125	115	2	5	15	361

est eco-units was attributed to vines and climbers, and the remainder to trees.

Some types of tree architecture appear to be more successful in competing with vines and climbers. *Schizolobium parahyba* (Vell.) Blake (Cesalpinaceae), for example, is a fast-growing species with a smooth bark and a high leaf turnover rate. Vines and climbers that manage to attach to these leaves fall with them as they senesce. The effect of tree architecture and leaf dynamics on the ability of trees to escape from vines and climbers deserves further study. There are some climbers that have very fast growth rates that seem to overcompete with some pioneer species. A noteworthy case is that of the climber *Celtis iguanae* (Ulmaceae), which dominates large patches of the fragment (up to 100 square meters) and has a competitive advantage over a member of the same family, *Trema micrantha* (a typical pioneer species). As a climber, it puts its weight on small trees, which are frequently found bent and broken. This may explain the low density of large-gap pioneers, which cannot occupy these habitats that are extremely favorable to them in physical terms. In conclusion, we hypothesize that these pioneer species are not regenerating well because they are losing the competition with vines and climbers that occupy these low forest eco-units.

Vine and climber competition is not an exclusive problem of large-gap pioneer species. Species of other guilds also show evidence of poor regeneration that can be associated with vine and climber competition. There are cases of large, emergent tree species that seem to have a barrier at the sapling stage. An example is *Cariniana legalis* (Mart.) Kumtze (Lecythidaceae), the largest tree of this ecosystem (up to 55 meters in height) and one that has a very diverse epiphyte flora and fauna. *Cariniana* has abundant saplings (24 individuals per hectare) (5–15 centimeters dbh) but it also has discontinuities in its population structure.

Other barriers for tree regeneration in forest fragments include lack of pollinators and seed dispersers, excessive predation and unfavorable microclimatic conditions (Nepstad et al., 1990; Viana 1990). These barriers often result in population structures with no juveniles present. *Cedrela fissilis* Vell. (Meliaceae), a gap opportunist canopy tree that was heavily exploited in the region for its valuable timber but supposedly not logged in this fragment, has no individuals smaller than 21 centimeters dbh. *Chorisia speciosa* St. Hil. (Bombacaceae), another gap opportunist species, has no individuals smaller than 13 centimeters dbh. The apparent lack of regeneration of some species, if not a result of sampling limitations, poses an important problem for regeneration ecology: Why are some species not regenerating? This is a question that requires population-level studies to analyze the factors influencing the demographic parameters at flowering, dispersal, predation, and recruitment phases.

Another factor that limits the prospects for self-sustainability of forest fragments is the small population size of several species (Soulé 1987; Viana 1990). This is particularly true for species that either had a low density in the nonfragmented forest ecosystem or became rare due to logging or to the ecological consequences of fragmentation. As shown in table 8-5, 34.7 percent of all tree species had an estimated total population of fewer than 15 individuals. The majority of the species sampled (67.3 percent) have estimated populations smaller than 45 individuals. This is likely to be a problem for long-term sustainability of these species since this may well be the effective population size due to the fact that these populations may be completely isolated from other populations. This isolation problem seems to be particularly serious in landscapes such as this, with low "porosity" (*sensu* Forman and Godron 1986). In this sugarcane-dominated landscape, movement of animals is restricted because of poor habitat quality and annual fires. Another factor is the probability that several species of seed dispersers and pollinators may have gone locally extinct in those fragments and landscapes, further reducing genetic flow and creating barriers for self-sustainability.

Table 8-5

Estimated total population size for species of different guilds in the Santa Rita forest fragment. Values are number of species.

	Population size				
Ecological group	1–15	15–30	30–45	45–60	> 60
Large-gap pioneer	1	1	1	0	0
Small-gap pioneer	14	5	2	1	7
Gap opportunist	5	2	6	2	5
Shade tolerant	1	2	1	1	4
Understory	3	0	0	1	8
Unknown	10	9	3	0	3
Total (%)	34.7	19.4	13.3	5.1	27.5

Conclusions

The results obtained from the Santa Rita case study indicate that this fragment is not likely to be self-sustainable. These findings are representative of a significant part of the remaining forest cover of the Atlantic moist forests, especially in the semideciduous forest of the plateau region, where fragments are smaller and more isolated and disturbed than those along the coastal hillsides (Viana et al., 1992). The problem of sustainability of small forest fragments is also very serious in the northern limits of the Atlantic forests (SOS Mata Atlântica 1990).

There is evidence that forest succession in small forest fragments in the plateau forests is not likely to recover forest diversity in low diversity eco-units, as evidenced by the low density of pioneers and overabundance of vines and climbers. The problem of vines in edges of forest fragments has been described elsewhere and seems to be a general problem (Lawrence 1991; Lovejoy et al., 1986; Waldorff and Viana 1993). Restoration practices should include vine control, especially vine cutting followed by plantings of fast-growing tree species (Jesus et al., 1989). Control of vines and climbers in fragments in the Atlantic moist forests would greatly benefit from a better understanding of the ecology of different species (Morellato 1991). Vine control should focus on forest edges and low forest eco-units.

A critical issue associated with forest fragmentation is reduction of population sizes (Soulé 1987). The demographic analysis of minimum viable populations indicates that the median extinction time depends on the initial population size. However, the rate of extinction will depend on the level of migration between populations (Ewens et al., 1990). Currently there is little data on seed dispersal between forest fragments in the Atlantic moist forests. The same is true for regeneration ecology of species, which is particularly important for species that show no regeneration in some forest fragments. Studies on the reproductive ecology of individual species in forest fragments give important information to design restoration programs (Amaral 1993). Restoration practices should include enrichment planting with species with overly small and isolated populations, as well as reintroduction of key pollinators and seed dispersers. Priority should be given to maintenance of populations of key tree species such as keystone species that play a critical role in maintaining animal populations and emergent species, which provide habitat for a large number of epiphyte fauna and flora.

At the landscape level, the consequences of forest fragmentation generally vary depending on the matrix characteristics (Forman and Godron 1986; Hudson 1991). In the Atlantic moist forest, landscape matrix varies significantly, with different degrees of isolation of forest fragments, with shape and size structures of fragments, and with types of neighborhood. There is still a poor understanding of the effects of landscape matrix on fragment dynamics in the Atlantic moist forests. Restoration practices such as corridors between fragments should probably be an important component for a conservation strategy for the region. It is too early to advocate their use in all cases because their benefits and drawbacks are still being debated. Ideally, each situation should be assessed individually (Shafer 1990). The assessment of individual cases of restoration of fragmented landscapes should include not only biological factors but also socioeconomic ones (Viana, in press). It is critical to remember that fragments in the Atlantic moist forests are mostly privately owned. Therefore, the future of these fragments will largely depend on a good understanding of how landowners perceive the value of forest fragments and how effective government policies and environmental campaigns can create a more positive attitude toward forest fragments.

In conclusion, the Atlantic moist forests in Brazil are even more threatened than deforestation statistics—although themselves alarming—indicate. Small, isolated, and disturbed fragments, such as the Santa Rita forest, are often not self-sustainable and require active restoration practices. These restoration practices are urgent in many cases, especially in the plateau region, where ecosystems are most threatened. The challenge is how to reconcile the urgent need for restoration with the limited amount of biological and socioeconomic information on the dynamics of fragmentation. The most promising approach seems to be the parallel development of restoration practices and research on conservation biology and socioeconomics. Restoration practices will have to be cautious because fragments are too few, too small, too isolated, and too vulnerable. Applied research would greatly benefit from increased interaction with restoration activities by yielding results that are relevant to the challenge of conservation of the fragmented landscapes of the Atlantic moist forests.

Acknowledgments

We would like to thank the financial support of Fundação Boticário para a Proteção da Natureza and FAPESP; the institutional support of IPE and IPEF; the generous collaboration of the owners of the Santa Rita forest fragment; the students and researchers of the Laboratory of Tropical Forestry at ESALQ, especially André Dias and Leandro Pinheiro; Dr. Ricardo Rodrigues for his assistance in botanical identifications; Dr. Julie Denslow for many fruitful discussions; Drs. Elizabeth Taylor, Jõao Luiz Ferreira Batista, and Hilton T. Z. do Couto, who reviewed an earlier version of this manuscript; and Liliana Correa Viana for her support in preparing the figures and revising one of the final versions of the manuscript.

References

Amaral, W. A. N. 1993. "Ecologia reprodutiva de cytarexylum myrianthum cham. (Verbenaceae) em mata ciliar no município de Botucat, SP." M.Sc. thesis, ESALQ, University of São Paulo.

Baur, B. N. 1964. "The ecological basis of moist forest management." Sydney, New South Wales, Australia, Forestry Commission.

Bernardes, A. T., A. B. M. Machado, and A. B. Rylands. 1990. "Fauna Brasileira Ameaçada de extinção." Belo Horizonte, Brazil, Fundação Biodiversitas.

Bertoni, J. E., F. R. Martins, J. L. Moraes, and G. J. Shepherd. 1988. "Composição florística do Parque Estadual de Vassununga, Sanat Rita do Passa Quatro." *Boletim Técnico do Instituto Florestal* 42: 149–170.

Blockhus, J. L., M. Dillenbeck, J. Sayer, and P. Wegge. 1992. *Conserving Biological Diversity in Managed Tropical Forests.* Gland, Switzerland: IUCN.

Brown, K. S., and G. G. Brown. 1992. "Habitat alteration and species loss in Brazilian forests." In *Tropical Deforestation and Species Extinction,* edited by T. C. Whitmore and J. A. Sayer, pp. 120–142. London: Chapman and Hall.

CIMA 1991. "Relatório nacional." Comissão Interministerial sobre Meio Ambiente e Desenvolvimento, Brasília.

Denslow, J. 1987. "Tropical treefall gap and tree species diversity." *Annual Review of Ecology and Systematics* 18: 431–451.

DEPRN. 1991. "Projeto Olho Verde." Departemento de Proteção aos Recursos Naturais do Estado de São Paulo.

Ewens, W., P. J. Brockwell, J. M. Gani, and S. I. Resnick. 1990. "Minimum viable population size in the presence of catastrophes." In *Viable Populations for Conservation,* edited by M. Soulé, pp. 59–68. Cambridge, UK: Cambridge University Press.

Fonseca, G. A. B. 1985. "The vanishing Brazilian Atlantic forest." *Biological Conservation* 34: 17–34.

Forman, R. T. T., and M. Godron. 1986. *Landscape Ecology.* New York: Wiley.

Gradwohl, J., and R. Greenberg. 1991. "Small forest reserves: Making the best of a bad situation." *Climatic Change* 19: 253–256.

Harris, L. D. 1984. *The Fragmented Forest.* Chicago: University of Chicago Press.

Hudson, E. W. 1991. *Landscape Linkages and Biodiversity.* Washington, DC: Island Press.

Janzen, D. 1988. "Management of habitat fragments in a tropical dry forest: Growth." *Annal Missouri Botanical Garden* 75: 105–116.

Jesus, R. M., G. B. N. Dias, E. M. Cardoso, and M. S. Menandro. 1989. "Enriquecimento em matas degradadas e em formação de menor potencial." Compania Vale do Rio Doce (unpublished paper).

Kapos, V. 1989. "Effects of isolation on water status of forest patches in the Brazilian Amazon." *Journal of Tropical Ecology* 5: 1735–185.

Lawrence, W. F. 1991. "Edge effects in tropical forest fragments: Application of model for the design of nature reserves." *Biological Conservation* 57: 205–219.

Leitão Filho, H. F. 1987. "Aspectos taxonômicos das florestas do Estado de São Paulo." *Silvicultura em São Paulo* 16: 197–206.

Lovejoy, T. E., R. O. Bierregaard Jr., A. B. Rylands, J. R. Malcolm, C. E. Quintela, L. H. Harper, K. S. Brown Jr., A. H. Powell, G. V. N. Powell, H. O. R. Schubart, and M. B. Hays. 1986. "Edge and other effects of isolation on Amazon forest fragments." In *Conservation Biology,* edited by M. E. Soulé, pp. 257–285. Sunderland, MA: Sinauer.

Martinez-Ramos, M. 1985. "Claros, ciclos vitales de los arboles tropicales y regeneracion natural de las selvas altas perenifolias." In *Investigaciones Sobre la Regeneracion de Selvas Altas en Veracruz, Mexico, Vol. II,* edited by A. Gomez-Pompa, pp. 191–239. Xalapa, Ver., Mexico: Instituto Nacional de Investigaciones Sobre Recursos Bioticos.

Morellato, L. P. C. 1991. "Estudo da fenologia de árvores, arbustos e lianas de uma floresta semidecídua no sudeste do Brasil." M.Sc. thesis, Universidade de Campinas.

Mori, S. A., B. M. Boom, and G. T. Prance. 1981. "Distribution patterns and conservation of eastern Brazilian coastal forest species." *Brittonia* 33(2): 233–245.

Nepstad, D., C. Uhl, and A. Serrão. 1990. "Surmounting barriers to forest regeneration in abandoned, highly degraded pastures: A case study from Paragominas, Pará, Brasil." In *Alternatives to Deforestation in Amazonia,* edited by A. Anderson, pp. 215–231. New York: Columbia University Press.

Oldeman, R. A. A. 1983. "Tropical rainforest architecture, silvigenesis and diversity." In *Tropical Rain Forest: Ecology and Management,* edited by S. L.

Sutton, T. C. Whitmore, and A. C. Chadwick, pp. 139–150. Oxford, UK: Blackwell.

Putz, F. E. 1992. "Silvicultural effects of lianas." In *The Biology of Vines,* edited by F. E. Putz and H. A. Mooney, pp. 93–501. Cambridge, UK: Cambridge University Press.

Shafer, C. L. 1990. *Nature Reserves*. Washington, DC: Smithsonian Institution Press.

Soulé, M. 1987. *Viable Populations for Conservation.* Cambridge, UK: Cambridge University Press.

SOS Mata Atlântica. 1990. "Workshop Mata Atlântica. Reunião Nacional sobre a Proteção dos Ecossistemas Naturais da Mata Atlântica." SOS Mata Atlântica, São Paulo.

SOS Mata Atlântica and IINPE. 1993. "Evolução dos Remanescentes Florestais e Ecossistemas Associados do Domínio da Mata Atlântica." São Paulo, SOS Mata Atlântica and Instituto de Pesquisas Espaciais.

Tabanez, A. J., V. M. Viana, and A. Dias. Submitted. "Efeito de borda e estrutura de um fragmento de floresta de planalto." Revista Brasileira de Botanica.

Viana, V. M. 1990. "Biologia e manejo de fragmentos de florestas naturais." VI Congresso Florestal Brasileiro. SBS/SBEF, Campos do Jordão, S.P.

Viana, V. M., A. J. Tabanez, and J. Aguirre. 1992. "Restauração e manejo de fragmentos de florestas naturais." II Congresso Nacional sobre Essências Nativas. Instituto Florestal, São Paulo, S.P.

Viana, V. M. In press. "Conservation of biodiversity in Neotropical forest fragments in intensively cultivated landscapes." In *Interdisciplinary Approaches to Biodiversity Conservation and Land Use Dynamics in the New World,* edited by G. Fonseca and M. Schmink. Washington, DC: Conservation International.

Waldorff, P., and V. M. Viana. 1993. "Efeito de borda na Reserva Florestal de Linhares." VI Congresso Florestal Brasileiro. SBEF/SBS, Curitiba, Brazil.

Wilcox, B. A., and D. D. Murphy. 1985. "Conservation strategy: The effects of fragmentation on extinction." *The American Naturalist* 125: 879–887.

Chapter 9

The Importance of Forest Fragments to the Maintenance of Regional Biodiversity in Costa Rica

Carlos F. Guindon

The current rate of conversion of primary forest to agricultural land is widely recognized as a threat to tropical biodiversity and sustainable land use. Costa Rica is no exception, despite its advances in land conservation efforts (Gamez and Ugalde 1988). With a 3.6 percent annual rate of deforestation, two-thirds of Costa Rica's forests have been lost or degraded since 1950 (Gamez and Ugalde 1988). As in other previously forested tropical areas, deforestation has resulted in a few isolated tracts of protected forest and many small unprotected forest fragments of various sizes, shapes, and degrees of isolation and disturbance (Lovejoy 1988; Wilcove et al. 1986).

Of particular concern is that most protected areas tend to be located in relatively inaccessible (mountaintops) or inhospitable (lowland rainforest) regions. In Costa Rica this has resulted in most protected areas being located either below 50 meters or above 1,000 meters. The area in between tends to be where most human settlement occurs, resulting in greater natural habitat fragmentation and alteration. Because changes in moisture and temperature along the altitudinal gradient result in high species diversity, many species are inevitably left out of existing protected area systems. These species may occur within a limited section along the altitudinal gradient or depend on tracking seasonal resources or conditions along the gradient. The uneven distribution and limited coverage of protected areas, combined with increasing human demands for the land outside these areas, means that habitat fragments will play an increasingly important role in maintaining biodiversity within the tropics.

Forest fragment "islands" may be particularly important to maintaining the biodiversity of tropical highland reserves (Stiles 1988; Wheelwright 1983). For example, up to 25 percent of the resident bird species in Costa Rica's upland forests make altitudinal movements on a regular basis (Stiles 1988). Most altitudinally migrating birds depend on fruit or nectar (Stiles 1988). The distribution of fragments along moisture or elevation gradients may be as important to maintaining tree and bird species diversity as their size, proximity, or configuration at any one point along the gradient.

The Tilarán mountain range in northwestern Costa Rica provides a good example of this situation. The top of the range, mostly above 1,500 meters, is protected; but at lower elevations, the forest, especially on the Pacific side, becomes increasingly fragmented. In this region, as in other tropical montane regions, trees in the avocado family (Lauraceae) form a particularly important component of tree species diversity, density, and basal area (Guindon 1988; Wheelwright 1983).

There are approximately 40 species of bird-dispersed lauraceous trees in the Monteverde region of the mountain range (Haber 1991). Several species, such as *Ocotea whitei* and *Ocotea monteverdensis,* provide valuable timber. Both of these species are endemic to Costa Rica and have distributions limited primarily to the Tilarán mountain range (Burger and van der Werff 1990), where they occur primarily in forest fragments outside existing protected areas due to their altitudinal distribution (pers. obs.).

A mutualistic relationship may exist between some altitudinally migrating fruit-eating birds and lauraceous tree species within these forest fragments. Lauraceous fruits do not have to be ingested by a bird to germinate (pers. obs.); however, as found for other plants, dispersal away from parent trees may increase the chances of seed survival by moving seeds beyond the densest portion of the seed shadow (Dirzo and Dominguez 1986). Dispersal may also deliver seeds to more favorable sites for germination and growth (Howe 1986). Birds with large gapes are the only known dispersers of the large fruit produced by all but two of the lauraceous tree species in the Monteverde region. Therefore the maintenance of Lauraceae within forest fragments may depend on birds in several ways. Birds may disperse seeds within fragments, between fragments, or between the extensive forest and fragments. Seed dispersal into fragments may be critical to maintain tree populations that are overharvested or experience catastrophic mortality from disease or storms.

Forest fragments are known to be important to the resplendent quetzal (*Pharomachrus mocinno*), which feeds largely on Lauraceae fruit. The quetzal migrates altitudinally in accordance with fruit availability (Wheelwright 1983; Powell and Bjork 1995). From June through September they move from the protected area to forest fragments on the Pacific slope up to 5 kilometers away and 600 meters lower in elevation.

Other birds in the Monteverde region also eat large lauraceous fruit and may migrate altitudinally. They include the keel-billed toucan (*Ramphastos sulfuratus*), the three-wattled bellbird (*Procnias tricarunculata*), the emerald toucanet (*Aulacorhynchus prasinus*), and the black guan (*Chamaepetes unicolor*) (Wheelwright et al., 1984). Even less is known about the importance of forest fragments to these species.

Several studies have shown that forest fragment size and distance to extensive forest may determine the number of bird species occurring in any given fragment (e.g., Ambuel and Temple 1983; Blake and Karr 1987; Galli et al., 1976; Howe 1984; Howe et al., 1981; Moore and Hooper 1975; Opdam et al., 1985; van Dorp and Opdam 1987). Few studies have determined if there is a similar relationship between bird abundance and fragment size and isolation. In a study looking at understory birds using mist nets, Rappole and Morton (1985) documented a reduction in bird abundance with the fragmentation of a 4.85 hectare section of a much larger forest in Mexico.

Between July 1990 and January 1993 I determined the abundance, diversity, and density of quetzals, toucans, toucanets, bellbirds, and guans in 30 forest fragments. I also censused the availability of Lauraceae fruit in each fragment. From this data, I predicted that fragment use by the five frugivorous birds would be (1) positively correlated with fragment size, Lauraceae fruit production, and percentage forest within a 200-meter radius of the fragment and (2) negatively correlated with fragment isolation as measured by distance from extensive forest and total open space between the fragment and extensive forest. Frugivore abundance, density, and diversity were used to identify the most important of the 30 forest fragments for these five species of frugivores.

I also predicted that seeds would be dispersed more evenly and in larger numbers in fragments most often used by the five frugivorous birds. Greater seed dispersal should translate into increased Lauraceae regeneration, as measured by increased seedling and sapling establishment. Determining which factors, individually or combined, best predict frugivore use, and quantifying the importance of frugivore use to the regeneration of the Lauraceae, allowed me to recommend forest fragment management options that can enhance local and regional species diversity while providing timber and nontimber forest resources for landowners and communities. I describe briefly how these forest fragment management options are currently being implemented in order to propose a system of corridors and protected areas.

Study Site

This research was conducted in 30 forest fragments located between 900 and 1,500 meters on the Pacific slope of the Tilarán mountain range (10°18'N, 84°48'W) (figure 9-1). The forest fragments were on small farms outside of the 28,000-hectare Monteverde reserve complex (MRC) comprised of the Monteverde Biological Cloud Forest Reserve, the Santa Elena Reserve, and Bosque Eterno de los Niños. The fragments ranged in size from 0.30 to 7.92 hectares and covered a total area of 72.56 hectares. The terrain is highly dissected by streams and rivers, and the vegetation changes rapidly along an altitudinal moisture gradient. In the short elevational drop within my study area there is an approximately 90 percent turnover in tree species (Guindon, unpublished data). The region supports a very diverse flora and fauna, including over 800 species of woody plants, 400 species of birds, and 490 species of butterflies (Lawton and Dryer 1980; Wheelwright 1986).

Research Design

Initial research was conducted from June 15 to September 15, 1990. During this time 12 forest fragments were selected, transects established, and initial bird censuses made. In the summer of 1991, 18 additional fragments were added in order to include a more complete range of sizes and altitudes (table 9-1). Most fragments were less than two hectares in size and within two kilometers of the Monteverde reserve complex. Censuses were conducted in all 30 fragments between July 1991 and January 1993. Potential fragments were initially selected from aerial photographs and then ground checked. Forest fragments were de-

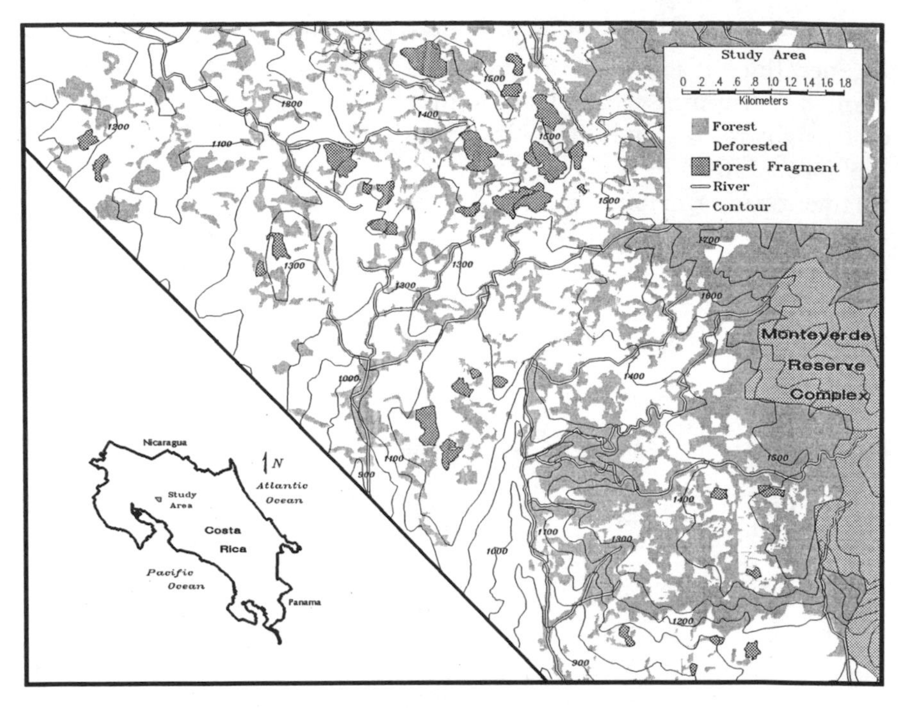

Figure 9-1
Location of the study site on the Pacific slope of the Tilarán mountain range in northwestern Costa Rica and the distribution of the 30 forest fragments.

fined as having a minimum width and length of 50 meters. Likewise transects did not extend into sections of woods greater than 50 meters wide. Transects were established along parallel compass lines running the full length of each fragment, starting 25 meters from one edge and repeated every 50 meters until reaching within 25 meters of the opposite edge.

Censuses were conducted twice a month during the three to four months when the five frugivores (resplendent quetzal, three-wattled bellbird, keel-billed toucan, emerald toucanet, and black guan) concentrate in the forest fragments. During the rest of the year, monthly censuses were conducted to detect the presence of species, or individuals, that do not leave the area. Census counts were made by walking the fixed parallel transects at a rate of 10 meters per minute and recording all individuals of the five bird species seen or heard within 25 meters to each side.

Table 9-1

Size distribution of forest fragments and distance distribution in relationship to the Monteverde Reserve Complex (MRC).

Size (ha)	Distance from MRC		
	≤2 km	> 2 km	Total
< 2	10	6	16
2–4	5	4	9
> 4	2	3	5
Total	17	13	30

All Lauraceae estimated to be of fruiting age and located within 2.5 meters of the transects were tagged with aluminum tags. Monthly observations were made on the phenologies of these trees. Fruit production and removal were estimated on all tagged Lauraceae by counting approximately 10 percent of the fruit on each major branch and extrapolating for the total tree. These counts were used to estimate total fruit crop for each fragment. Focal tree observations were conducted on a biweekly or monthly basis for all tagged trees with mature fruit. These observations consisted of spending 10 minutes at the tree and recording the presence of dispersers, fruit removal, and seed dispersal. These observations were complimented by focal bird observations, where individuals of each of the five species of dispersers were observed for a period of time to record fruit intake, fruit processing time, dispersal distance, and dispersal location.

Lauraceae density and diversity were determined for each fragment by using the modified 10 percent strip method (Cain and Castro 1959; Guindon 1988), which involved using the previously described transects for the bird censuses and measuring trees falling within 2.5 meters to each side of the transect. These 5-meter-wide transects were divided into 25-meter-long sections. Within each section, all Lauraceae larger than 1.3 meters in height were measured and their diameter at breast height (dbh) determined. In each fragment a series of 30 randomly placed 1-meter by 5-meter-plots, running perpendicular to the transects, were used to determine the frequency and density of Lauraceae seedlings (less than or equal to 1.3 meters in height).

Fragment size was determined from the number and lengths of transects. Total distance from and total open space between extensive intact forest and forest fragments were determined from aerial photos combined with ground checks. The percentage of forest within a 200-meter radius was determined using the computer-aided mapping and resource inventory system, CAMRIS 3.46 (Ford 1989).

Results

All fragments were used by at least one of the five species and all five species used forest fragments for at least part of the year. In 14 fragments three of the five species of frugivore were observed, in 7 four were observed, in 5 two were observed, in 2 only one was observed, and in 2 all five were observed. The emerald toucanet was observed in all of the forest fragments, accounting for 62.50 percent (1,421) of the total observations. Bellbirds were observed in 17 fragments, accounting for 15.52 percent (353) of observations; followed by toucans observed in 25 fragments, accounting for 9.63 percent (219) of the total observations; guans in 9 fragments, accounting for 8.88 percent (202) of the total observations; and quetzals in 11 fragments, accounting for 3.47 percent (79) of the total observations. The toucanet was the most abundant of the five frugivores in 23 fragments, the bellbird in 6 fragments, and the toucan in 1 fragment.

Seasonality

Although frugivore use of the fragments was seasonal for all species, there were at least two of the five species present in the fragments during any given month. Only the toucanet was present in the fragments during every month of the year. Birds were most abundant June through September, as measured by mean number of individuals observed per census. Species diversity, as measured by mean number of species observed per census, was also greatest June through September.

The overall abundance of frugivores was greater in 1991 than in 1992, which was mostly due to greater numbers of toucanets and quetzals in 1991.

Fragment Size

When looking at all five species of avian frugivores together, frugivore abundance was most highly correlated with fragment size (r_s 0.781, $P < 0.001$; where r is the Spearman rank correlation coefficient and P is the probability of this correlation coefficient occurring by chance, table 9-2). Frugivore diversity was also most highly correlated with fragment size (r_s 0.799, $P < 0.001$). When looking at the five species individually, fragment size was the highest correlate of frugivore abundance only for the toucanet (r_s 0.754, $P < 0.001$). Guan abundance and density (mean number per census per hectare) were significantly positively correlated with fragment size (r_s 0.462, $P < 0.01$; and r_s 0.340, $P < 0.05$). Quetzals were also significantly more abundant in larger fragments (r_s 0.419, $P < 0.05$).

Forest fragment size was significantly positively correlated with the percentage forest within a 200-meter radius (r_s 0.325, $P < 0.05$) and altitude (r_s 0.461, $P < 0.01$). The larger fragments also tended to have more Lauraceae fruit (r_s 0.405, $P < 0.05$).

Lauraceae Fruit

There was a significant tendency for more Lauraceae fruit not only in the larger fragments but also in those more distant from extensive forest (r_s 0.478, $P < 0.01$) and with the greatest amount of open space between them and the extensive forest (r_s 0.437, $P < 0.01$). After frag-

Table 9-2

Spearman rank correlation coefficients between frugivore abundance and diversity, expressed in mean number of birds or species observed per census (20–25 censuses), forest fragment Lauraceae fruit abundance, size, total distance and open distance to extensive forest, percentage forest within a 200-m radius, and altitude for 30 forest fragments (June 1991–January 1993).

Species	Fruit	Size	Total distance	Open distance	Percentage forest	Altitude
All Species	0.683***	0.781***	0.384*	0.226	0.173	0.256
Bellbird	0.634***	0.239	0.575***	0.739***	−0.089	−0.236
Guan	−0.077	0.462**	−0.263	−0.486**	0.380*	0.770***
Toucan	0.739***	0.320*	0.718***	0.662***	−0.204	−0.312*
Toucanet	0.577***	0.754***	0.135	−0.043	0.173	0.248
Quetzal	0.255	0.419*	−0.053	−0.284	0.308*	0.536**
Number of species	0.712***	0.799***	0.355*	0.153	0.247	0.409*

*, $P < 0.05$; **, $P < 0.01$; ***, $P < 0.001$

ment size the highest correlate of frugivore abundance and diversity was Lauraceae fruit abundance (mean number of Lauraceae fruit per census—r_s 0.683 and 0.712, $P < 0.001$; table 9-2). All species but the guan and quetzal showed significant positive correlations with Lauraceae fruit abundance. Toucans showed the highest correlation with Lauraceae fruit abundance (r_s 0.739, $P < 0.001$); bellbird abundance showed the second highest correlation (r_s 0.634, $P < 0.001$) followed by toucanets (r_s 0.577, $P < 0.001$). All five species of frugivores were observed feeding on Lauraceae fruit within the forest fragments.

Due to the size of their fruit, 16 of the 23 species of Lauraceae observed depend primarily on the 5 species of frugivores for their seed dispersal. Like the birds, fruiting individuals of Lauraceae were distributed unequally between the fragments. The unequal distribution of Lauraceae species and bird species among the fragments may result in many of the Lauraceae being dependent on only a few of the frugivore species for their seed dispersal.

When looking at the correlation between frugivore abundance and the fruit abundance of each Lauraceae species, only the bellbird showed a significant correlation with *Beilschmiedia* sp. 1. The quetzal showed the highest significant correlation with *B. brenesii* fruit abundance. The toucan showed the highest significant correlation with *B.* sp. 2. The guan and the quetzal showed the highest, and similar, significant correlations with *B. ovalis, B. pendula, Cinnamomum neurophylla, Ocotea* sp. 1, *O. insularis, Pleurothyrium* sp. 1, and *P. palmanum*. The guan showed the highest, and only significant, correlation with *Licaria* sp. 1. The bellbird and toucan showed the highest significant correlations with *Nectandra* sp. 1, *N. salicina*, and *Ocotea floribunda*. The guan showed the highest significant correlation with *Nectandra smithii*. The toucanet and quetzal showed the highest, and similar, significant correlations with *Ocotea membranacea*. Toucanets showed the highest significant correlation with *Ocotea tenera*. The guan showed the highest significant correlation with *Ocotea nicaraguensis* and *O. valeriana*. The quetzal showed the highest significant correlation with *Ocotea whitei*.

Distance from Extensive Forest

The only two species that showed the predicted negative correlation with distance from extensive forest were the guan and the quetzal (table 9-2). In both cases the correlation was not significant. Only guan density was significantly negatively correlated with total distance from extensive forest (r_s –0.309, $P < 0.05$). Toucan and bellbird abundance and density were significantly positively correlated with distance from extensive forest. Both guan abundance and density were significantly negatively correlated with total open space (r_s –0.486, –0.475, $P < 0.01$). Distance from extensive forest was highly correlated with total open space (r_s 0.845, $P < 0.001$); there also tended to be more Lauraceae fruit in the fragments further away from the extensive forest (r_s 0.478, $P < 0.01$). The fragments furthest away from extensive forest tended to have less forest within 200 meters of their boundaries (r_s –0.422, $P < 0.05$) and tended to be at lower elevations (r_s –0.382, $P < 0.05$).

Percentage Forest within a 200-Meter Radius

Only guans and quetzals were significantly more abundant in fragments with a higher percentage of forest within a 200-meter radius (r_s 0.380 and r_s 0.308, $P < 0.05$; table 9-2). Guan density was also significantly positively correlated with percentage forest within a 200-meter radius of the fragments (r_s 0.333, $P < 0.05$). There was more forest within a 200-meter radius of the forest fragments at higher elevations (r_s 0.597, $P < 0.01$) and less forest within 200 meters of the fragments with more open space between them and the extensive forest (r_s –0.313, $P < 0.05$).

Altitude

Fragment altitude was the highest correlate of guan and quetzal abundance (r_s 0.770, $P < 0.001$; and r_s 0.536, $P < 0.01$) and density (r_s 0.735, $P < 0.001$; and r_s 0.495, $P < 0.01$). Toucan abundance was significantly negatively correlated with altitude (r_s –0.312, $P < 0.05$). Both toucan and bellbird densities were significantly negatively correlated with altitude (r_s –0.484, $P < 0.01$; and r_s –0.318, $P < 0.05$).

Identification of the Most Important Fragments

A one-way ANOVA found that the abundance of frugivores differed significantly between fragments (F-ratio = 13.19; df = 29; $P < 0.001$). Fragments AC2, JW, and RLC had significantly higher abundances of frugivores than 23 of the other fragments (figure 9-2). Fragment AC1 had a significantly higher abundance of frugivores than 19 of the other fragments, fragment MA than 18 of the other fragments, and fragment RA2 than 14 of the other fragments. None of the remaining fragments showed a significantly greater abundance of frugivores than more than 4 other fragments.

The mean number of frugivore species observed per census also differed significantly between fragments (F-ratio = 10.77; df = 29; P < 0.001). Fragment RLC had a significantly higher diversity of frugivores than 21 of the other fragments, AC2 than 19 of the other fragments, JW than 17 of the other fragments, MA than 13 of the other fragments, AC1 than 12 of the other fragments, RA2 than 11 of the other fragments, AF1 than 10 of the other fragments, and CM than 8 of the other fragments. None of the remaining fragments showed a significantly greater diversity of frugivores than more than 3 other fragments. Frugivore abundance and diversity were highly positively correlated (r_s 0.929, $P < 0.001$).

Research Findings

The results show that forest fragments are important to all five species of frugivores for at least part of the year. The peak in fragment use, June through September, coincides with the peak in Lauraceae fruit production. Although I analyzed frugivore use of the fragments just during the months June through September, the significant relationships did not change. All of the most mobile species, bellbird, toucan, and toucanet, appear to select the forest fragments with the most Lauraceae fruit.

The guan and quetzal appear to be limited by distance from extensive forest and, in particular, the amount of open space they must cross to reach fragments. This may have prevented them from reaching the fragments with the higher Lauraceae fruit production, which tended to be further from extensive forest (r_s 0.479, $P < 0.01$) and with more open space to cross (r_s 0.438, $P < 0.01$). The guan and quetzal also may be limited by fragment size; both not only were more abundant in larger fragments but also occurred in higher densities in larger fragments (table 9-2). The negative correlation between guan abundance and Lauraceae fruit production may be due to a greater dependence on other species of large fruit. Guans were observed in particular to feed on the fruit of species in the Symplocaceae. Guan use of forest fragments is probably influenced as well by hunting pressure, which may be greater in the fragments further away from the protected forest and nearer to the larger population centers.

Besides differences in the willingness or ability to cross open space and greater distances, there may be physiological reasons for the differences in forest fragment selection. Bellbirds in captivity were observed not only to regurgitate the large seeds but also to regurgitate the skins of Lauraceae fruit (Guindon and Powell pers. obs.). This may allow them to key on Lauraceae fruit such as *Nectandra salicina,* which are relatively regular annual fruiters with relatively low reward (pers. obs.). In the middle to lower fragments, quetzals, which are not known to regurgitate fruit skins, were observed to feed primarily on the large fruit of *Beilschmiedia brenesii* and B. sp. 2.

Frugivore abundance was positively correlated with fragment size as predicted. However, bellbird abundance and density appear to depend more on Lauraceae fruit abundance and density than on fragment size. This may be due to their mobility across relatively long distances and open areas. Bellbirds may view the landscape at a much larger scale, selecting areas with groups of forest fragments relatively near to one another over areas where fragments are more dispersed or fruit is less abundant. They are also highly mobile and may view the landscape in a similar way. Toucans nest in lower areas where the forest is already fragmented, moving seasonally to the upper fragments or extensive forest to feed. Their use of the fragments is the reverse of guans and quetzals, which nest in the upper areas, mostly within extensive forest, migrating seasonally down into the forest fragments. The high correlation of toucanet abundance with fragment size is probably due to the fact that they nest in the forest fragments. They were also the only species present year-round and observed in all of the fragments. Interestingly, their low numbers between November and December suggests that a portion of the population either migrates out of the region or into different habitat, such as edges or second growth during this time of the year. Riley and Smith (1992) found that toucanets in the Monteverde region fed on smaller and lighter fruits during the dry season and at a lower foraging height. This would suggest that toucanets may shift in their habitat use during the time of year that their abundance is lowest in the forest fragments.

In order to maintain all five species of frugivores in the region it will be necessary to provide forest along the altitudinal gradient between 900 and 1,500 meters, as well as beyond. The only two fragments where all five species were observed, JW and AC2, occurred at 1,500 meters. Although relatively large and containing 12 percent of the Lauraceae fruit, JW and AC2 only accounted for 7 percent of the bellbird observations and 1 percent of the toucan observations, but 37 percent of the guan observations, 35 percent of the quetzal observations, and 25 percent of the toucanet observations. Quetzals and guans may benefit the most from forest left in the upper part of the gradient, while bellbirds and toucans may benefit the most from forest left in the lower part of the gradient.

The seasonal use of the fragments suggests that all species require forest outside of the region for at least part of the year. Radio telemetry data have shown that quetzals stay within the Tilarán mountain range but migrate over to the Caribbean slope part of the year (Powell and Bjork 1995). Bellbirds using the fragments migrate as far as the Caribbean lowlands of Nicaragua and the Talamanca mountain range in Panama (Powell and Bjork 1995). One keel-billed toucan with a transmitter stayed on the Pacific slope in the region of the study area but moved downslope for part of the year (Powell and Bjork 1995). At least part of the toucan population may migrate further, as they were observed to aggregate into groups of 15 to 20 in September and October just prior to the time their numbers decreased in the study area. The same behavior was observed in the toucanets.

The distribution of forest along the altitudinal gradient will also be critical for maintaining plant diversity. This is made evident by a comparison between forest fragment AC2 at 1,500 meters and forest fragment MA at 1,250 meters (figure 9-1). These fragments are of similar size, are on the same mountain slope, and are separated by only two kilometers. Of the 106 tree species detected in a 10 percent sample of the two fragments, only 11 species were shared and none of the 10 species with the highest importance values were held in common. Likewise, none of the 10 species of bird-dispersed Lauraceae occurring in these fragments were held in common.

In the case of large fruit-eating birds, maintaining plant diversity, and in particular Lauraceae diversity, may be critical due to differences in fruiting cycles. Maintaining significant populations of each tree species will also be important. Only a few species of Lauraceae fruit consistently every year (Wheelwright 1986), and many species fruit every two years or irregularly; within populations of a given species there are individuals that fruit at different times. This means that the fewer species there are, and the fewer individuals, the greater the probability that in any given year there will be a shortage of fruit.

Application of Research Findings

The initial findings on where the five species of birds occur in greatest numbers, densities, and diversity is being integrated into current conservation efforts to protect and interconnect key forest fragments under a regional biodiversity conservation plan. The fragments considered to be key for the maintenance of frugivore and Lauraceae diversity include JW (1,482 meters), AC2 (1,535 meters), RLC (1,488 meters), AC1 (1,370 meters), MA (1,235 meters), RA2 (2,400 meters) and FA4 (1,295 meters) (figure 9-1). These seven fragments accounted for 54 percent of the total forest fragment area surveyed, 71 percent of the Lauraceae fruit produced, 58 percent of total frugivore abundance, 49 percent of toucanet abundance, 56 percent of bellbird abundance, 41 percent of toucan abundance, 58 percent of guan abundance, and 72 percent of quetzal abundance. Unfortunately no fragments in the larger size range were included at the lower end of the elevational transect. This was due to a lack of large fragments at this elevation with the exception of forest on extremely steep slopes, where frugivores could not be easily censused. These forests are known to be used by most of the five species of frugivores (pers. obs.) and should also be considered critical based on their size and location in a region of the elevational gradient where little forest is left. This information, along with data from George Powell and Robin Bjork's telemetry work, was utilized to propose a system of microcorridors (10–100 meters wide) connecting "critical" habitats with the MRC (figure 9-2). Support is also being provided for an initiative to establish a Pacific slope macrocorridor (kilometers wide).

The microcorridor proposal is now being implemented through the Forest on Farms and Corridors Project administered by the Asociación Conservacionista de Monteverde. The project proposes to unite the efforts of researchers, extensionists, educators, and landowners to design and implement regional plans for the protection and sustainable use of natural forest resources within the buffer zone ("benefit zone") of the privately and publicly owned protected areas within the Tilarán mountain range. The project addresses the problems of most buffer zones: (1) land-use practices do not recognize and take full advantage of the potential goods and services (water, seed dispersal, pollination, wildlife diversity) provided by their proximity to extensive protected areas, and (2) components of the protected areas are being lost due to the reduction and degradation of the surrounding critical habitat.

The research component of the project is based on the previously described methodology. Its primary function is to identify the forest fragments outside of the protected areas with the greatest diversity and abundance of forest-dependent species.

The education component is directed primarily at promoting the establishment of a land ethic that recognizes (1) the importance of natural forest within the context of other land uses and (2) the interdepen-

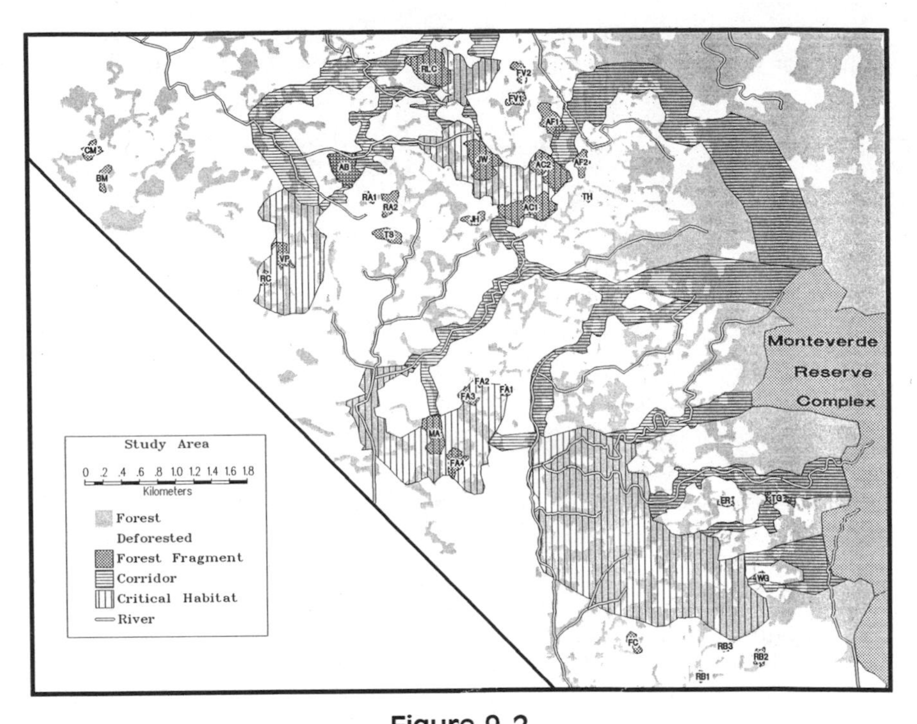

Figure 9-2

Proposed corridor system for the Monteverde region showing critical habitat, corridors, forest fragments included in the study, and protected area.

dence of habitats at a regional level. Education is viewed as a two-way process; therefore an important part of this activity is learning about the present and past uses of natural forest resources. This information is gathered from current landowners as well as previous or present hunters and sawmill operators. All members of the project team are viewed as educators. This means that information must be shared between the different project components and that activities are designed to involve landowner family members or other forest resource users directly.

The extension component is directed primarily at integrating information on the biology and the present and previous uses of natural forest areas into the socioeconomic context of a given region in order to come up with alternative land-use practices that strive for a sustainability of the forests. Tree abundance data are utilized, along with landowner knowledge of what trees have been most heavily exploited, to select species for reforestation within the corridors. Species that are found to be rare in the forest fragments, and that are known to have been extracted due to their commercial value, are give priority in reforestation efforts. Particular emphasis is directed at halting current degradation of forest remnants from cattle by providing fencing materials. Concrete posts interspersed with living fence posts are being promoted in order to reduce the impact of the traditional extraction of trees for split or round posts. The protection of water-caption areas and stream and river margins is also emphasized, as it often compliments the establishment of forest corridors and the protection of critical forest habitat.

When timber extraction is necessary, assistance is provided to the landowner to minimize the impact. Trees are bored to determine if they are sound before cutting permits are processed. Once permits are obtained for extracting timber, assistance is provided on how to fell and process the logs to minimize damage and waste.

The project is also promoting and providing assistance to landowners interested in establishing *servidumbres ecológicas* which are comparable to conservation easements. This means of providing long-term protection of areas, an approach that is relatively new in Costa Rica (Atmetlla 1992), is being implemented to establish a corridor, through the Monteverde community, along the Guacimal River. Within the Monteverde community, there is considerable interest in the use of *servidumbres ecologicas* as a tool to help control development as well as to set aside green areas. At this point, the participation is totally voluntary, with the only incentive being legal and logistical assistance. The easements are written into the property title and can only be modified by consensus of all signatories. It is hoped that, if they can be successfully used in Monteverde, this will create regional interest for their application in land-use planning. The easements will be more permanent than the landowner agreements being signed in return for fencing and reforestation incentives, which serve to establish more of a moral than a legal obligation and look only at a 10-year horizon. These agreements are dependent on landowner awareness and understanding of the objectives, which require an ongoing educational effort to assure their implementation.

Conclusions

Research has shown that forest fragments on the Pacific slope of the Tilarán mountain range are critical for maintaining regional biodiversity. Of primary importance is the distribution of the fragments along the altitudinal gradient, followed by their size and proximity or connection to other forested areas. Similar situations are probably true for other regions where protected areas do not adequately incorporate the full range of environmental gradients.

Understanding the dynamics of habitat fragments in relationship to extensive protected areas is key to being able to propose land-use alternatives that will minimize the loss of species and processes such as seed dispersal and pollination, which are critical for the regeneration and maintenance of the fragments. Once the components and processes are better understood, many approaches and tools can be used to achieve the desired land-use changes. Key to the success of such initiatives is agreement by local landowners and land users with the need for the proposed changes. An integrated, multidisciplinary team approach, comprised primarily of local residents, has been an effective way of identifying and implementing land-use changes in the Tilarán mountain range buffer zone.

Acknowledgments

I thank Danilo Brenes for field assistance and the many landowners who gave me permission to work on their property. My research would not have been possible without generous financial assistance from AID (through the WWF–US Biodiversity Support Program), the New York Zoological Society, Lincoln Park Zoological Society, and the National Audubon Society. Funding for the initial phase of the Forest on Farms and Corridor Project has been provided by AID through the National Fish and Wildlife Foundation and the RARE Center for Tropical Research. The support and experience of the Asociación Conservacionista de Monteverde has been essential to the project's success.

References

Ambuel, B., and S. A. Temple. 1983. "Area-dependent changes in the bird communities and vegetation of southern Wisconsin forest." *Ecology* 64: 1057–1068.

Atmetlla, A. 1992. "Servidumbres ecológicas." *Ivstitia* 68: 13–17.

Blake, J. G., and J. R. Karr. 1987. "Breeding birds of isolated woodlots: Area and habitat relationships." *Ecology* 68: 1724–1734.

Burger, W., and H. van der Werff. 1990. "Flora costaricensis: Family 80: Lauraceae." *Fieldiana (Botany) New Series* 23: 1–29.

Cain, S. A., and G. M. Castro. 1959. *Manual of Vegetation Analysis.* New York: Harper and Brothers.

Dirzo, R., and C. A. Dominguez. 1986. "Seed shadows, seed predation and the advantages of dispersal." In *Frugivores and Seed Dispersal,* edited by A. Estrada and T. H. Fleming, pp. 237–249. Dordrecht, Netherlands: Dr. W. Junk Publishers.

Ford, R. G. 1989. "CAMRIS: Computer-aided mapping and resource inventory system." Unpublished document, 124 pp.

Galli, A. E., C. F. Leck, and R. T. T. Forman. 1976. "Avian distribution patterns in forest islands of different sizes in central New Jersey." *Auk* 93: 356–364.

Gamez, R., and A. Ugalde. 1988. "Costa Rica's national park system and the preservation of biological diversity: Linking conservation with socioeconomic development." In *Tropical Rainforests: Diversity and Conservation,* edited by F. Almeda and C. M. Pringle, pp. 131–142. San Francisco: California Academy of Sciences.

Guindon, C. F. 1988. "Protection of habitat critical to the resplendent quetzal, (*Pharomachrus mocinno*), on private land bordering the Monteverde Cloud Forest Reserve." Master's thesis. Ball State University, Muncie, Indiana.

Haber, W. A. 1991. "Lista provisional de las plantas de monteverde, Costa Rica." *Brenesia* 34: 63–120.

Howe, H. F. 1986. "Seed dispersal by fruit-eating birds and mammals." In *Seed Dispersal,* edited by D. R. Murray, pp. 123–190. Australia: Academic Press.

Howe, R. W. 1984. "Local dynamics of bird assemblages in small forest habitat islands in Australia and North America." *Ecology* 54: 1585–1601.

Howe, R. W., T. D. Howe, and H. A. Ford. 1981. "Bird distribution on small remnants in New South Wales." *Australian Wildlife Research* 8: 637–651.

Lawton, R., and V. Dryer. 1980. "The vegetation of the Monteverde Cloud Forest Reserve." *Brenesia* 18: 101–116.

Lovejoy, T. E. 1988. "A celebration of life on earth." In *Tropical Rainforests: Diversity and Conservation,* edited by F. Almeda and C. M. Pringle, pp. 1–12. San Francisco: California Academy of Sciences.

Moore, N. W., and M. D. Hooper. 1975. "On the number of bird species in British woods." *Biological Conservation* 8: 239–250.

Opdam, P., G. Rijsdijk, and F. Hustings. 1985. "Bird communities in small woods in an agricultural landscape: Effects of area and isolation." *Biological Conservation* 34: 333–352.

Powell, G. V. N., and R. D. Bjork. 1995. "Implications of intratropical migrations on reserve design: A case study using Pharomachrus mocinno." *Conservation Biology* 9: 354–362.

Rappole, J. H., and E. S. Morton. 1985. "Effects of habitat alteration on a tropical avian forest community." In *Neotropical Ornithology,* edited by P. A.

Buckley, M. S. Foster, E. S. Morton, R. S. Ridgely, and F. G. Buckley, pp. 67–95. Washington DC: American Ornithologists Union.

Riley, C. M., and K. G. Smith. 1992. "Sexual dimorphism and foraging behavior of emerald toucanets *Aulacorhynchus prasinus* in Costa Rica." *Ornis Scandinavica* 23: 459–466.

Stiles, F. G. 1988. "Altitudinal movements of birds on the Caribbean slope of Costa Rica: Implications for conservation." In *Tropical Rainforests: Diversity and Conservation,* edited by F. Almeda and C. M. Pringle, pp. 243–258. San Francisco: California Academy of Sciences.

van Dorp, D., and P. R. M. Opdam. 1987. "Effects of patch size, isolation and regional abundance on forest bird communities." *Landscape Ecology* 1: 59–73.

Wheelwright, N. T. 1983. "Fruits and the ecology of resplendent quetzals." *Auk* 100: 286–301.

Wheelwright, N. T. 1986. "A seven-year study of individual variation in fruit production in tropical bird-dispersed tree species in the family Lauraceae." In *Frugivores and Seed Dispersal,* edited by A. Estrada and T. H. Fleming, pp. 19–25. Dordrecht, Netherlands: Dr. W. Junk Publishers.

Wheelwright, N. T., W. A. Haber, K. G. Murray, and C. Guindon. 1984. "Tropical fruit-eating birds and their food plants: A survey of a Costa Rican lower montane forest." *Biotropica* 16: 173–192.

Wilcove, D. S., C. H. McLellan, and A. P. Dobson. 1986. "Habitat fragmentation in the temperate zone." In *Conservation Biology: The Science of Scarcity and Diversity,* edited by M. E. Soulé, pp. 237–256. Sunderland, MA: Sinauer.

Chapter 10

Islands in an Ever-Changing Sea: The Ecological and Socioeconomic Dynamics of Amazonian Rainforest Fragments

Richard O. Bierregaard Jr. and Virginia H. Dale

Introduction

Our ever-growing human population is modifying the very structure of the planet's biosphere at a pace perhaps unprecedented in evolutionary history. Large-scale anthropogenic alteration of the environment has been occurring since the advent of agriculture. However, in the latter half of this century, human activities have reached the most remote regions of the earth and are having truly global repercussions. Global climate change, loss of species and habitat, and pollution now represent the unprecedented impact of humans on the environment.

Tropical rainforests now cover between eight and nine million square kilometers, or slightly less than 60 percent of our best estimates of their former extent (Myers 1988). Much of this deforestation has happened in the past several decades. Only a few large tracts of rainforest, such as western Amazonia and the Zaire basin, remain at least structurally intact, while others, such as the Atlantic coastal forests of Brazil and the rainforests of Madagascar, have been reduced to less than one-tenth of their original extent (Myers 1988).

The most striking alteration of the environment is the conversion of continuous habitat into a landscape mosaic of remnants of the native ecosystem surrounded by a habitat often dramatically different from the typical climax vegetation of the region. Because the rapid transition from continuous forest to a landscape mosaic of forest remnants will undoubtedly affect plant and animal species and their interactions, the ecosystem itself will change in ways that are difficult to predict.

Most forest remnants do not exist in isolation from humans. Humans create forest islands and continue to interact with them in a myriad of ways, many of which are described in other chapters in this book. In fact, changes in ecosystem structure and function in "habitat islands" often depend as much on what happens in the sea of habitat around islands as they do on changes that take place within the forest patches themselves. Most important, we have learned that conditions around forest remnants are very dynamic and usually driven by socioeconomic factors—road construction, colonization projects sponsored by local governments, and the economics of a fluctuating marketplace. While ecosystem change and socioeconomic factors are thus inextricably linked, they are rarely studied in concert.

There is a need for a predictive model to understand the biological dynamics of forest fragments and the factors driving human activities surrounding and, in fact, creating the remnants. In this chapter, we will review recent research into the ecological and socioeconomic changes that are associated with forest remnants in Amazonia and discuss ways these two approaches may interrelate.

Ecosystem Changes

The logical place to begin searching for clues to the ecological changes that can be expected in habitat remnants is on naturally occurring islands. In particular "land-bridge" islands, which were connected to nearby mainland before the most recent sea-level rise, are compelling analogs to the habitat islands that remain in the wake of human colonization.

Empirical evidence from many studies of island biotas demonstrates that islands typically support fewer species than comparable areas on nearby mainland (see references in MacArthur and Wilson 1967; Shafer 1990). MacArthur and Wilson (1963, 1967) proposed a simple equilibrium model that accounts for the observed patterns, arguing that the number of species on an island should be an equilibrium between the input of species through colonization and their "outflow" via extinction. Immigration rates should be dependent on both the size of an island as well as its distance from the nearest colonizing source. Extinction rates would depend on the size of the island (principally because larger islands can support larger populations, which are more likely to survive disease outbreaks, habitat destruction through fires or storms, or similar pressures).

This paradigm appealed to conservation biologists, who were quick to apply the principle to the design of wildlife reserves. Using MacArthur and Wilson's model, they argued that nature reserves should be as large as possible. However, Simberloff and Abele (1976) pointed out that MacArthur and Wilson's theory by itself did not necessarily imply that "bigger is better," because the model treats all species equally. When comparing a single large reserve to several smaller reserves of similar total area, the model does not predict that the large reserve will have more species than the smaller reserves in aggregate. Because the species in each of the smaller reserves are not necessarily the same (in the model all species have equal immigration and extinction probabilities), the total number of different species in all the smaller reserves might, in sum, be the same as or even greater than that in the single large reserve.

A very spirited debate erupted almost immediately in the biological literature (Diamond 1976; Terborgh 1976; Whitcomb et al., 1976; and others). Of particular concern was the fear that unscrupulous developers could use Simberloff and Abele's arguments to subdivide existing reserves or block the creation of the large, protected areas that many biologists believe are necessary to maintain regional biodiversity. For almost two decades researchers have addressed the issues raised by Simberloff and Abele without ever reaching a consensus (see review by Shafer 1990).

Of the many examples of island biota cited to quantify the relationship between extinction rates and island size, very few (e.g., Karr 1982; Leck 1989; Willis 1974) relied on accurate estimations of the numbers of species present when the forest tract in question was separated from continuous forest. Extinction rates in almost all published studies were estimated by assuming that the number of species on the island in question was once equal to that of the closest adjacent continuous habitat. However, the number of species in the small areas that later became either habitat or real islands would have been only a subset of the regional fauna, and hence the denominator in calculations of extinction rates (usually presented as a proportion of the local flora or fauna) cannot be known (Bierregaard and Lovejoy 1989).

With this shortcoming especially in mind, in 1979 the World Wildlife Fund and Brazil's National Institute for Research in Amazonia (INPA) launched the Biological Dynamics Forest Fragments Project (BDFFP) in the forests north of Manaus, Brazil (2°30' S x 60°W) (Lovejoy and Bierregaard 1990). The BDFFP, now jointly run by the Smithsonian Institution and INPA, takes advantage of a Brazilian law requiring landowners in Amazonia to leave the forest standing on 50 percent of any land to be developed.

Working in accord with local ranchers, BDFFP scientists marked off areas of 1, 10, 100, and 1,000 hectares of virgin forest slated for clearing for cattle pasture. Researchers then began censuses of a wide range of plants and animals in the very plots of forest that would later become habitat islands. From 1980 to 1990, the ranchers felled the forest around about half (11) of the 1-, 10-, and 100-hectare plots, isolating them from adjacent continuous forest by 100–1,000 meters (figure10-1). Continued censuses in the now-isolated plots have provided the only quantitative, replicated data (of which we are aware) anywhere in the world on population changes after the "insularization" of tropical forest remnants.

Soon after isolation of the experimental reserves, researchers detected several surprising effects of habitat fragmentation. Plants and animals reacted quickly to the dramatic changes in their local environment.

As soon as the ranchers began to fell the forest around the reserves and for several months after the study plots were separated from adjacent forest, we witnessed an increase in numbers of species of birds and population densities as individuals fled the felled forest for the sanctuary of the small forest remnants (Bierregaard and Lovejoy 1988). We speculate that these elevated population densities may be related to the subsequent disappearance of species from the small plots (Bierregaard and Lovejoy 1989).

Remarkably, understory birds and euglossine bees (important pollinators for many plant species) rarely crossed an open area of as little as 100 meters (Bierregaard and Lovejoy 1988, Powell and Powell 1987). The reluctance to cross such a narrow gap of nonforest habitat suggests that tropical rainforests may be particularly susceptible to fragmentation effects.

It came as no surprise that adult trees suffered along the windward edge of all isolated reserves (Lovejoy et al., 1984, 1986). In contrast, a

Figure 10-1

Photograph of 1- and 10-hectare forest reserves of the Biological Dynamics of Forest Fragments Project, north of Manaus, Brazil. Eleven such reserves, from 1 to 100 hectares, were isolated from continuous forest by cattle ranchers felling forest to implant pasture. (Photo by R. O. Bierregaard.)

dramatic increase in the mortality of canopy trees in the center of a 10-hectare reserve within two years of isolation was unexpected (Lovejoy et al., 1984). Litterfall, seedling recruitment, and seedling mortality were also affected near reserve edges (Sizer 1992). These edge effects were created by clearing, which exposed the sides of the reserves to desiccation by wind and sun and were strongest in the smaller reserves. Direct physical measurements (Kapos 1989; Kapos et al., 1993) as well as inferences from censuses of butterflies (Lovejoy, et al., 1984) indicate that a core area of only 25–60 hectares maintained microclimatic conditions in the understory typical of undisturbed forest in 100-hectare reserves.

While most avian guilds showed dramatic changes in 1- and 10-hectare isolates, most small understory birds were relatively unaffected in 100-hectare fragments. However, birds that require large territories, especially those that follow army ants, disappeared (Bierregaard and

Lovejoy 1989; Harper 1989) in the 100-hectare plot that was fully isolated.

In the smaller isolates, both species number and population densities of understory birds had begun to decline dramatically by a year after isolation, with no compensatory invasion by birds of the surrounding pastures (Bierregaard and Lovejoy 1988, 1989). Canopy species, such as parrots, hummingbirds, and some frugivores, which are more likely to cross gaps in the forest, continued to use the forest remnants after the understory birds suffered their population declines. These species were not censused quantitatively, so no accurate measure of changes in their activity in the isolated reserves is available.

After isolation of the reserves, the ranchers were not always able to maintain the area around the reserves in pasture, and a thick, weedy, second-growth forest began to grow up. Within three years the second growth attained a height of several meters and maintained its own shaded understory. In some cases this successional forest reconnected the reserves to adjacent, continuous forest and some, but not all, bird species that had disappeared from isolates returned (Stouffer and Bierregaard, in press a, b). We learned that for small mammals (Malcolm 1991) and birds (Stouffer and Bierregaard, in press a) the conditions around the reserve were a more important determinant of what species could persist in the reserves than was the size of the reserve.

These data lead us to believe that the absolute minimum forest patch size that could be considered viable for a substantial percentage of the species in the Amazon forest is 100 hectares. Clearly such small patches will not offer sanctuary to large vertebrates, whose loss may have a substantial destabilizing effect on the ecosystem (Terborgh 1992). Remnants of 100 hectares surrounded by pasture would probably lose enough species to disrupt many interspecific interdependencies, especially between plants and their pollinators or seed dispersers; and some or even many, species that persisted might be too small to be genetically viable in the long term. However, since the vast majority of animals in the forest are small, 100-hectare reserves should harbor a significant number of species. Furthermore, our experience suggests that the human-dominated landscapes in Amazonia often succeed to a mosaic of forest remnants, second-growth forest, and land kept clear for agriculture. Work at the BDFFP site with birds (S. Borges, pers. comm.) and small mammals (Malcolm 1991), and snakes in Belém, Brazil (Cunha and Nascimento 1978) suggests that many primary-forest species can persist in, or at least use second-growth forests. In such a matrix, complete devastation of the ecosystem is not to be expected.

Offerman et al. (1995) have reviewed available knowledge of habitat fragmentation effects on selected tropical animal groups. Their goal was to identify those groups most at risk for extirpation due to fragmentation. The review emphasizes the need for more information on fragmentation effects on organisms of the sort provided by experiments such as the BDFFP. The replication of the BDFFP experimental design in other tropical forests or in temperate forests would greatly improve understanding of the effects of habitat fragmentation on species survival. Also, Offerman et al. (1995) identify gaps in knowledge both in groups of species (e.g., bats and reptiles) and in life-history characteristics (e.g., minimum habitat size, mobility between habitat patches, and response to edge effects). Detailed knowledge of species' relationships to the spatial distribution of their habitat will link theoretical analyses of forest fragmentation and field data on changes in species abundance and distribution.

Relating Land-Use Alterations to Ecosystem Changes

Human land-use activities are the most common cause of habitat fragmentation. Meyer and Turner (1992) suggest that global land-use activities are affected by six factors: human population, affluence, technology, socioeconomic organization, level of economic development, and culture. Turner et al. (1994) emphasize the need for a conceptual framework and common methods for case studies of the interactions between human activities and land-cover changes.

Simulation experiments using species with different life-history patterns on heterogeneous landscapes have shown that natural disturbance and forest management practices interact with existing landscape patterns to dramatically affect the risk of species loss (Gardner et al., 1993). The most vulnerable species are restricted to specific habitat types and become isolated as a result of landscape fragmentation. Simulation results also show that land management practices that change the degree of landscape fragmentation instigate a change in the competitive balance between species, further complicating the maintenance of native species diversity (Gardner et al., 1993).

A critical challenge for improving understanding of the relationship between land-use management alternatives and ecosystem changes is developing useful tools. Models that link socioeconomic processes to ecological impacts offer such a tool. For example, the Man and the Biosphere project in the southern Appalachians and Olympic forests of the

United States is developing a model that focuses on land uses as road and second-home developments (Lee et al., 1992).

Description of a Model of Land-Use Change in the Amazon

More pertinent to the issue of dynamics in the Amazon is a model developed for central Rondônia, Brazil that analyzes the effects of small farmer settlement on deforestation (Southworth et al.,1991; Dale et al., 1993a,b). The Dynamic Ecological–Land-Use Tenure Analysis (DELTA) model is designed to improve understanding of the socioeconomic factors that drive the transition from continuous forest to a habitat mosaic. By tracking not only the history of each farm lot (which average 100 hectares), but also the history of colonists on the land, greater care can be taken to ensure that the resulting (aggregate) land-use patterns appropriately reflect the human settlement process. The tracking of each lot also allows DELTA to become a prescriptive tool with which settlement-policy options may be evaluated.

DELTA employs a suite of data and programs to analyze the impacts of land-use change (figure 10-2; major steps in the program elements are described in more detail below). The spatial data are organized in the ARC\INFO geographic information system (GIS) running on a Microvax 3500. Spatially explicit data used to initiate the simulation model include lot size and location with respect to neighboring lots, distance to market and to primary and feeder roads, soil conditions, and original vegetation types. Data on roads (alignments, lengths, type of pavement), agricultural and pasture suitability of the soils, and vegetation types are entered into the GIS that interfaces with the model. Each lot is assigned a set of numerical values corresponding to road distances, soil suitability, and original vegetation type. As the model is run, data on the land-use history are recorded for each lot every year.

Information from the literature is used to set the parameter values for the DELTA model (table 10-1). The nonspatial parameters that feed into the model include information about the size of the area, lot occupancy decisions (such as the proportion of a lot's area to be cleared each year), conditions for coalescing lots into pastures, criteria for changes over time in a tenant's mix of land uses on a lot, and choice variables that deal with conditions under which colonists may move to a new lot. A user interface to the model requests values for each parameter and assists the user by providing example inputs. Simulation results are based on the specific initial conditions, highway building patterns, and popu-

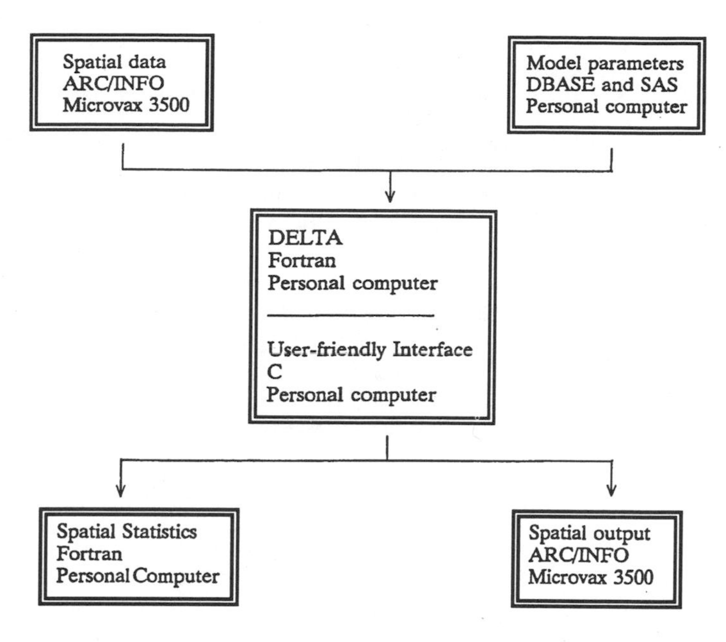

Figure 10-2
Interface between data and models requires a diversity of software and hardware systems. In each box, the type of data or models is on the first line, the software system on the second line, and the computer on the third line.

lation immigration and emigration conditions set by the user. Numbers of immigrants entering the region per year is currently an exogenous input.

DELTA is an integrated settlement diffusion, land-use change, and ecological change model. The simulation model is written in FORTRAN, and runs on a personal computer. Run times for a 3,000-lot system (300,000 hectares) and for a planning horizon of 50 years require only a few minutes. The model tracks the land-use and migrant status of each lot over a period of years. Farmers can move between lots dependent on lot conditions and distance to market. Agricultural lots can be coalesced into pasture. Lots can be fragmented by division and during sales with two or more colonist families occupying a lot that was originally designated for one family. Coalescing of adjacent lots or fragmentation can occur randomly if a set of conditions has been met. For coa-

Table 10-1

Parameters in DELTA.

Category	Parameter
System size	Number of land lots Number of time periods (years) Number of new settlers in each time period
Lot selection and occupancy period	Maximum number of consecutive occupancies Minimum time between consecutive occupancies Number of lot attraction variables Lot occupancy option Weights on lot attraction variables Probability of colonist leaving after each time period Logit choice model sensitivity (distance decay rate)
Coalescing	Minimum joint level of clearing Probability that lots will coalesce Profit per unit of coalesced land Minimum proportion of a lot that must be developed Maximum number of neighboring lots
Land-use mix	Rate of change to a given land use type Percent of a lot's land used in a given land use type for a given tenure
Land development	Land quality effect Tenure time effect Transportation cost Minimum area that must be developed Ratio of land productivity to land quality Unit transportation cost by land use
Carbon loss	Initial carbon on forested lot (mg/ha) Rate of carbon return Minimum carbon (proportion of initial carbon) Rate of decrease in carbon

lescing, the conditions include that a combined percentage of the cleared land area exceed a user-specified value, for large cleared areas are frequently combined under a single owner (Millikan 1988). Colonists leaving a lot either purchase a new lot, emigrate from the area, or become part of the local labor pool of landless workers.

Model outputs include spatial statistics and maps, as well as net regionwide impacts. Specifically, these are a time series of deforestation maps, changes in land degradation tied to the temporal and spatial pattern of deforestation, and aggregate statistics recording the effects of a particular settlement pattern on the spatiotemporal pattern of forest clearing and, in turn, the results of this land development on carbon release. Comparison of the spatial implications of particular land-use scenarios includes potential effects on the broad-scale pattern of forests, cleared areas, and secondary vegetation.

The model has been tested by comparing projections to farmers choices and to patterns of land-use changes as observed from remotely sensed images. Responses to surveys of the farmers in central Rondônia regarding their land-use decisions compare well to model projections, with the exception that the effects of diseases such as malaria on farmers are not included (Dale et al., 1993b). Furthermore, Frohn et al. (in review) show that LANDSAT imagery for central Rondônia indicate similar spatial and temporal patterns of land cover changes as the model projection based on the percentage of area cleared of forest, a contagion index, and the fractal dimension of the landscape.

Description of Land-Use Change in Rondônia and Application of the DELTA Model

The Brazilian state of Rondônia, located in the central Amazon Basin, is dominated by mature Neotropical forests. Government initiatives produced an extensive network of roads (an 18-fold increase in the total length of roads occurred between 1979 and 1988—(Frohn et al., 1990), which opened the interior forest areas to colonization. On the family-owned farms in central Rondônia, human encroachment into continuous forest typically results in a landscape mosaic of deforested areas and isolated forest fragments on the order of 10 to 100 hectares. The land for small family farms has been parceled into lots averaging 100 hectares in size (Leite and Furley 1985). Because of the difficulty of clearing land, each lot is now always entirely cultivated (the 50 percent law is usually not complied with in Rondônia).

Colonists used slash-and-burn techniques to clear the forest for agriculture, producing a dynamic mosaic of agricultural fields, pasture, regrowth, and mature forest, with most of the clearing originating along roads. Between 1978 and 1988, 17,717 square kilometers of Rondônia's forest were cleared, and an additional 1,417 square kilometers of forest were isolated from the contiguous forest into small (less than 100-square-kilometer) patches (Skole and Tucker 1993).

In areas undergoing colonization, a suite of socioeconomic factors, including government immigration policies, land tenure practices, mean clearing size, road development scenarios, and local market economies, affects the size and distribution of forest fragments scattered through deforested areas. The DELTA model incorporates these factors to assess the ecological impacts of deforestation as projected by the amount of land cleared and carbon released in colonization projects under different values of the model's parameters (Dale et al., 1994a). Changing patterns of forest clearance and isolation can also be simulated by the DELTA model (Dale et al., 1993a, 1994a; Southworth et al., 1991). The number of families remaining on the lots over time is projected by the model as a measure of the sustainability of the land management practices (Dale et al., 1993b).

DELTA model simulations suggest that different scenarios of land management result in unique land-cover patterns (Dale et al., 1994a). Land-use activities that are typical for colonists in Rondônia (Coy 1987; Dale and Pedlowski 1992; Leite and Furley 1985) involve rapid clearing of the forest and almost complete deforestation within 18 years. Model projections for the typical case compare well to changes over time in the patterns of land-cover change depicted from LANDSAT imagery of the region (Frohn et al., in review). The worst-case scenario (taken from the extreme of the Transamazon Highway experience as reported by Moran 1981 and Fearnside 1980, 1984, and 1986) results in total clearance in the first 10 years. On the other hand, a best-case scenario can be simulated in which forest clearance stabilizes at about 40 percent by year 20. The best-case scenario involves some clearing, but no burning, of the virgin forest and planting of perennial trees. Using the model to simulate different scenarios of land management permits evaluation of causes of specific land-cover changes. The worst- and best-case model projections are hypothetical, but the typical model scenario is meant to replicate recent land management activities in central Rondônia.

The basic insight underlying the DELTA model is that deforestation is a socioeconomic process that has ecological implications which may feed back to socioeconomic processes. The model provides the potential to predict future trends in deforestation.

Relating Ecosystem Changes to Species Distribution

When the DELTA model is combined with the ecological data from the Forest Fragments Project, it is possible to evaluate the impact of management strategies resulting from particular socioeconomic policies on the distribution of tropical forest species. For example, knowledge of different species' home range requirements allows prediction of the extinction rates for species in different-sized forest remnants. Data on tendencies to cross gaps of deforested land also provide insights into probable rates of recolonization. Note that the data on species-specific gap-crossing abilities contradict the assumption of the MacArthur and Wilson model, discussed earlier, that all species have an equal probability of colonizing an area. Together, data from the BDFFP on the maximum gap width between habitat patches that an animal is physically able to cross, and the minimum patch area required to maintain normal behavioral patterns (e.g., including special habitats for breeding), were used to calculate the area available for nine taxonomically diverse groups of animals (Dale et al.,1994b).

The area of forest habitat suitable for each animal group was projected for each year. First, "connected" clusters of habitat cells were identified. A cluster is connected if an animal in one cell can move to any other cell in that cluster (i.e., gaps between cells in a cluster are not wider than the maximum gap width that the animal is able to cross). Next, clusters with areas less than the minimum area required by an individual or group (for those that only occur in groups) were discarded. Further discussion of this technique can be found in Pearson et al. (in press).

This analysis showed that changes in available habitat are similar for animals that have their gap-crossing ability proportional to area requirements (figure 10-3), regardless of taxonomic affiliation (Dale et al., 1994b). For instance, the model suggests that species with large gap-crossing abilities and large area requirements (e.g., jaguars) respond in a similar fashion as do species with small gap-crossing abilities and smaller area requirements (e.g., sloths). In contrast, animals with gap-crossing ability disproportionately small in comparison to their area requirements (e.g., scarab beetles) decline more rapidly (figure 10-3). Few animals larger than insects seem to fall into this latter group; therefore, landscape-level analysis using simple gap-crossing ability and area requirements may provide a swift preliminary identification of the animals most susceptible to rapid decline and possible extirpation.

Once sensitive species have been identified, additional spatial data may be incorporated to improve the accuracy of the assessment. For example, when possible edge effects and breeding habitat requirements are included in the assessment of suitable habitat available for the tropical frog (*Chiasmocleis shudikarensis*), the amount of suitable habitat is decreased to 39 percent of the original area defined by gap-crossing and area requirements alone (Dale et al., 1994b).

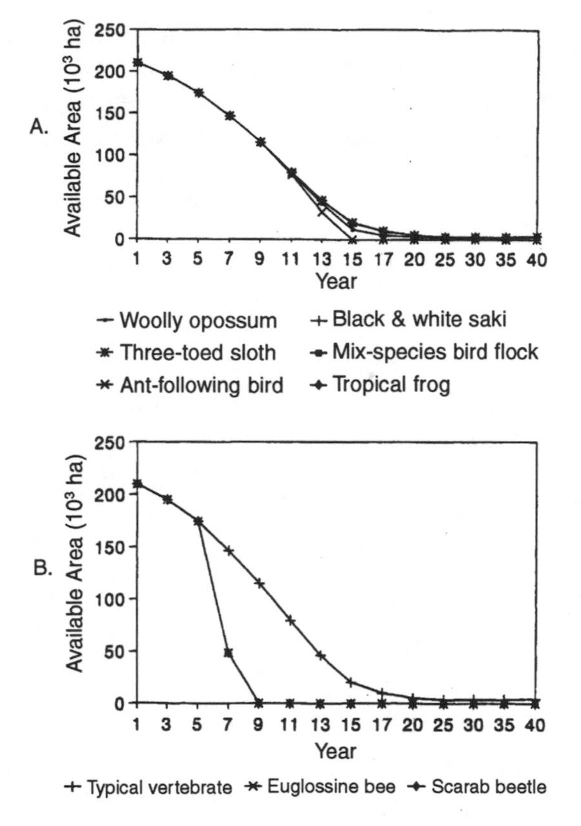

Figure 10-3

Simulated changes in an area of suitable habitat for the typical land-use scenario projected with the DELTA model over 40 years for fauna species whose gap-crossing ability is (a) proportional to their area requirements and (b) less than their area requirements.
Source: Dale et al., 1994b.

Conclusions

A coherent regional development scheme will depend in part on an integration of our knowledge of ecological processes operating within forest remnants and socioeconomic forces at work around them. The Biological Dynamics Forest Fragments Project extinction and immigration data will enable the DELTA model to predict losses in biodiversity associated with different suites of policy and socioeconomic factors.

Combining these analyses, we will be able to offer guidelines that will maximize the compatibility of development and ecosystem preservation of forest islands that remain after fragmentation of the Amazon rainforest.

Acknowledgments

Reviews of the manuscript by Rebecca Efromyson and Bill Hargrove were most helpful. Research was sponsored by the ORNL Director's R&D Fund and by the Carbon Dioxide Research Program, Atmospheric and Climatic Change Division, Office of Health and Environmental Research, U.S. Department of Energy, under contract DE-AC05-84OR21400 with Martin Marietta Energy Systems Inc. This is ESD publication no. 111 of the Biological Dynamics of Forest Fragments Project.

References

Bierregaard, R. O. Jr., and T. E. Lovejoy. 1988. "Birds in Amazonian forest fragments: Effects of insularization." In *Acta XIX Congress Internationalis Ornithologici, Volume II,* edited by H. Ouellet, pp. 1564–1579. Ottawa: University of Ottawa Press.

Bierregaard, R. O. Jr., and T. E. Lovejoy. 1989. "Effects of forest fragmentation on Amazonian understory bird communities." *Acta Amazonica* 19: 215–241.

Coy, M. 1987. "Rondônia: Frente pioneira e programa Polonoroeste. O processo de diferenciação sócio-econômica na periferia e os limites do planejamento público." *Tübingen Geographische Studien* 95: 253–270.

da Cunha, O.R., and F. P. do Nascimento. 1978. *Ofídios da Amazônia.* Pub. Avulsas no. 31. Museu Paraense Emílio Goeldi.

Dale, V. H., and M. A. Pedlowski. 1992. "Farming the forests." *Forum for Applied Research and Public Policy* 7(4): 20–21.

Dale, V. H., F. Southworth, R. V. O'Neill, A. Rose, and R. Frohn, 1993a. "Simulating spatial patterns of land-use change in Rondônia, Brazil." In *Some Mathematical Questions in Biology,* edited by R. H. Gardner, pp. 29–56. Providence, RI: American Mathematical Society.

Dale, V. H., R. V. O'Neill, and F. Southworth. 1993b. "Causes and effects of land-use change in central Rondônia, Brazil." *Photogrammetric Engineering & Remote Sensing* 59: 997–1005.

Dale, V. H., R. V. O'Neill, F. Southworth, and M. A. Pedlowski. 1994a. "Modeling effects of land-use change in central Rondônia, Brazil." *Conservation Biology* 8: 196–206.

Dale, V. H., H. Offerman, S. Pearson, and R. V. O'Neill. 1994b. "Effects of forest fragmentation on Neotropical fauna." *Conservation Biology* 8:1027–1036.

Diamond, J. M. 1976. "Island biogeography and conservation: Strategy and limitations." *Science* 193: 1027–1029.

Fearnside, P. M. 1980. "Land-use allocation of the Transamazon highway colonists of Brazil and its relation to human carrying capacity." In *Land, People and Planning in Contemporary Amazonia,* edited by F. Barbira-Scazzocchio, pp. 114–138. Cambridge, UK: Centre of Latin American Studies Occasional Paper No. 3.

Fearnside, P. M. 1984. "Land clearing behavior in small farmer settlement schemes in the Brazilian Amazon and its relation to human carrying capacity." In *Tropical Rain Forests: The Leeds Symposium,* edited by A. C. Chadwick, and S. L. Sutton, pp. 255–271. Leeds, UK: Leeds Philosophical and Literary Society.

Fearnside, P. M. 1986. *Human Carrying Capacity of the Brazilian Rainforest.* New York: Columbia University Press.

Frohn, R. C., V. H. Dale, and B. Jimenez. 1990. "The effects of colonization on deforestation in Rondoinia, Brazil." Oak Ridge, TN: *ORNL\TM 11470.*

Frohn, R. C., K. C. McGwire, V. H. Dale, and J. E. Estes. In review. "Using satellite remote sensing analysis to evaluate a socioeconomic and ecologic model of land-use change in Rondônia, Brazil." *International Journal of Remote Sensing.*

Gardner, R. H., A. W. King, and V. H. Dale. 1993. "Interactions between forest harvesting, landscape heterogeneity, and species persistence." In *Modeling Sustainable Forest Ecosystems,* edited by D. C. LeMaster and R. A. Sedjo, pp. 65–75. Washington, DC: American Forests.

Harper, L. H. 1989. "Birds and army ants (*Eciton burchelli*), observations on their ecology in undisturbed forest and isolated reserves." *Acta Amazônica* 19: 249–263.

Kapos, V. 1989. "Effects of isolation on the water status of forest patches in the Brazilian Amazon." *Journal of Tropical Ecology* 5: 173–185.

Kapos, V., G. M. Ganade, E. Matsui, and R. L. Victoria. 1993. "$\delta^{13}C$ as an indicator of edge effects in tropical rain forest reserves." *Ecology* 81: 425–432.

Karr, J. R. 1982. "Avian extinction on Barro Colorado Island, Panama: A reassessment." *American Naturalist* 119: 220–239.

Leck, C. F. 1989. "Avian extinctions in an isolated tropical wet-forest preserve, Ecuador." *Auk* 96: 343–352.

Lee, R. G., R. Flamm, M. G. Turner, C. Bledsoe, P. Chandler, C. DeFerrari, R. Gottfried, R. J. Naiman, N. Schumaker, and D. Wear. 1992. "Integrating sustainable development and environmental vitality: A landscape ecology approach." In *Watershed Management,* edited by R. J. Naiman, pp. 499–521. New York: Springer Verlag.

Leite, L. L., and P. A. Furley. 1985. "Land development in the Brazilian Amazon with particular reference to Rondônia and the Ouro Preto colonization project." In *Change in the Amazon Basin,* edited by J. Hemming, pp. 119–140. Manchester, UK: Manchester University Press.

Lovejoy, T. E., and R. O. Bierregaard Jr. 1990. "Central Amazonian forests and the Minimum Critical Size of Ecosystems Project." In *Four Neotropical Rainforests,* edited by A. Gentry, pp. 60–74. New Haven, CT: Yale University Press.

Lovejoy, T. E., J. M. Rankin, R. O. Bierregaard Jr., K. S. Brown Jr., L. H. Emmons, and M. van der Voort. 1984. "Ecosystem decay of Amazon forest remnants." In *Extinctions,* edited by M. H. Nitecki, pp. 295–325. Chicago: University of Chicago Press.

Lovejoy, T. E., R. O. Bierregaard Jr., A. B. Rylands, J. R. Malcolm, C. E. Quintela, L. H. Harper, K. S. Brown Jr., A. H. Powell, G. V. N. Powell, H. O. R. Schubart, and M. Hays. 1986. "Edge and other effects of isolation in Amazon forest fragments." In *Conservation Biology,* edited by M. Soulé, pp. 257–285. Sunderland, MA: Sinauer.

MacArthur, R. H., and E. O. Wilson. 1963. "An equilibrium theory of insular zoogeography." *Evolution* 17: 373–387.

MacArthur, R. H., and E. O. Wilson. 1967. *The Theory of Island Biogeography.* Princeton, NJ: Princeton University Press.

Malcolm, J. R. 1991. "The small mammals of Amazonian forest fragments: Pattern and process." Ph.D. dissertation, University of Florida.

Meyer, W. B., and B. L. Turner II. 1992. "Human population growth and global land-use and land-cover change." *Annual Review Ecology System* 23: 39–61.

Millikan, B. H. 1988. *The dialectics of devastation: Tropical deforestation, land degradation, and society in Rondônia, Brazil.* M.A. thesis, University of California, Berkeley.

Moran, E. F. 1981. *Developing the Amazon.* Bloomington, IN: Indiana University Press.

Myers, N. 1988. "Tropical forests and their species: Going, going...?" In *Biodiversity,* edited by E. O. Wilson, pp. 28–35. Washington, DC: National Academy Press.

Offerman, H. L., V. H. Dale, S. M. Pearson, R. O. Bierregaard Jr., and R. V. O'Neill. 1995. "Effects of forest fragmentation on Neotropical fauna: A review of current research." *Environmental Reviews* 3: 191–211.

Pearson, S. M., M. G. Turner, R. H. Gardner, and R. V. O'Neill. In press. An organism-based perspective of habitat fragmentation. In *Biodiversity in Managed Landscapes: Theory and Practice,* edited by R. C. Szaro. Oxford, UK: Oxford University Press.

Powell, A. H., and G. V. N. Powell. 1987. "Population dynamics of euglossine bees in Amazonian forest fragments." *Biotropica* 19: 176–179.

Shafer, C. L. 1990. *Nature Reserves: Island Theory and Conservation Practice.* Washington, DC: Smithsonian Institution Press.

Simberloff, D. S., and L. G. Abele. 1976. "Island biogeography theory and conservation practice." *Science* 191: 285–286.

Sizer, N. 1992. "The impact of edge formation on regeneration and litterfall in a tropical rain forest fragment in Amazonia." Ph.D. dissertation, Cambridge University, England.

Skole, D., and C. Tucker. 1993. "Tropical deforestation and habitat fragmentation in the Amazon: Satellite data from 1978 to 1988." *Science* 260: 1905–1910.

Southworth, F., V. H. Dale, R. V. O'Neill. 1991. "Contrasting patterns of land use in Rondônia, Brazil: Simulating the effects on carbon release." *International Social Sciences Journal* 130: 681–698.

Stouffer, P. C., and R. O. Bierregaard Jr. 1995. "Use of Amazonian forest fragments by understory insectivorous birds: Effects of fragment size, surrounding secondary vegetation, and time since isolation." *Ecology Monograph* 76: 2429–2445.

Stouffer, P. C., and R. O. Bierregaard Jr. 1995. "Effects of forest fragmentation on understory hummingbirds in Amazônia, Brazil." *Conservation Biology* 9: 1086–1095.

Terborgh, J. 1976. "Island biogeography and conservation: Strategy and limitations." *Science* 139: 1028–1029.

Terborgh, J. 1992. "Maintenance of diversity in tropical forests." *Biotropica* 24: 283–292.

Turner, B. L. II, W. B. Meyer, and D. L. Skole. 1994. "Global land-use/land-cover change: Towards an integrated study." *Ambio* 23: 91–95.

Whitcomb, R. F., J. F. Lynch, P. A. Opler, and C. S. Robbins. 1976. "Island biogeography and conservation: Strategy and limitations." *Science* 193: 1030–1032.

Willis, E. O. 1974. "Populations and local extinctions of birds on Barro Colorado Island, Panama." *Ecology Monograph* 44: 153–169.

Chapter 11

Modification of Tropical Forest Patches for Wildlife Protection and Community Conservation in Belize

Jonathan Lyon and Robert H. Horwich

Introduction

Tropical forest landscapes are undergoing rapid and dramatic alterations due to increasing human pressure to utilize and convert existing forests. Because of the high rate of conversion, traditional forest-protection strategies that emphasize human exclusion are no longer effective in protecting and maintaining biodiversity in disturbed tropical landscapes. Over the past two decades, there has been growing awareness that maintenance of biological diversity in the tropics is inescapably linked to the protection, management, and restoration of forest patches (Bierregaard et al., 1992; Ewel 1979; Gómez-Pompa and Vásquez-Yanes 1974; Hough 1988; Janzen 1988; Laurance 1991; Lovejoy et al., 1986; Lugo et al., 1993; Schwarzkopf and Rylands 1989; Terborgh 1992). Given this new focus for conservation, conservation planners have the task of identifying appropriate strategies and conservation goals for fragmented landscapes. Most small-scale forest patch or patch mosaic environments can meet and sustain the niche requirements of a limited number of wildlife species and provide only a finite amount of resources for human users. Protecting and sustaining forest patch viability for wildlife conservation may thus require modification and management to meet conservation goals.

Assessing the structure, composition, and unique ecological dynamics of individual forest patches has long been heralded as an essential building block in formulating long-term management plans (Pickett and Thompson 1978). Forest patches are dynamic entities with their constituent biota being controlled and influenced by a host of interacting biotic and abiotic factors and historic/present land use patterns (Brown and Lugo 1990; Gómez-Pompa and Vásquez-Yanes 1981; Redford 1992; Uhl et al., 1990). Within a given landscape they can range from relatively undisturbed forest, to highly disturbed secondary forest, to scrub, or to managed monoculture plantations. Wildlife conservation and management planning in patchy landscapes should be based on sound ecological assessments that include analysis of the existing forest structure and composition, the existing wildlife habitat, and the existing impacts of human utilization on the residual forest ecosystem (Mwalyosi 1991; Noss and Harris 1986). The development of conservation goals must also be bounded within the limitations of the existing ecological landscape (Noss 1987). Utilizing highly disturbed forest patches for wildlife conservation offers new challenges and also holds the promise of new opportunities.

In this chapter we discuss two projects in Belize that have targeted highly disturbed forest landscapes for wildlife conservation. These projects represent attempts to integrate wildlife and wildlife habitat protection within the cultural and socioeconomic environment of rural communities. The projects have been developed under two very different milieus and at different landscape scales. The main focus of the discussion is the Community Baboon Sanctuary (CBS).

The CBS is a small-scale (47 square kilometers), community-based wildlife conservation initiative developed on privately owned or leased lands. The CBS conservation mission is centered on community-based protection and enhancement of black howler monkey (*Alouatta pigra*) habitat in a lowland forest area encompassing several small, rural communities. In this chapter, we discuss how ecological assessments of forest patches were conducted and how the assessment results relate to howler conservation, forest habitat protection, and the development of a community-based management program. We also discuss the importance of utilizing local knowledge of land-use practices and forest dynamics in tandem with academic research efforts, and how combining these two sources of information can lead to successful efforts to protect, modify, and improve existing habitat. We conclude the discussion of the CBS by describing the management methods employed to promote simple, low or no-cost habitat preservation and enhancement, and the importance of choosing methods that fit the socioeconomic and cultural norms of the local participants.

The second project, the Manatee Special Development Area Biodiversity Project (MSDABP), is a landscape-level effort (46,760 hectares) encompassing both public and private lands superimposed on a complex matrix of forest, savanna, and aquatic ecosystems. The MSDABPs conservation mission is the protection of numerous biotic communities and their constituent flora and fauna in a multiple-use landscape. Unlike the CBS, management in the MSDABP is landscape based and not focused exclusively on individual land holdings. The brief discussion of the MSDABP is centered on the application of the habitat protection and community conservation programs implemented at the CBS to an ecologically diverse, large-scale, public, multiple-use landscape.

The discussion of these two projects focuses on three aspects of wildlife conservation planning in disturbed landscapes, namely (1) integrating wildlife conservation strategy for a given fragmented and patchy landscape within both the ecological and human contextual setting, (2) the importance of using ecological research in partnership with local knowledge in habitat modification planning, and (3) the importance of involving local communities in all aspects of the conservation effort and maintaining a strong level of local control in terms of land-use management, resource monitoring, and potential economic returns.

Community Conservation and Wildlife Protection on Private Lands: The Community Baboon Sanctuary

The Community Baboon Sanctuary is an experimental, community-based conservation project created in 1985 to protect the habitat of the vegetarian black howler monkey (*Alouatta pigra*). The CBS gets its name from the Belizean name for the black howler—"baboon." The CBS project area is located in north-central Belize along the Belize River (see figure 11-1). The CBS includes eight villages, representing approximately 450 people. The Belize River site was chosen for research and conservation based on reconnaissance work that showed the area sup-

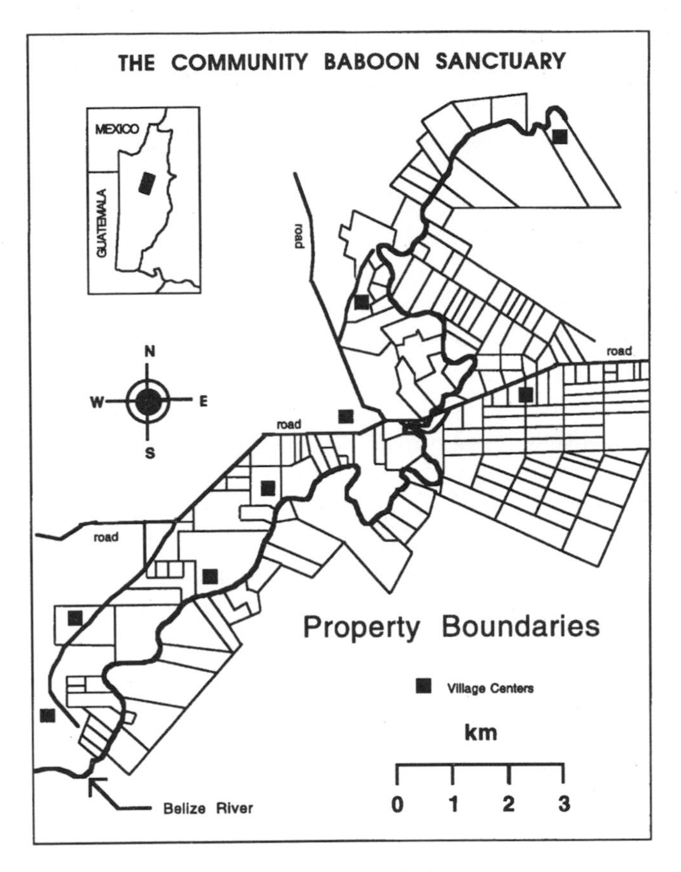

Figure 11-1
Map of the Community Baboon Sanctuary (CBS) in Belize. Individual property boundaries are shown as well as major roads and the Belize River.

ported an especially large population of black howler monkeys (Horwich and Johnson 1986). The howler population along the Belize River is concentrated in areas of highly disturbed lowland forest and parallels the occurrence of other howler species in disturbed habitats (Bernstein et al. 1976; Schwarzkopf and Rylands 1989). Three unique aspects of the Belize River site have been noted: (1) One of the largest black howler populations in the species range is found in a populated region of widespread and continuous land use; (2) howlers and humans have been coexisting at the site for many years; and (3) the howlers are not hunted to any large extent by local people, despite abundant hunting of other game animals. These characteristics provided a unique opportunity for conservation efforts.

Land Use within the CBS

The CBS is located in a riparian corridor of lowland rainforest in an area of relatively fertile alluvial soils (Birchall and Jenkin 1979). Rainfall in the region ranges from 150 to 175 centimeters per year, with a pronounced dry season occurring from February through May. Many hardwood trees shed their leaves during the dry season, giving the forest a semideciduous character. The conservation–research area consists of over 110 individual land holdings ranging in size from 2 to 202 hectares. Land tenure status of CBS residents consists of three categories: freehold titled (16 percent), leased and part titled (41 percent), and part leased (43 percent) (Hartup 1994). The vast majority of residents in the CBS have lived in the area for most of their lives.

Crop and livestock farming are the primary occupations of most CBS residents. The local cropping system is small-scale (0.1–1-hectare slash-and-burn plots). Land clearings for crops (milpas) are undertaken almost exclusively in the dry season (February to mid-May). Milpas are typically located along riparian corridors due to the fertility of alluvial soils and the ease in reaching riparian areas by land or canoe (Robinson 1985). Milpas are generally maintained for 2–4 years then left fallow for 15–40 years. The fallow period is relatively long for two main reasons: (1) the human population in the CBS has been more or less stable for many years, and (2) properties are sufficiently large such that not all forest land is cleared at any one time. The main food crops grown are rice, beans, and corn, although root crops, vegetables, and fruit trees are also grown. Hartup (1994) noted that over 72 percent of landowners in the CBS reported producing sufficient yields of staple crops from their milpas to meet the needs of their families. Many small-scale cattle operations (herds of 5–40 head) are also functioning within the CBS. Pasture clearings range from 1 to 40 hectares with a mean of 4.2 hectares (based on 1988 data). The vast majority of pastures are unimproved and maintained without the use of machinery.

Due to recent extensive agricultural activity in the region, approximately 41 percent of the 47-square-kilometer CBS area is currently cleared of forest cover. Assessments of historical patterns of land usage gleaned from a sequence of aerial photographs of the area are summarized in figure 11-2. In the period prior to the 1960s, most agricultural activity in the region was in the form of small-scale, slash-and-burn mil-

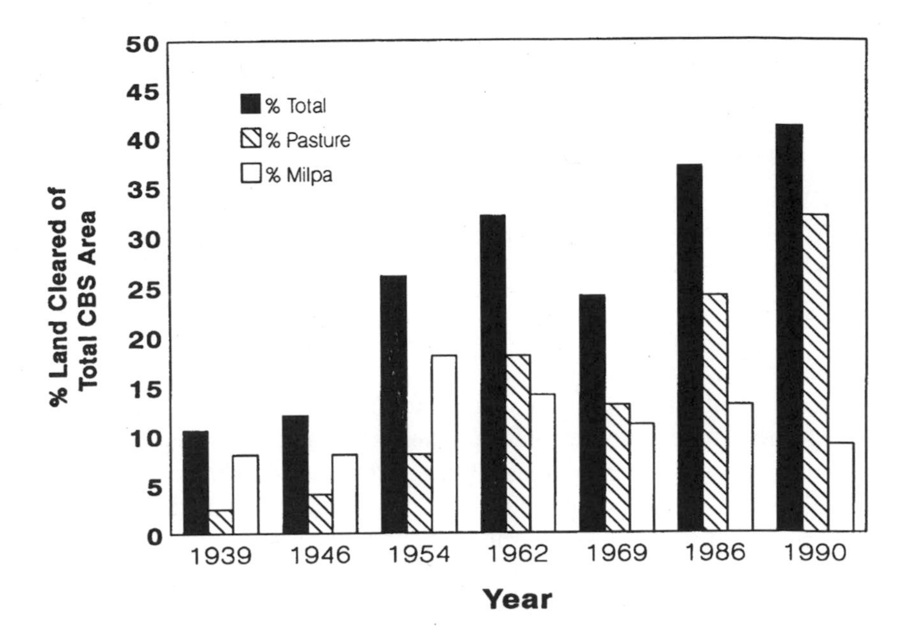

Figure 11-2

Land-clearing history summary for the CBS. Percentage and type of forest clearings within the CBS are shown based on aerial surveys from seven different years.

pas. With construction of a road into the area from the east in the early 1960s, land clearing substantially increased and shifted toward pasture clearing and away from milpa clearing. This latter trend continues to the present day and is likely to go on. Viewed as a whole, CBS land holdings form a fragmented and patchy landscape consisting of a mosaic of small, slash-and-burn clearings, variable-sized pastures, and successional forests of various stages of development (5–75 years).

Existing Patch Structure and Composition

Influenced by both historic land-use patterns and current land usage, the forest patch mosaic comprising the CBS is complex and dynamic. The CBS consists of numerous, mixed-age forest patches (5–75 years) interspersed with various-sized agricultural clearings. Highly variable, multiple successional pathways exist in CBS forests. These pathways can produce very different forest patch composition, depending on fac-

Table 11-1

Vegetative reproduction capacity and relative dominance of the 25 most common tree species in the CBS. All plots were 0.2 ha in size and were located in fragments of similar size and age. Percentage basal area of each tree is shown relative to the total basal area on the plot.

Tree species	Vegetative reproduction	Percentage of total basal				
		Plot 1	Plot 2	Plot 3	Plot 4	Plot 5
Orbigyna cohune	no	94	42	15	—	2
Guazuma ulmifolia	yes	1	8	26	8	1
Luehea seemanii	yes	2	3	15	5	5
Spondias mombin	yes	1	12	18	12	4
Bactris major	no	—	2	—	13	1
Cecropia obtusifolia	no	—	—	8	—	12
Lonchocarpus guatemalensis	yes	—	1	8	8	—
Inga edulis	yes	—	1	2	11	—
Ficus glabrata	yes	—	2	2	4	2
Coccoloba belizensis	yes	—	8	3	7	3
Cassia grandis	yes	1	3	—	5	9
Enterolobim cyclocarpum	yes	—	4	—	—	11
Ficus spp.	yes	1	1	—	—	5
Bursera simaruba	yes	—	4	1	2	3
Tabebuia rosea	yes	—	—	1	—	5
Ceiba pentandra	yes	—	1	1	2	8
Zanthoxylum kellermanii	yes	—	—	—	—	2
Schizolobium parahybum	yes	—	2	—	4	3
Bucida buceras	no	—	—	—	7	—
Faramea occidentalis	yes	—	2	—	2	2
Erythrina fusca	yes	—	—	—	8	—
Vochysia hondurensis	no	—	—	—	—	8
Pithecellobium brownii	yes	—	3	—	2	—
Lonchocarpus hondurensis	yes	—	1	—	—	—
Cordia diversifolia	no	—	—	—	—	15

tors such as soil type; season of land-clearing activity (dry versus rainy season); previous forest composition; and the type, intensity, and duration of various land-use practices. Most forest patches have partially broken canopies, show little stratification, and reach maximum heights of only 20–25 meters. Research on the composition of 24 forest patches within the CBS has shown that most patches contain 2–5 dominant tree species, 20–25 codominant tree species, and 10–15 rarer tree and shrub species. Total basal area of trees in the majority of patches is dominated by only a handful of species. There are, however, striking differences in the ranking in dominance and relative importance of tree species between patches. Table 11-1 provides a summary of the variation in relative dominance of the 25 most common trees in the CBS within five different forest patch research plots. It also contains information on the vegetative reproductive capacity of those same species.

The tree-age structure within patches shows two distinct patterns. The first type, a typical reverse-J curve indicative of highly disturbed forest environments, is displayed in a majority of plots. In these patches, most trees are in smaller-diameter classes, although a substantial number of larger-diameter trees may also be present. The larger-sized trees are typically remnants from past land-clearing activities that were left standing because of difficulties in felling and/or the tree's value as a source of canoe wood, fiber, food, or construction materials. The second type of age structure pattern, much less common, exhibits a more equitable distribution of age classes and tree species, and a higher tree species richness. This pattern is indicative of a maturing secondary forest experiencing minimal disturbance. Interestingly, the more disturbed and less species-rich forest patches appear to be the preferred habitat of the black howlers.

Influence of Disturbance History on the Ecological Dynamics of Forest Patches

The existing highly disturbed secondary forests and nonforest communities within the CBS have been shaped by the interrelationships between natural and human factors. For centuries, hurricane damage has been documented as a recurring disturbance in the region (Hartshorn 1984). It is also possible that historic Mayan land use has altered the vegetation in the region as it has been reported to have done in other areas of Belize and the Yucatan (Folan et al., 1979; Gómez-Pompa and Vásquez-Yanes 1981; Lambert and Arnason 1978). It is likely that many tree species used by the Maya now exist in the landscape as secondary successional species in disturbed and regenerating areas (Turner and Miksicek 1984). In addition to Mayan influences, some 300 years of European-based timber exploitation and related land use have had dramatic impacts on the forests of the region. These impacts include the removal of nearly all the marketable timber species (such as, mahogany *Swietenia macrophylla,* and cedar, *Cedrela odorata*), clearing of forest for the growth of animal fodder, and extensive hunting of game. Logging-related disturbances were particularly acute along the Belize River (Leslie 1987). High levels of disturbance can have a strong influence on the life-history parameters of plant species (Denslow 1987) and on the composition of the remaining secondary forests (Brown and Lugo 1990). For example, many forests previously thought to be relatively undisturbed primary forests in Central America have been recategorized as human-modified secondary or late secondary forests (Gómez-Pompa and Vásquez-Yanes 1981; Lanly 1982). The high frequency, intensity, and duration of disturbance in the area that now is the CBS have contributed to the present manifestation of several disturbance-induced forest characteristics.

There are three dominant, identifiable, disturbance-induced characteristics within the mosaic of CBS forests: (1) the widespread dominance of a single tree species, the cohune palm (*Orbigyna cohune*); (2) the preponderance of stump sprouting and vegetative reproduction; and (3) the capacity of tree species typically classified as climax species to behave similarly to pioneer species in their ability to occupy disturbed sites. Each of these characteristics has a strong impact on the composition and regeneration potential of forest patches and each has to be assessed and considered in the formulation of howler habitat conservation strategies.

Cohune palms have historically been left in milpa and pasture clearings. Difficult to fell with machetes, they provide local people with usable materials, including inexpensive and abundant sources of roof thatch, cooking oil, and charcoal. Even with the recent use of chain saws for land clearing, cohunes are still commonly left in pasture clearings because they provide shade for cattle. Cohunes are also capable of surviving in slash-and-burn clearings; the genus *Orbigyna* has meristems that can survive defoliation and fire and can persist and expand in highly disturbed landscapes (Hecht et al., 1988). Research on 38, 0.2-hectare forest plots in the CBS showed that the basal area represented by cohunes ranged from 10 to 94 percent (mean: 42 percent). Cohune seedlings and saplings also typically comprise a large portion of understory vegetation within many forest patches. Interestingly, despite the fact that no part of the cohune is eaten by the howlers, the prevalence of cohune palms can nonetheless be beneficial to howler feeding ecology. Surveys of strangler fig growth on remnant cohune palms in pastures and milpas have shown fig infestations ranging from 42 to 86 percent in palm canopies. This is significant because fig trees are an important component of the howler diet. Furthermore, Furley (1975) reported that in Belize cohune palms may be beneficial to the entire forest community because of the important role they play in soil formation and fertility.

Following natural disturbance, forest clearing, fire, and land abandonment, many tree species exhibit the capacity to resprout. Of the 102 tree and treelet species found on CBS forest patch research plots, 67 species (66 percent) have been observed to resprout following cutting and/or fire. Of the 25 most common tree species, 19 have the capacity for vegetative reproduction. The widespread presence of multiple boles of many tree species in the present landscape is also indicative of the prevalence of vegetative reproduction. Similar to stump sprouting is the capacity of portions of damaged tree stems or branches to propagate (Kinsman 1990). Propagation experiments using freshly cut stem and branch fragments have demonstrated that many tree and shrub species in the CBS possess this capacity. The existence of "living fence posts" provides additional evidence of plant fragment propagation. The importance of vegetative reproduction and sprouting in determining forest successional patterns and forest composition have been noted in other disturbed tropical forests such as the dry forests in Costa Rica and Puerto Rico (Ewel 1977), forests on Barro Colorado Island (Putz et al., 1983; Putz and Brokaw 1989), and disturbed Amazonian forests (Uhl et al., 1981). The prevalence of vegetative reproduction means that many tree species are able to maintain their occupancy on a site without seed recruitment. Thus, vegetative reproduction has enormous potential as a management tool to aid in the retention of specific tree species on a given site, lessening the reliance on seed dispersal for colonization of disturbed sites and for reintroducing species into forest patches.

The high frequency of sprouting and vegetative reproduction among tree species in the CBS following disturbance can also promote unique successional assemblages of tree species. Many tree species, which in undisturbed environments typically occupy different positions on successional seres and rely on seed recruitment, are found growing in the same successional sere within the same-aged forest patch. This phenomenon results in the establishment of forest patches that contain a mixture of early-, mid-, and late-successional tree species. An important consequence of this atypical tree species mixing is that howlers (and other wildlife) often have access to a wide successional range of tree species growing in a relatively small area. Thus, despite the intense disturbance history in the region, many of the disturbance-induced vegetative characteristics of tree species may actually benefit the howlers and partially explain their abundance in CBS forests.

Tree Phenology within Patches and Howler Feeding Ecology

Phenological research (assessment of the seasonal dynamics of leaf, flower, and fruit production) has provided important information and insight into the capacity of CBS forest patches to support a large population of howlers. Phenological data were collected from 1985 to 1988 on the 42 most common tree species found in the CBS. Of these 42 species, 25 made up 95 percent of the total number of trees found along the 10 kilometers of phenological transects. Three aspects of the data-collection process itself were particularly significant and should be noted. First, the phenology data collection was conducted by salaried CBS staff. This provided research training and ensured a level of local involvement in resource assessment. Second, phenological investigations provided a formal framework into which local informal knowledge on the seasonal aspects of forest dynamics could be placed. The result was a linkage between basic ecological research and local knowledge. Third, the compiled phenology data provided detailed information that could be directly applied in forest and howler research.

Figure 11-3, which summarizes the phenological research results, illustrates the availability of new leaves, flowers, and fruits (the three key foods in the howler diet) for the 42 most common tree species in the CBS. The intensity index in this figure refers to the relative proportion of leaves, flowers or fruits occupying a hypothetical canopy (the index runs from 0 to 2.5). While it by no means represents the inclusive diet of the howlers (which also utilize a number of shrub, liana, and herbaceous species), it is indicative of the general forest environment with respect to howler feeding ecology. The figure reveals that, despite disturbance history of the area, relatively low number of tree species, pronounced dry season, and different phenological patterns of individual species, a relatively abundant supply of new leaves, flowers, and fruits is available year-round for the howlers. Although only a small number of tree species dominate the overall landscape, the phenological patterns within the CBS provide an all-season food source for the howlers. The local success of the vegetarian black howler is clearly linked to its preferential use of the suite of tree species found in the existing forest patches. The small troop sizes and home ranges of the black howler are also well suited to the highly fragmented forests of the re-

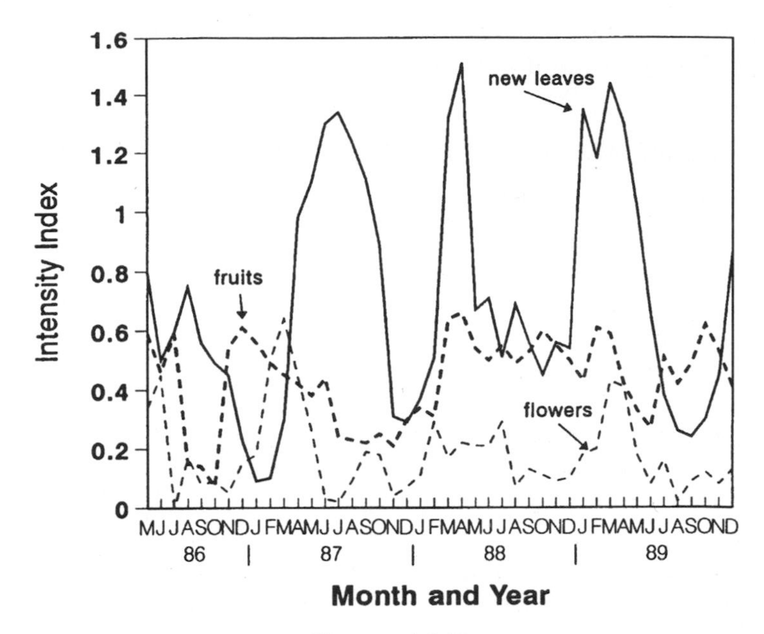

Figure 11-3

Phenological summary for the 25 most common tree species in the CBS from the period May 1986 to December 1989. The intensity of new leaf, flower, and fruit production is shown based on an intensity index ranging from 0 to 2.5. The intensity index refers to the proportion of a hypothetical canopy that contains a full complement of new leaves, flowers, or fruits.

gion (Horwich 1983). Thus, phenological research provided valuable and insightful information on the seasonal dynamics of howler habitat and the data also provided practical information that could be incorporated into management plans. The CBS phenological research effort is a prime example of how valuable habitat data can be collected inexpensively while providing local people with income, training, and insight into forest dynamics.

Significance of Seed-Dispersing Animals

The majority of the 25 most common tree species in the CBS have animal-dispersed seeds. The maintenance of the existing tree species in the riparian landscape is dependent, at least in part, on the presence of seed-dispersing animals. In addition to the howler monkeys, a wide variety of mammal- and bird-disperser species are present within the CBS (Horwich and Lyon 1990). The importance of maintaining animal seed dispersers in disturbed landscapes has been emphasized by Redford (1992), who noted that forest patches devoid of animal dispersers are susceptible to the long-term elimination of animal-dispersed tree species from the landscape. However, the prevalence of vegetative reproduction among trees found in the CBS reduces dependence on seed-dispersing animals to maintain the current assemblages of trees. Within the CBS, disturbance-induced vegetative characteristics in trees may act to maintain tree species diversity, which in turn may encourage the return of locally extinct or diminished animal populations.

The Influence of Human Factors on Forest Patch Composition

Individual land-clearing practices account for much of the structural complexity of patches in the CBS. For generations, many local farming families have unknowlingly been promoting a regional diversity of tree species and maintaining adequate black howler habitat. For example, large-diameter trees are frequently left in slash-and-burn and pasture clearings (remnants). Which particular species is retained on a site is typically landowner specific and may reflect potential use of the tree (e.g., canoe making or construction materials), difficulty in felling the tree, or historic family patterns of leaving certain tree species. This human selection of remnant species, in addition to distinct patterns of land use leading to different successional pathways, have, in part, created and maintained a level of forest patch heterogeneity between properties within the CBS. Although tree species richness and diversity on a given land holding may be relatively low, on a larger, sanctuary-wide scale, the species and structural diversity of the forests is much higher. Based on the variety of land-use practices in the CBS, the resultant vegetation patterns apparently have been capable of sustaining a relatively large population of howlers for a number of decades.

The importance of landowners leaving remnant trees in milpas and pastures within the CBS cannot be overstated. These remnant trees serve as emergent links in otherwise denuded forests, act as seed sources, and provide a structural nucleus (Janzen 1988). Remnant trees also provide shade, bird habitat, cover for wildlife, and platforms for epiphytes. Within the CBS, remnant trees act as platforms for the growth of several species of strangler figs (*Ficus* spp.), which are an important part of the howler diet. Most milpas and pastures within the

CBS are also adjacent to remnant patches of various-aged secondary forest. While the forest patches are themselves typically disturbed, their constituent tree species do act as seed sources, provide perches for seed-dispersing birds, and often expand by root suckers and other forms of vegetative reproduction. These regenerative processes facilitate various forms of recruitment into the clearing and encourage animal movement to remnant trees. Promoting the practice of leaving remnant trees in clearings within the CBS has been important in maintaining adequate structural and tree species stocks in forest clearings.

Patch Variability, Habitat Protection, and Community Conservation

The composition of any given forest patch is controlled by a host of factors, including dry season effects (Ewel 1977), disturbance (Brokaw 1985), and forest succession (Gómez-Pompa and Vásquez-Yanes 1981). Study of the interrelationships between these factors and human-mediated forest alteration is an essential step in planning a restoration strategy for patches (Purata 1986). Hence, detailed composition and structural analysis of fragments should be interpreted in light of interactions between ecological dynamics and human utilization. Janzen (1988) noted the important ecological distinctions and management ramifications of "shrinking relicts" of forest versus forest fragment "restoration." However, these two categories are not always mutually exclusive. For example, the fragmented landscape in the CBS includes characteristics of both types of fragments which often overlap spatially on the same limited landscape and even within a given forest patch. Given this complexity, individual land holdings have been used as the basic management unit in the CBS.

In an effort to protect howler habitat along the Belize River site, an innovative approach was required. Due to the existing land tenure system in the region, and the strong desire of local people to maintain control of their property rights, the establishment of a voluntary, community-based howler sanctuary (the CBS) was essentially the only viable option (Horwich and Lyon 1988). The CBS was formed by obtaining voluntary, and nonlegally binding pledges from local landowners who agreed to protect the existing howler habitat on their properties. Under the sanctuary's pledge system, individual landholders had the crux of responsibility for stewardship and long-term maintenance of howler habitat. This community-based conservation and management approach was pursued as an experiment to test our contention that rural landowners in the area possessed the knowledge, ability, and commitment to voluntarily protect howler habitat. In formulating management suggestions for landowners, local knowledge of forests and land-use practices were solicited as a significant resource that was publicly and positively linked to habitat-protection planning. Land-use management and habitat protection were viewed as compatible concepts that could be melded under a voluntary, participatory management system (Horwich and Lyon 1988; Horwich 1990; Hartup 1994). More extensive discussion of this approach and of the establishment, expansion, and history of the CBS can be found in other publications (Horwich and Lyon 1987, 1988; Horwich 1990).

Howler habitat management in the CBS is linked to both professional research programs and a locally controlled management structure. The management structure includes an oversight council, composed of representatives from each of the eight participating villages, which oversees the CBS staff. CBS staff are responsible for the day-to-day management of the CBS. All council members and employees of the CBS are local residents. The original mission of the CBS, basic wildlife habitat protection, has expanded and now consists of four main components: forest and wildlife research, land-use management, natural resource education centered around a small natural history museum in the sanctuary, and a locally controlled ecotourism program. The first two components are the substance of the present discussion. Further details on education and ecotourism programs in the CBS are discussed in detail elsewhere (Bruner 1993; Horwich et al. 1993; Horwich and Lyon, in press a). It is important to note, however, that the economic benefits received from the ecotourism component of the CBS are closely linked to the conservation of the howlers, protection of howler habitat, and the continued success of the project. The ecotourism effort has placed the efforts of CBS participants in the national and international spotlight. This outside recognition of the CBS as an important conservaton advance and a model for other rural communities has instilled in the local people a sense of pride and achievement (Horwich and Lyon, in press b).

Protection, Modification, and Enhancement of Howler Habitat through Community Cooperation

The habitat management strategy of the CBS has emphasized protection of existing forests rather than restoration of existing forests to some earlier, more pristine state. CBS management plans are based on present habitat conditions and provide suggestions for modest modifications of existing land use within the framework of local land-use practices. The value and potential of recognizing the capacity of local peoples to protect habitat within existing cultural frameworks remains unrealized (Unruh 1994). Under the CBS habitat management program, individual landownership parcels are used as the working management unit, and each landowner is responsible for voluntary adherence to a mutually agreed upon habitat management plan. This approach promotes individual stewardship, maintains local control, and clarifies the role of each land user in the overall management strategy. Under the CBS structure, the land user maintains complete control over which specific areas of forest or individual trees are cut. However, CBS staff encourage landowners to retain specific food trees, trees that the landowner has identified as important resting stations and/or transportation routes for howlers, and forest patches and corridors along waterways and property boundaries.

The CBS's overall habitat management and forest patch enhancement strategy employs three basic nature reserve design concepts: (1) establishment and protection of forest corridors between existing forest patches, across large clearings, and along property boundaries; (2) protection and renewal of forest corridors along the main river course and large tributaries; and (3) retention and/or establishment of remnant trees to provide food for howlers in slash-and-burn and pasture clearings. The balance of forest and other vegetation on properties remains under indigenous crop and pasture rotation. It should be noted, however, that disturbed environments also typically provide additional forest habitat and food sources for the howlers. Although Belizean law requires a 20-meter buffer strip along each side of all waterways, the law is rarely observed or enforced. In the CBS, exact dimensions of buffer strips are not formally defined because the dimensions of buffer strips and corridors are matched to individual land users and properties. Nonetheless, corridors of 10–20 meters width have been encouraged and are thought to be sufficient to maintain a stable howler population. The minimum conservation objective of the CBS is protection of a skeletal framework of secondary forests and forest patches along the Belize River and along all property boundaries. A comparison of the minimum conservation objective and the current forest cover is provided in figure 11-4.

The methods outlined in the individualized management plans are simple, and they require no capital outlay and little, if any, added physical labor by participating landowners. The potential costs to landown-

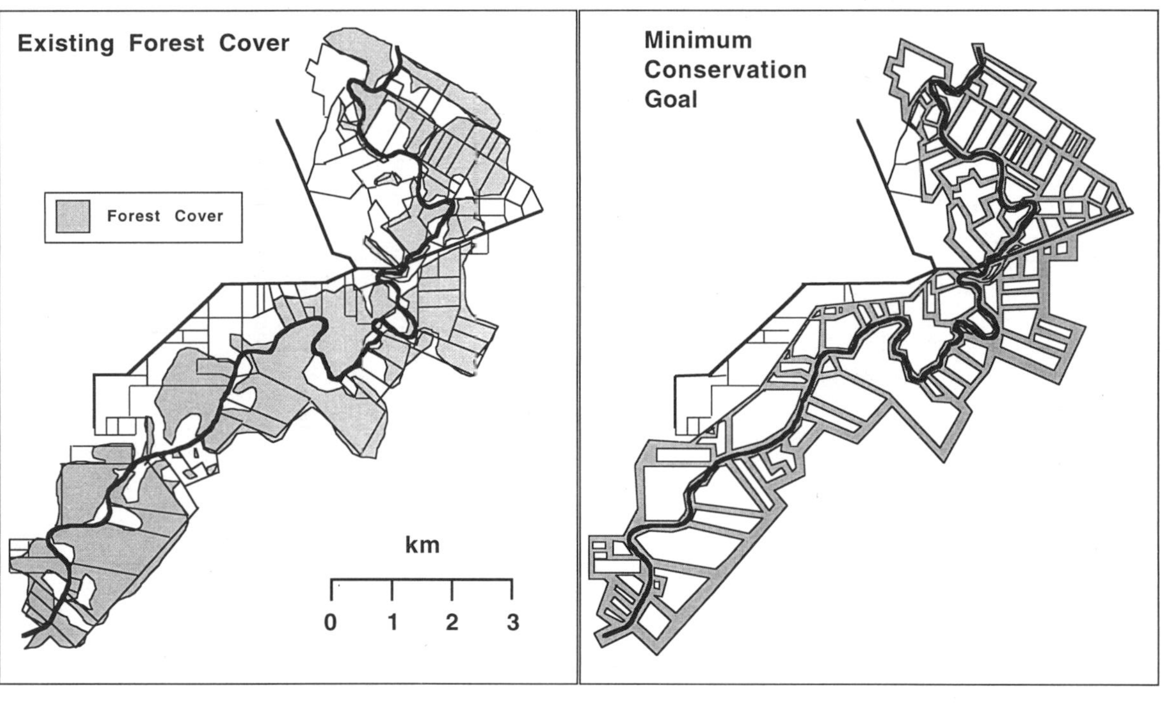

Figure 11-4

A comparison of the existing forest cover (*left*) with the minimum conservation goal (*right*) for maintenance of black howler habitat in the CBS is shown. The minimum objective is based on the protection of a continuous, interconnected corridor of forest and forest fragments along the Belize River, property boundaries, and roadways.

ers are in the form of slightly smaller slash-and-burn clearings or pastures. Reductions in land area under cultivation or in pasture may result in decreased agricultural yields in the short term. But in the long term, the management methods introduced provide benefits through reduced soil erosion, faster forest regeneration potential during fallow periods, and slower rates of nutrient depletion on clearings. Protection of howler habitat also generates income from ecotourism. Additionally, although the black howler is the focal conservation species, the management plans and forest patch habitat maintenance effort are likely to benefit other wildlife and maintain populations of game animals (Polisar and Horwich 1994). The following list represents a sample of the management tools and techniques that have been or could be employed to enhance and modify existing forest patches in the CBS. This list, while not meant to be inclusive, highlights the types of no-cost or low-cost management options that can be used in habitat modification efforts.

1. Construct artificial nesting, feeding, or habitat sites in forest patches
2. Selective removal of nonnative tree species from patches for construction or fuelwood
3. Plant specific "living fence posts" to facilitate reforestation and to create corridors and buffer strips
4. Plant locally extinct tree species in existing habitat patches
5. Promote a diverse set of animal-dispersed remnant trees in agricultural clearings
6. Build inexpensive aerial bridges to facilitate wildlife movement across roads and other obstacles
7. Establish low-cost, locally operated tree propagation centers to grow and distribute fruit/hardwood trees
8. Plant nurse tree crops in highly disturbed areas to promote forest regeneration

Howler populations in the CBS have steadily increased since the initiation of the sanctuary. Currently, over 80 percent of CBS landholders are adhering to their voluntary management plans. The habitat management success of the CBS is due in large part to the integrative research and innovative management philosophy initiated at the outset of the project. Basic research on the ecology of the howlers and on the structure and composition of existing forest fragments was undertaken before, during, and after the initial development of individual management plans. Furthermore, research on forest patch composition and phenology was conducted with local involvement and input. Local people were directly involved in the research effort as paid assistants, consultants, and interested volunteers. Local knowledge of forest dynamics, plant identification, land use, and howler ecology was treated as an important information resource and the starting point for management planning. Through this process, a large number of residents became aware of the nature, and in many cases the details, of the various research efforts and considered themselves active participants. This working arrangement also encouraged discussion between researchers and land users about the management of the local forests. All project researchers lived within the villages comprising the CBS, which facilitated communication, understanding, respect, and friendship between researchers and local people and allowed for a smooth transition from research to management. Community participation in all aspects of the CBS effort proved a successful and fruitful approach, and similar efforts have been shown to be effective elsewhere (Cox and Elmqvist 1991; Dhar 1994; Nimlos and Savage 1991; Saunders 1993).

The CBS experiment has shown that much can be achieved in maintaining and enhancing disturbed forest habitats with simple techniques. The continued success of the CBS as a wildlife conservation effort ultimately rests on the cooperation of landholders and land users, who remain the key players in the management effort. The impact of their voluntary commitment to pursue habitat management and forest patch modification in the CBS can not be overstated. The capacity of rural land users to function as managers and stewards of their own lands remains an untapped resource in many regions of the tropical landscape.

Promoting Wildlife Conservation and Habitat Protection in a Multiple-Use Landscape: The Manatee Special Development Area Biodiversity Project

The Manatee Special Development Area (MSDA) was established in 1991 by the government of Belize to provide interim protection for a region of unique biological diversity. The 46,760-hectare MSDA encompasses public and private lands that are superimposed on a complex matrix of biotic communities, including coastal beaches, saline marshes, freshwater wetlands, mangrove forests, pine forests, savannas, brackish lagoon systems, broadleaf semideciduous forests, riparian forests, and karst hill forests. The only permanent human settlement in the area is a single village, Gales Point, with about 200 residents. The biotic characteristics of the MSDA are more complex than in the CBS, in that the MSDA landscape is comprised of fragmented and distributed biotic communities.

The first research/management proposal for what is now the MSDA was drawn up by Horwich and Lyon (1991). The plan called for forest and wildlife research targeted within each of the biotic communities and promoted implementation of some of the land-use management aspects of the Biosphere Reserve concept (MAB 1984). With the formal establishment of the MSDA, the Belize government mandated the development of a management and zoning plan for the area. The MSDA was to be partitioned into zones for farming, citrus development, residential and commercial use, tourism, forestry, wildlife reserve, and environmental protection (Horwich et al. 1993; Horwich and Boardman 1993; Greenlee 1994). A brief summary of the efforts being undertaken within the MSDA to characterize forest vegetation and wildlife habitats, and to integrate wildlife conservation into all aspects of zoning and management is given below.

Biotic Communities as Individual Research and Management Units

To meet the zoning mandate, the MSDA Biodiversity Project (MSDABP) initiated a research/management program slated for completion in 1995 or 1996. The MSDABP involves numerous individuals and organizations, including Gales Point villagers, scientists, agricultural and fishing cooperatives, land planners, Belizean government personnel, and community conservation specialists. The conservation goals of the MSDABP are two-fold: (1) to catalog and characterize the biota of each community type, and (2) to use this information to formulate an overall management plan aimed at optimizing biological diversity. The emphasis on characterizing individual biotic communities is a critical prerequisite to comprehensive forest and wildlife management planning for the entire MSDA.

MSDABP personnel are in the process of collecting basic ecological information on forest patch species composition, physiography, age structure, and phenology. To facilitate this process, permanent vegetation plots have been established in each of the representative forest types found in the MSDA. The vegetation plots are the focal point of forest patch characterizations as well as for a host of other mammal, bird, insect, and soil studies. The permanent plots are also used for assessment of wildlife utilization and hunting pressure in each forest type. Overall, the permanent plots and the research associated with them provide basic and important information on the entire ecological community and a common reference point for researchers, local people, land-use planners, and government workers.

The implementation of the MSDABP represents an expansion of the experimental conservation approach pioneered at the CBS into an ecologically diverse, public–private landscape. An integral component of the MSDABP is the involvement of local people, researchers, and government and nongovernment persons and agencies in all aspects of the project, including biological assessment of the various biotic communities, the design of research programs, and the integration of research results into a management format. Despite the challenges of effectively utilizing a wide array of ecological information in the management and mediation of a host of competing land use interests, the conservation approach taken by the MSDABP and the CBS represents an opportunity to revitalize promising aspects of the Biosphere Reserve concept and to promote wildlife conservation in a new context. By working to empower local people who utilize the public lands for agriculture and hunting, the MSDABP is working to insure that those persons who have a vested interest in protecting the resource are included in the formulation of land-use planning and management of those lands.

Summary and Conclusions

Targeting highly disturbed landscapes for wildlife conservation holds great promise and many challenges. The physical, ecological, and human dynamics of disturbed landscapes are intertwined and complex. Direct application of theoretical, ecological, and conservation concepts to these landscapes without specific assessment of their inherent ecological structure and dynamics may not be appropriate or useful. The CBS provides an example of how basic ecological analysis of a fragmented and patchy forest landscape can provide essential information for management and conservation of a focal species (the black howler). The CBS also provides an example of how human land-use practices can be successfully molded to meet wildlife conservation goals and protect and enhance habitat. CBS management plans call for relatively simple, low- or no-cost techniques. A key ingredient of the CBS is that local input and concerns about the management and scope of wildlife conservation are continually addressed so that forest management remains an active conservation force in the community. Wildlife conservation in disturbed areas may require the forging of new alliances and cooperation between researchers, conservationists, and local people. The MSD-ABP represents a new venture in this direction on a large-scale, multiple-use landscape.

Given the inherent limitations of using highly disturbed forests for wildlife conservation, it is likely that many wildlife species will not be appropriate conservation targets for these landscapes. However, future research and experimentation may provide fragment configurations that will prove suitable for a host of wildlife, including deep forest or large territory species. Efforts to modify patches and/or simulate later successional stages or more complex environments in existing fragmented landscapes can provide additional tools in the arsenal of species conservation. The value and potential of utilizing vegetative propagation as a conservation tool/technique to modify and enhance forest patches and corridors also needs further exploration.

Like many developing nations, Belize does not have the necessary resources and staff to monitor and enforce adherence to land-use restrictions and/or zoning regulations. Education of local people about the consequences of improper land use, combined with local empowerment for active participation in decision making, can help bridge the gap between rural land-use pressure and wildlife conservation. In those situations in which the people utilizing the existing forest landscape have a vested interest in maintaining the forest and its resources to maintain their standard of living, community-based conservation may be a viable alternative. Integration of wildlife conservation within the social, cultural, and socioeconomic context of local peoples has great potential to diffuse some of the present conflicts between conservationists and rural communities. Providing local land users with specific management training and employment opportunities in ecological research helps to involve local people directly with the conservation effort. Such an integrative approach can lead to tangible conservation results at the local level, where they are most needed. The ability of rural land users to function as managers and stewards of their own lands, as well as of public lands, remains an untapped resource in many regions of the tropical world.

References

Bernstein, I. S., P. Balcaen, L. Dresdale, H. Gouzoules, M. Kavanagh, T. Patterson, and P. Neyman-Warner. 1976. "Differential effects of forest degradation on primate populations." *Primates* 17: 401–411.

Bierregaard, R. O., T. E. Lovejoy, V. Kapos, A. Augusto dos Santos, and R. W. Hutchings. 1992. "The biological dynamics of tropical rainforest fragments." *BioScience* 42(11): 859–866.

Birchall, C. J., and R. M. Jenkin. 1979. *The Soils of the Belize Valley, Belize.* Tolworth: Land Resources Development Centre.

Brokaw, N. V. L. 1985. "Treefalls, regrowth, and community structure in tropical forests." In *The Ecology of Natural Disturbance and Patch Dynamics,* edited by S. T. A. Pickett and P. S. White, pp. 53–69. New York: Academic Press.

Brown, S., and A. E. Lugo. 1990. "Tropical secondary forests." *Journal of Tropical Ecology* 6: 1–32.

Bruner, Y. G. 1993. "Evaluating a model of private-ownership conservation: Ecotourism in the community baboon sanctuary in Belize." M.S. thesis, Georgia Institute of Technology.

Cox, P. A., and T. Elmqvist. 1991. "Indigenous control of tropical rain-forest reserves: An alternative strategy for conservation." *Ambio* 20(7): 317–321.

Denslow, J. S. 1987. "Tropical rain forest gaps and tree species diversity." *Annual Review of Ecology and Systematics* 18: 431–451.

Dhar, S. K. 1994. "Rehabilitation of degraded tropical forest watershed with people's participation." *Ambio* 23(3): 216–221.

Ewel, J. J. 1977. "Differences between wet and dry successional tropical ecosystems." *Geo-Eco-Trop* 1(2): 103–117.

Ewel, J. J. 1979. "Secondary forests: The tropical wood resource of the future." In *Simposio Internacional Sobre las Ciencias Forestales y su Contribucion al Desarrollo de la America Tropical,* edited by M. Chavarria, pp. 53–60. San Jose, Costa Rica: Cocit/Interciencia.

Folan, W. J., L. A. Fletcher, and E. R. Kintz. 1979. "Fruit, fiber, bark and resin: Social organization of a Mayan urban center." *Science* 204: 697–701.

Furley, P. A. 1975. "The significance of the cohune palm, *Orbigyna cohune* (Mart.) Dahlgren, on the nature and in the development of the soil profile." *Biotropica* 7(1): 32–36.

Gómez-Pompa, A., and C. Vásquez-Yanes. 1974. "Studies on secondary succession of tropical lowlands: The life cycle of secondary species." In *Proceedings First International Congress of Ecology,* pp. 336–342. Netherlands: The Hague.

Gómez-Pompa, A., and C. Vásquez-Yanes. 1981. "Successional studies of a rain forest in Mexico." In *Forest Succession: Concepts and Applications,* edited by D. C. West, H. H. Shugart, and D. B. Botkin, pp. 246–266. New York: Springer.

Greenlee, D. 1994. "Community based ecotourism in Gales Point, Belize." *Belize Currents* April: 10–11.

Hartshorn, G. 1984. *Belize Country Environmental Profile: A Field Study*. Belize City: Robert Nicolait and Associates with USAID.

Hartup, B. K. 1994. "Community conservation in Belize: Demography, resource use, and attitudes of participating landowners." *Biological Conservation* 69: 235–241.

Hecht, S. B., A. B. Anderson, and P. May. 1988. "The subsidy from nature: Shifting cultivation, successional palm forests and rural development." *Human Organization* 47: 23–35.

Horwich, R. H. 1983. "Breeding behaviors in the black howler monkey (*Alouatta pigra*) of Belize." *Primates* 24: 222–230.

Horwich, R. H. 1990. "How to develop a community sanctuary: An experimental approach to the conservation of private lands." *Oryx* 24: 95–102.

Horwich, R. H., and B. Boardman. 1993. "Community conservation and ecotourism—Gales Point, Manatee." *Belize Review* April: 14–20.

Horwich, R. H., and E. D. Johnson. 1986. "Geographic distribution of the black howler, *Alouatta pigra,* in Central America." *Primates* 24: 290–296.

Horwich, R. H., and J. Lyon. 1987. "Development of the 'Community Baboon Sanctuary' in Belize: An experiment in grass roots conservation." *Primate Conservation* 8: 32–34.

Horwich, R. H., and J. Lyon. 1988. "Experimental techniques for the conservation of private lands." *Journal of Medical Primatology* 17(3): 169–176.

Horwich, R. H., and J. Lyon. 1990. *A Belizean Rain Forest: The Community Baboon Sanctuary.* Gays Mills, WI: Orangutan Press.

Horwich, R. H., and J. Lyon, 1991. "Proposal for a multiple land use system for the community manatee reserve." Community conservation consultants manuscript GP-2. Gays Mills, WI: Orangutan Press.

Horwich, R. H., and J. Lyon. 1995. "Multi-level education and conservation at the community baboon sanctuary, Belize." In *Conserving Wildlife: International Education and Communication Approaches,* edited by S. K. Jacobson, pp. 235–253. New York: Columbia University Press.

Horwich, R. H., and J. Lyon. In press. "Rural ecotourism as a conservation tool." In *Development of Tourism in Critical Environments,* edited by T. V. Singh. Lucknow, India: Centre for Tourism Research and Development.

Horwich, R. H., D. Murray, E. Saqui, J. Lyon, and D. Godfrey. 1993. "Ecotourism and community development : A view from Belize." In *Ecotourism: A Guide for Planners and Managers,* edited by K. Lindberg and D. E. Hawkins, pp. 152–168. North Bennington, VT: The Ecotourism Society.

Hough, J. L. 1988. "Obstacles to effective management of conflicts between national parks and surrounding human communities in developing countries." *Environmental Conservation* 15: 129–136.

Janzen, D. H. 1988. "Management of habitat fragments in a tropical dry forest: Growth." *Annals of the Missouri Botanical Garden* 75: 105–116.

Kinsman, S. 1990. "Regeneration by fragmentation in tropical montane forest shrubs." *American Journal of Botany* 77(12): 1626–1633.

Lambert, J. D. H., and T. Arnason. 1978. "Distribution of vegetation on Maya ruins and its relationship to ancient land-use at Lamanai, Belize." *Turrialba* 28(1): 33–41.

Lanly, J. P. 1982. *Tropical Forest Resources.* FAO Forestry Paper 30. Rome: Food and Agricultural Organization of the United Nations.

Laurance, W. F. 1991. "Edge effects in tropical forest fragments: Application of a model for the design of nature reserves." *Biological Conservation* 57: 205–219.

Leslie, V. 1987. "The Belize River boat traffic." *Caribbean Quarterly* 33: 1–28.

Lovejoy, T. E., R. O. Bierregaard Jr., A. B. Rylands, J. R. Malcolm, C. E. Quinetella, L. H. Harper, K. S. Brown Jr., A. H. Powell, G. V. N Powell, H. O. R. Schubart, and M. B. Hays. 1986. "Edge and other effects of isolation on Amazon forest fragments." In *Conservation Biology: The Science of Scarcity and Diversity,* edited by M. E. Soulé, pp. 257–285. Sunderland MA: Sinauer.

Lugo, A. E., J. A. Parrotta, and S. Brown. 1993. "Loss in species caused by tropical deforestation and their recovery through management." *Ambio* 22(3): 106–109.

MAB (Man and Biosphere). 1984. "Action plan for biosphere reserves." *Nature and Resources* 20(4): 1–12.

Mwalyosi, R. B. B. 1991. "Ecological evaluation for wildlife corridors and buffer zones for Lake Manyara National Park, Tanzania, and its immediate environment." *Biological Conservation* 57: 171–186.

Nimlos, T. J., and R. F. Savage. 1991. "Successful soil conservation in the Ecuadorian highlands." *Journal of Soil and Water Conservation* September–October: 341–343.

Noss, R. F. 1987. "From plant communities to landscapes in conservation inventories: A look at the Nature Conservancy (USA)." *Biological Conservation* 41: 11–37.

Noss, R. F., and L. D. Harris. 1986. "Nodes, networks, and MUMs: Preserving diversity at all scales." *Environmental Management* 10: 299–309.

Pickett, S. T. A., and J. N. Thompson. 1978. "Patch dynamics and the design of nature reserves." *Biological Conservation* 13: 27–37.

Polisar, J., and R. H. Horwich. 1994. "Conservation of the large, economically important river turtle *Dermatemys mawii* in Belize." *Conservation Biology* 8: 338–342.

Purata, S. E. 1986. "Floristic and structural changes during old-field succession in the Mexican tropics in relation to site history and species availability." *Journal of Tropical Ecology* 2: 257–276.

Putz, F. E., and N. V. L. Brokaw. 1989. "Sprouting of broken trees on Barro Colorado Island, Panama." *Ecology* 70(2): 508–512.

Putz, F. E., P. D. Coley, K. Lu, A. Montalvo, and A. Aiello. 1983. "Snapping and uprooting of trees: Structural determinants and ecological consequences." *Canadian Journal of Forest Research* 13: 1011–1020.

Redford, K. H. 1992. "The empty forest." *BioScience* 42(6): 412–422.

Robinson, G. M. 1985. "Agricultural change in the Belize River Valley." *Caribbean Geography* 2: 33–44.

Saunders, D. A. 1993. "A community-based observer scheme to assess avian response to habitat reduction and fragmentation in south-western Australia." *Biological Conservation* 64: 203–218.

Schwarzkopf, L., and A. B. Rylands. 1989. "Primate species richness in relation to habitat structure in Amazonian rainforest fragments." *Biological Conservation* 48: 1–12.

Terborgh, J. 1992. "Maintenance of diversity in tropical forests." *Biotropica* 24: 283–292.

Turner, B. L., and C. H. Miksicek. 1984. "Economic plant species associated with prehistoric agriculture in the Maya Lowlands." *Economic Botany* 38: 179–193.

Uhl, C., K. Clark, H. Clark, and P. Murphy. 1981. "Early plant succession after cutting and burning in the upper Rio Negro region of the Amazon basin." *Journal of Ecology* 69: 631–649.

Uhl, C., D. Hepstad, R. Buschbacher, K. Clark, B. Kauffman, and S. Subler. 1990. "Studies of ecosystem response to natural and anthropogenic disturbances provide guidelines for designing sustainable land-use systems in Amazonia." In *Alternatives to Deforestation: Steps toward Sustainable Use of the Amazon Rain Forest,* edited by A. B. Anderson, pp. 24–42. New York: Columbia University Press.

Unruh, J. D. 1994. "The role of land use pattern and process in the diffusion of valuable tree species." *Journal of Biogeography* 21: 283–295.

Part III

Human Dimensions

Chapter 12

Forest Use and Ownership: Patterns, Issues, and Recommendations

Janis B. Alcorn

Introduction

This volume on forest patches achieves two very important objectives. First, it brings greater recognition to the existence of local natural forests. These real, local forests have largely been invisible in discussions of tropical forests in the abstract. Second, it recognizes that action is needed to alter current policies undermining the existence of these forests. People who live in, and own the forests of, Central America, South America, Africa, South Pacific, and Asia would probably be pleased with the topic of this book, although it is unlikely they will ever see it. I hope that all readers of this book will work to ensure that forest owners have increased opportunities to share their insights with others and participate in decisions about forest futures, both locally and at the global level.

In this chapter I argue that forest ownership is a critical conservation issue that has received insufficient attention. Security of tenure is critically important for conservation, because it provides incentives for people to forgo short-term gains in order to maintain forest or other habitat into the future. Secure tenure means they are ensured access to benefits from their long-term investments (Lynch and Alcorn 1994). If it were possible for a satellite image or aerial photo to show not only the forest structure at a given place but also the tenure structure and the community organizational structures and institutions that exist in forested areas, then we would have a better picture of the world's forests. If on top of the community organizational layer could be overlain the past and present forest users (including indigenous and outsider users, and the final end-users in other countries), the de jure and de facto tenurial rights exercised over the areas in question, and the status of the common property management systems in place, then we would have an even better starting point for understanding forest dynamics and the factors behind the loss of biodiversity. But such overlays do not exist. In their absence, analyses of forest depletion and standard solutions are based on incomplete information about forest users, forest owners, and faulty models of deforestation dynamics (Alcorn 1995).

In the following sections I sketch a tenurial structure and user overlay for satellite imagery based on my own fieldwork and discussions with those involved in conservation on the ground, as well as case studies and analyses done by others. I draw generalizations, a venture always fraught with dangers inherent in glossing over local and regional differences. At this stage, however, it is useful to identify patterns in existing case studies and draw generalizations to guide research to refine and revise the generalizations and better understand local variations.

Local Forests and Local Peoples

Forest patches in the landscape, the subject of this book, may either be forest remnants or remnant forests. The term *remnant forest* retains the meaning of forest as a naturally regenerating community, as opposed to a *forest remnant,* which may be patches of "living dead" or even single trees. Remnant forests are forests of any size that have a lower number of species, especially large mammal species, and ecological relationships than are found in the most biodiverse forests under similar ecological conditions. The term *remnant* isn't a size distinction as much as a function of the existing biological community's capacity to maintain itself or evolve while retaining biodiversity.

In this chapter, I focus on forests in the sense that rural people use the term—to refer to local, naturally wooded areas that are large or small blocks or narrow, fingerlike corridors along rivers or ridges. In some places, these forests are highly diverse biotic communities. In other places, forests may be large in areal extent but disturbed or even, as in many of India's forests, arrested at the earliest stages of succession. Sahelian West African remnant forests are vast, brushy open woodlands. The mangroves of the world's tropical coasts are now highly fragmented remnant forests. At one end of the size scale stand the forests of Papua New Guinea, Indonesia, Central Africa, and Amazonia, where large swathes of forest occupied by indigenous peoples are being fragmented by loggers, oil companies, mining operations, and resettlement programs. At the other end of the size scale are the community forests of Ghana, Thailand, and Mexico, which range in size from a few hectares to hundreds of hectares, and stand as islands in land under nonforest use. These forests comprise most of the world's forest cover.

An estimated 200 to 500 million people live in the world's forests (Lynch 1990). They have long used forests and forest plants, buried their ancestors in these forests, and named the hills, rivers, and valleys where the forests are found. It is their forests that are under discussion in this volume. These forest owners and their remnant forests, which generally cover 20 to 70 percent of the communities' lands, are primarily found in "marginal" or "peripheral" areas in relation to the global economy.

Most forest-dwelling people are farmers (figure 12-1). They grow cereals such as rice and maize, or root crops, such as manioc, sweet potato, taro, and yams. But regardless of their main crop, they all depend on forests for material goods, food, and the definition of their identity. Indigenous forest people's dependence on forest and field is evident in ritual exchanges of field and forest products from these areas. For example, when two different communities of the Bora of the Peruvian and Colombian Amazon have a ceremony or meeting, manioc bread and pineapples from one community's garden are ritually exchanged for forest products such as fish, insects, and game brought by the other community. This exchange signifies recognition of mutual interdependence between Bora communities, as well as the recognition of mutual interdependence between the human and forest communities.

The remaining regions of robust, intact forest often overlap with areas used and claimed by indigenous communities. The term *indigenous* is used here to refer to long-term residents of a given place. Historically, the term *indigenous peoples* has referred to peoples being ruled by colonizers. In Africa, Europe, and Asia, one indigenous group often

Figure 12-1
Forest farmers, such as this Bora farmer in the Upper Amazon of Peru, use swidden farming methods to manage forest processes, produce crops, and maintain forests. (Photo by Janis B. Alcorn.)

controls the state, and that group prefers to view other indigenous peoples as ethnic "populations" of their state. In the Americas, indigenous forest-dwelling communities are usually remnants of the original indigenous groups that survived conquest by Europeans and subsequent marginalization by neocolonial states. In Latin America, there are also indigenous peasant communities of mixed ancestry that rely on forests and have developed methods for managing them, often by learning from their indigenous neighbors or relatives.

Aside from the political connotations of the term, there are important ecological aspects of being indigenous. Indigenous peoples often have personal and ancestral relationships with the lands on which they reside (or regularly migrate across), including the forests, waters, sky, and mountains associated with those lands. For people accustomed, as we are, to extracting resources from distant areas belonging to others, land and resources are commodities. Our identities are defined by the commodities we consume, and this relationship does not engender any sense of stewardship for the resources that were the source of the commodity. Indigenous peoples' relationships with one another and with their past, on the other hand, are tightly linked to their rights to resources within the lands and waters owned by their communities. Indigenous peoples' relationship to a particular place supports a strong ethical sense of responsibility toward stewardship of those lands and waters, even as internal and external stresses work against that relationships (Lynch and Alcorn, 1994; Ventocilla et al., 1995).

Forest Uses, Users, and Owners

There are thousands of articles about commercially valuable forest products (e.g., see the journal *Economic Botany*). In some cases, it is the harvest of these products that increasingly leads to depletion of forests, particularly extraction of timber and nontimber products by large companies. The benefits from harvest of these products do not usually flow to local people who serve as collectors or laborers in forests that once belonged to their communities.

There are also thousands of articles, in many different languages, describing the local uses of forest products. The general patterns of forest use are similar around the world: Communities rely on forests for food, medicine, construction materials, craft materials, fuel, insurance, and products for sale and trade—in other words a significant portion of their livelihoods. These uses of the forest often go unrecognized by outsiders who equate use with commodities for sale.

Indigenous residents use the forest and its products for multiple purposes. For example, the Mexican Huastec Maya name nearly 90 percent of plants available in their environment (Alcorn 1984). Two-thirds of those plants (forest and nonforest) have uses. One might assume that when people have need of a plant they just go out, find it, and bring it home. But forest-dwelling communities do not just harvest useful plants; they manage their ecosystems to ensure that a wide variety of useful species are available. In the communal lands owned by the Huastec, one-third of available plant species are "managed for"—in the sense that those species are planted or protected in order to maintain them within the habitats created by general agricultural and forestry management practices (Alcorn 1981). In the Huastec case, about one-third of available plant species are found in forests, and 90 percent of forest species have specific uses (Alcorn 1983). This level of plant knowledge and interest in maintaining useful species is generally typical of indigenous forest dwellers around the world (e.g., Balée 1994; Conklin 1954; Messerschmidt 1993; Warner 1991).

Forests provide food in the form of mushrooms, edible greens and roots, fruits, river products such as fish and reptiles, and game that can be shot in the forest or trapped in gardens scattered in the forest. Studies of forest dwellers around the world show that it is common for forest foods to make up between 50 and 80 percent of nutritional intake, with the higher percentages going to women, children, and poor people (Scoones et al. 1992). Salt can even be made by burning certain forest leaves and roots together. Forests also provide forage for forest-dwellers' animals and forage for the animals of pastoralists that move through forested areas.

Beyond food, forest plants have numerous other, local uses. The forest itself creates homes—both as backyards and playgrounds for children and as workplaces for their parents. Forest plants are used to make the skeleton of the house, to lash house parts together, for thatching roofs, and for the walls. Forest plants provide fibers used for producing household items such as bags, carrying baskets, and bark cloth. Bark cloth still has ritual importance in many forested areas of the world. Large forest trees are used to make drums for communication, as well as to make canoes, which serve for communication and transport. Forest dwellers rely on their forests for access to fresh medicines from living plants and animals on sudden notice when someone falls ill.

Local communities generally value forests beyond the products that are immediately harvested. They value forests as living factories from which products can be harvested, for their natural regenerative processes that subsidize agriculture, and for their ecological services in protecting watersheds and coasts and providing habitat for game and fish reproduction. Forest dwellers consciously incorporate the forest into their land-use systems. While outsiders often think that local forests are still standing because people haven't gotten around to cutting them down yet (it is only a matter of time before they will be gone), these forests often remain as the result of conscious land-use decisions made by households and communitites to keep these patches of forests (it is unlikely they will be cut down in the near future).

For example, swidden agriculture (aka slash and burn) was long vilified as destructive by those who think of forest as a permanent group of standing trees. This attitude toward swidden is changing as more observers realize that forests are really processes, communities in flux, not permanent objects (Warner 1991). These processes form the basis for slash-and-burn farming (Alcorn 1989), which has played a role in traditional forest management. Swidden spatially and temporally integrates complex agroforestry systems into agriculture systems around the world (Alcorn 1990; Messerschmidt 1993; Olofson 1983; Shepherd 1992; Warner 1991).

In addition to swidden, or in communities not practicing swidden, there may be several types of more permanent forest patches being managed for different products. For example, Indonesians who manage forests for the commercial production of illipe, durian, or damar also maintain more natural and biodiverse patches of community forest that have not been enriched with any particular commercially valuable species. In sum, all the forest patches seen from air, even in a small geographic area, are usually not equal.

Aware of their dependence on the forest, communities that are able to modify forest and land use respond to reduced availability of needed forest goods and services. They have altered land-use and livelihood strategies in order to retain forests—including intensification of agriculture on nonforest lands, increased reliance on income from outside jobs, and dedication of increased land to natural forests. In some cases, agricultural intensification and increased wage labor opportunities have enabled farmers to retain, enrich, and even expand permanent forest (e.g., Alcorn 1984; Pinedo-Vasquez and Padoch, chapter 16, this volume; Chayan Vaddhanaphuti, pers. comm. 1990). On the other hand, in situations where local communities do not have the right to manage their forests, they must suffer the consequences of the state forest departments' mismanagement.

Traditionally, forests were managed for trade and local use, as well as for ecological services. Increasingly, however, communities are moving to exploit their forests for monetary gains. Medicinal plants and trees that produce salable fruits, resins, and other products are managed for sale or trade into local or regional markets; animal skins and meat are sold to buyers seeking products for outside markets; and orchids and other plants are sold for ornamental uses in urban settings. High demand for such nontimber forest products can severely degrade forests, especially in cases where communities are denied the right to police extraction from their resource base.

Although commercial operations extract the same products used by local people, they generally exploit the products for short-term profits without concern for long-term sustainability. Commercial harvests, particularly extraction of timber and nontimber products by large companies, increasingly leads to depletion of forests. Commercial extraction has also led to depletion or local extinction of valuable medicinal plants (e.g., Lewington 1993).

The level of exploitation by outsiders without a long-term dependence on the local resource often has immense impact and is highly unsustainable. Sometimes outsiders just come and gather bags of medicinal plant samples for phytochemical and pharmacological testing for the international market. If the medicinal plant becomes commercially valuable, however, laborers hired by collecting firms can severely reduce the populations of that species to the point where local people no longer have access to the medicine (e.g., Cunningham and Mbenkum 1993).

While community-based logging for local use has had minimal impacts, community-based logging for profit has a mixed record. There are indications that it has the greatest potential to be sustainable if involved communities are concerned for long-term productivity and have strong tenurial rights over the resource (e.g., Bray 1991). On the other hand, logging by state-sponsored concessionaires is usually carried out with minimal concern for long-term productivity and leaves the land in a condition of open access for settlement after the original tenurial rights have been broken. Such logging often continues even after local communities decry the ecological degradation and the loss of biodiversity. In Thailand, for example, a logging ban was enacted after widespread community protests against overlogging by government-sponsored concessionaires in community forests (Lohmann 1991). Yet logging continues in Thailand—logs are taken into Burma to be stamped as Burmese for return across the border into Thailand for sale as imported logs. The Thai military elite also finance and carry out logging in Burma, Cambodia, and Laos to profit from Thai market demand. This relationship between the military and logging is typical in Asian and Central African forests where logging is usually done by powerful interests with links to the government and the military.

Most often, outsiders come into forests, claim the land, and create orange groves, plantations, or pastures where forest once stood. While degraded natural forests often have the potential for regeneration, the levels of biodiversity plummet after conversion to plantation or pasture, and regeneration of forest is unlikely due to loss of seed sources and the alteration of soils and waterways that attend the conversion. It is this radical land-use change accompanied by severe depletion of biodiversity that must be compared against the land use of marginalized indigenous communities that live in relatively biodiverse areas.

Finally, in addition to local and outsider users of forests discussed above, a third user class has recently been asserting its rights—the global users. The global community is now claiming rights to manage forest for conservation of biodiversity and for carbon sequestration to mitigate against global climate change. During the 1992 conference on which this book is based, for example, North American biologists suggested that they should write "owners' manuals" for tropical forests.

Conflict and Tenure

As the paragraphs above make clear, forests and forest species are useful. But who owns these forests and who benefits from using and conserving this biodiversity? Conflicts have arisen over who has the right to make decisions about forest use and management—that is, who has "ownership" of the forest. While we may think of ownership as the right to sell something, tenurial specialists refer to ownership or tenure as a bundle of rights and obligations as recognized or distributed by some (local or state) authority. Tenure defines relationships between people and mediates their use of natural resources (Crocombe 1971), but it does not define relationships between people and property. Those with tenurial rights and responsibilities have a specific relationship to others in terms of who can and cannot do what, and under what circumstances, with the property in question.

Local conflicts over resources have led to the evolution of local tenure systems and supportive social mores appropriate to the culture and sociopolitical organization of the community (e.g., Berkes 1989). What are some of the characteristics of these local tenure systems, and why are they resource conservative? Traditional tenure systems are in effect a partnership between individuals and the community to maintain a community's resource base by limiting access and imposing restrictions on forest use. They are extremely variable, complex mixtures of private and community rights and responsibilities (Crocombe 1971). Community members do not simply share joint ownership; rather, different community members have different rights and obligations, and different kinds of rights are exercised over different kinds of resources. Agricultural lands, for example, are often held by individual families, while rights to forest or pasture lands are more likely to overlap. These overlapping rights result in a sharing of benefits across a broad range of the community. For example, rights to a given patch of forest or even particular trees may be held by an individual household, several different people, or different groups of people at different times of the year. Overlapping claims work to exclude noncommunity members and protect the community from easy acquisition by outsiders or from exclusive use by any single individual who might destroy the community's forest resource base.

Customary tenure systems vary widely from place to place, yet they are similar to one another in that they derive their legitimacy from the community's authority. With rights come certain responsibilities toward the community and toward maintenance of the resource. Resource rights are inherited in complex ways, and bundles of rights often get reorganized in each new generation. Customary rights and obligations evolve with the changing availability of the resource, changing demands for the resource, and changing relationships between user communities. If forests are under stress, local regulations are often established and then enforced by local forest-protection committees. The effectiveness of these common property systems depends on widespread acceptance and adherence to rules governing access, strong institutions that administer local justice, and guidance by local leaders committed to the values of the system. Some have described traditional tenurial systems as a form of "institutional capital" (Field 1984), because compliance is sustained with low investment on enforcement.

State Responses to Traditional Tenure

The institutions, authority, and rights of communities to forests are almost invariably ignored when decisions about forests are made at the national level or in the context of colonial expansion. Such national-level decisions lead to conflicts—both between the state and the community, and within and between communities. These conflicts are very old. For example, in the late 1500s, Guatemalan communities argued in Spanish colonial courts that their forests should not be given over to Spaniards for conversion to pasture, because these forests belonged to them and were an integral part of their communities' productive base (M. J. MacLeod, pers. comm. 1992). The same arguments can be found in the historical literature from Europe (Westoby 1989), Africa (Porter et al., 1991), and Asia (Guha 1983; Scott 1976). Today, communities make the same arguments in the face of forest alienation for state-sanctioned logging, mining and petroleum ventures, plantations, and colonization or resettlement programs. In 1993, the Embera, Kuna, and Waunaan communities in the Darién region of Panama mapped their forests (figure 12-2) in an effort to register their rights to ancestral lands in the face of settler encroachment and the threatened completion of the last segment of the Panamerican Highway (Denniston 1994). The forest-protection committees established by communities to exclude their neighbors and regulate community forest use are usually unable to keep out powerful outsiders.

The historical trend is quite similar around the world. Initially, despite the fact that forests and adjacent agricultural lands were being administered under traditional tenurial systems, named in local languages, and claimed by local peoples, colonial authorities and/or the

Figure 12-2

Panamanian government cartographers assist Embera, Kuna, and Waunaan communities to map the land-use patterns in their Darien territories. (Photo by Nicanor Gonzales, Center for the Support of Native Lands.)

neocolonial state limited communities to certain specified forest product "usufruct rights," which were then downgraded to "privileges" (Guha 1983 lucidly describes this process in India). Eventually, states deny resident communities any rights or privileges in the state's forest, except perhaps temporary employment as laborers. In other cases, the forest is turned directly over to government-sanctioned concessionaires, ranchers, or plantation/reforestation companies; communities then immediately lose rights to their forest and become labeled "squatters" instead of being recognized as holders of property rights (figure 12-3). When community authority is usurped by the state, community property is transformed into "no one's property" (open access), no one takes responsibility for long-term forest management, and forest degradation ensues. Hardin's (1968) famous "tragedy of the commons" model was really describing the tragedy of "open access" (Berkes et al., 1989). Many forests were managed under common property regimes, but they were transformed into open access forests after logging or development projects ignored, and thereby invalidated, local systems of forest control. Once under open access, their biodiversity has been depleted by all and sundry, locals and outsiders.

States have generally ignored customary property rights. Although states often allow people to continue to exercise traditional rights, they generally refuse to acknowledge those rights when they impinge on the interests of the state (e.g., the rights to decide who logs and who gets profits from the sale of timber). In many places, states are effectively "executive committees of elites" who make policies and laws for their own personal benefit. States have historically assumed rights over forests in order to generate revenues. They commonly exercise rights to decide, for example, harvest schedules, royalties and distribution of profits, who can harvest (timber permits), who can transport logs, and export taxes. While states claim authority to replant or manage forests, they seldom exercise that right or otherwise take responsibility for it. Few states are held publicly accountable for meeting their obligations.

Some states grant indigenous peoples almost no rights of any kind. In other cases, indigenous peoples' rights to forests and land may be written into the constitution or in special policies or rulings, yet these rights are often undermined by other laws or ignored. For example, the state can retain the right to award logging concessions or mining permits. In other cases, indigenous peoples' land is classified as "public land." Where national laws or constitutions do recognize indigenous rights, they rarely recognize the legitimacy of indigenous social and political institutions that regulate forest use and land rights. This limits the abil-

Figure 12-3

Karen communities have maintained watershed forests as part of their terraced rice system in western Thailand (near Tak) for several hundred years. The rights of the Karen to control their common property are being undermined, with negative effects on forests. The Karen community has reported Hmong commercial farmers' clearing of their forest to Royal Forest Department authorities, who responded that the Karen were squatters because the forest belongs to the state and they had no right to report illegal clearing of forest. The allegation was made that the local government officer had no incentive to protect the forest due to kick-backs from commercial vegetable growers who were clearing forests. (Photo by Janis B. Alcorn.)

ity of communities to interact with national institutions because of the lack of recognized organizational linkages.

Nonetheless, attempts are made to work within the system to achieve recognition of rights. Sometimes indigenous groups choose to define themselves as cooperatives or corporations in order to receive recognition and action on tenure; but because these are not indigenous institutions, incorporation can create unexpected conflicts within communities. Sometimes forested areas designated as "community property" are actually controlled by a few powerful families who know how to work the larger system. Codification of existing tenure at a moment in time is another common attempt to validate traditional rights for incorporation into modern systems; however, codification freezes a situation and fails to allow the traditional system of conflict resolution to function and evolve.

Likewise, efforts to ensure resource tenure through individual titling have actually resulted in increased resource degradation and loss of forest. The state usually awards title to only one of the parties that have claims on the property in question. Those seeking titles, however, are often entrepreneurs who then sell the forest to outsiders who have no long-term interest in maintaining the forest. Yet, unfortunately, individual titling is the option chosen most often by those who want to address the perceived tenurial problem.

The alternative with the best track record is demarcation of community lands and award of community titles, leases, or special status (such as *comarcas* in Panama, *comunidades* in Mexico, and ancestral domain in Philippines) accompanied by recognition of traditional authority to resolve resource rights disputes. This option does not require the state to understand the complexities of the communities' tenurial system. Community-based tenure offers a protective and enabling shell, in the computer jargon sense, that allows locally derived management institutions to flourish and adapt to their ecological and social environment (Alcorn and Toledo 1995). Strong local tenurial shells protect the community and its institutions against powerful, resource-hungry entities with different cultural and ethical values that threaten the community and its resources. The state has the obligation to defend the rights of the community against outsiders. Therefore, in order to maintain the shell, the state can restrict the rights of communities to sell, lease, or rent community properties. It also retains the right to place broad zoning restrictions on the use of the property (land-use zoning, etc).

Donors and multilateral developments bands (MDBs) are aware of the importance of tenurial issues. The Asian Development Bank's (ADB) new policy on forestry, explicitly highlighting the critical importance of addressing tenure issues, states that long-term security of tenure is "imperative" for successful forest protection and management. The ADB acknowledges the loss of forest-dwellers' traditional rights as a major cause of deforestation and will support enforcement of tenurial rights of forest-dwelling and forest-dependent communities. But accomplishing what is laid out in policy has proven difficult for other MDBs. For example, a World Bank study (Wali and Davis 1992) reviewed the legal framework for land tenure, the resource tenure situation of indigenous peoples in specific Latin American countries, and World Bank–financed efforts to regularize tenure and demarcate indigenous lands for community-based tenure. The study found that even when indigenous lands have been demarcated and recognized by the government, they are still subject to exploitation by settlers and logging operations. In other words, although one of the primary functions of the state is to defend private property, it is not defending indigenous peoples' property rights.

In addition to direct state-sponsored deforestation, in many countries there is what Gerald Murray (unpublished) has called "institutional legitimization of deforestation," where multiple incentives for other sectors promote deforestation despite the claims of state interest in maintaining forest.

The bottom line is that while on paper ownership of forest resources may be held by the state, management of forests may be formally vested in a particular department, and indigenous people may have rights, in actuality the people who live in or use a given local forested area determine what happens there. Locally, decisions are influenced by state policies—not the formal policies, but the shadow or de facto policies as they are interpreted and implemented locally.

Many different sectoral and financial policies contribute to the loss of remnant forests, but policy reform that addresses the rights of communities to own forest and the obligations of the state to defend those community rights is a fundamental first step necessary for other policy reforms to have a significant effect. Policy analysts, nongovernmental organizations, and indigenous organizations have identified community tenure rights as critical for conservation of forests (International Alliance of the Indigenous-Tribal Peoples of the Tropical Forests 1992; Lynch and Alcorn 1994). Analysts looking at options for securing parks and reserves have suggested that instead of focusing on what kind of vegetation should be in buffer zones, conservationists should be trying to create appropriate ownership structures in buffer zones and focus on social processes instead of physical structures (Brown and Wyckoff-Baird 1992).

Options for Conserving Remnant Forests

Given the disturbing, repeated trend from community property to open access property that I have described, the issue that emerges can be captured in one question: What can be done to prevent the same story from happening in the world's remaining remnant forests?

Two answers offer contrasting options: (1) the state should assume a greater regulatory role over state property, or (2) local communities should be given greater authority to regulate forests as community property in partnerships with the state. Many states, and recently even some conservation biologists, cling to the first answer. India is an example of a country that followed the first option until it had a large, paramilitary forest department. India's forests still shrank and disappeared (Alcorn and Molnar 1996), however, and now it is switching to the second option. In view of the evidence, other countries are also seeking partnerships with communities in order to save forests.

The state will always be an important player in forest management—from serving an essential role as the last resort for justice to claiming the mantle of forest owner and adjudicator. It is possible for the state to play a positive role in conserving remnant forests, but general indications are that state regulation alone usually fails. Historically, many states have a bad record in terms of forest management and human rights; what they bring to the table varies widely. In some cases, states are strong, but their power decreases with distance from the capital; in other places, the state is so weak that local elites or outside businesses take on the functions of government.

Communities are also heterogeneous and bring various strengths to the table. It is important to recognize, however, that local forest dwellers are stakeholders with prior rights, which should not be abused when seeking solutions to forest management issues. Community rights are a given, regardless of the communities' track record in forest management. The challenge is to develop incentives and mechanisms for outsiders to negotiate with communities to achieve a mutually desirable goal of forest management.

The high degree of local, sometimes unexpected, variation in communities and governments cannot be overemphasized. In some cases where local communities are well organized and have demonstrated their management of forest resources, the state and its military engage them in bloody struggle. In other cases where states are weak, local conflict over ambiguous land rights among well-armed private communities discourages outsider logging companies from entering and deforesting an area (e.g., tribal areas of Pakistan). In other cases, communities and the fate of their forests are under the control of outside money lenders or drug lords aligned with interests that oppose national government.

Just as there is no typical community or local representative of the state, there are no packaged solutions for the loss of forests. Options for management depend on the state's commitment and capacity. Commitment and capacity can be determined by investigating the state's past performance on the ground, not by accepting a government's rhetoric or using budgets and manpower as indicators. But options for management cannot be imposed from outside. What is needed is for those with global interests to recognize that they must think locally, not just globally. Outsiders, including agencies of the state, should engage local communities as stakeholders with prior rights, together with those who claim stakeholder status, and initiate a long-term process of negotiations. Local negotiations are necessary to determine to what extent objectives are shared as well as for assessing the capacities of both parties to meet their obligations under any proposed partnership. The processes pioneered by the "joint forest management" programs of India for the past two decades (Malhotra and Poffenberger 1989; Raju et al., 1993) offer an example of such an approach.

Partnerships: Recommendations

The partnership option recognizes the authority of both parties' (state and community), uses the power of the state for defending the public good, and creates mechanisms for feedback to reach the central government. The options for partnerships are different in different places, even within one country, and must be worked out locally.

The option most likely to be successful is to stimulate appropriate partnerships between the state and local communities. In some cases, this may functionally be community partnerships with provincial or even municipal governments. The partnership option gives both parties authority, uses the power of the state for defending the public good, and creates a mechanism for local ecological feedback to reach the central government.

This option requires strengthening local organizations and providing communities with information to enable them to request useful research and assistance. It requires creating communication networks that provide the state with real information about the status of forests and threats to them. Building functional partnerships may require new legislation, policies, institutional linkages, and processes. Building viable partnerships requires that the state and outside agencies promoting forest conservation carry out the following steps:

1. Recognize the presence and rights of the (often invisible) people living in forests.
2. Recognize traditional (customary) forest tenure systems and defend communities' traditional rights against outsiders; ensure that benefits from intellectual property rights to valuable biodiversity and traditional knowledge are distributed to communities that live where useful biodiversity is being maintained.
3. Commit to a long-term negotiation process in good faith and by principles of due process, including ensuring that all parties are fully informed of their rights and options.
4. Legally define community-based tenurial instruments appropriate to safe guarding community-based rights.
5. Recognize "forest" as a valid land use, so that forest clearance is not a prerequisite to acquire tenure or loans.
6. Develop and assess comanagement options. At a national level, this would require exploring the existing legal and policy structures and looking for opportunities to build partnerships within existing policy frameworks.
7. Strengthen local organizations and provide communities with information about their rights and how to exercise them. This may include mapping and land-use planning through participatory methods (see Poole 1995).
8. Create communication networks so that communities can provide the state with real information about the status of forests and threats to them.
9. Encourage research and NGO partnerships with government and community organizations. This will serve to strengthen the state's commitment to the partnership as well as the chances for successful forest management through technical assistance.
10 Monitor impacts of policy and process changes for feedback. Experiment with pilot partnerships to give governments an opportunity to collaborate with communities in frameworks that are experimental and "illegal" under current law.

India and Nepal, for example, have pioneered some unique forest "comanagement" partnerships appropriate for South Asia under the name "joint forest management." There has recently been discussion of starting new joint protected area management programs in India. Another option is found in the forest leases being implemented and considered in Indonesia, the Philippines, and Thailand (Fox 1993). In the Philippines, the state has prevented migrant farmers from entering forests that are formally leased to indigenous peoples. Other options include the extractive and indigenous reserves of Latin America. Mexico's *ejidos* and *comunidades* provide communal tenure and a haven for biodiversity (Toledo 1992). Fully 70 percent of Mexico's remaining forest cover is on communally owned lands (Bray 1991), but those forests are now in jeopardy under new Mexican privatization policies being enacted in support of the North American Free Trade Agreement (NAFTA) (DeWalt and Rees 1994; Toledo 1994; Toledo and Alcorn 1995).

Scientists and universities can offer their partners assistance in many areas (figure 12-4), from designing ecological monitoring systems to assessing forest health (e.g., Peters 1994). In some cases, communities are

Figure 12-4

In West Kalimantan, Indonesia, forest farmers have been collaborating with social and biological scientists from Yale University, the New York Botanical Garden, and provincial forest officers to document and assess regeneration in their managed forests, such as the one pictured here.

seeking external assistance to ensure natural forest regeneration after realizing that decisions to market forest products have the potential to deplete their resource base. In other cases, forestry specialists at national universities have certified that community forests are not "degraded," keeping the state from awarding logging and reforestation licenses to concessionaires who would cut down community forests.

It is important to remember that, while the advantages of partnerships seem obvious, there are good reasons why some communities might not want their forest tenure to be recognized by the state. Recognition opens up communities to further danger of manipulation by the administrative arm of the state. In some cases, local communities are exercising many of their rights even when those rights are not recognized by the state. But in most places, after commercial ventures and speculators have entered forests and ignored the rights of residents, the ensuing problems have been sufficient to push communities to fight for recognition of their rights. There is always, however, a certain element of desirability to retaining an ambiguous identity vis-à-vis the state in the face of present danger from a state that moves to dominate, or extract resources from, those it has clearly identified. The rent-seeking behavior of local bureaucracies and departmental functionaries is notorious, and the weak systems of justice in most countries make rural people vulnerable to exploitation by distant or local powers. These forces can undermine proposed partnerships.

Any partnership should only be formulated and proposed after close consultation with all the stakeholders, including local communities and government. The Baguio Declaration (Legal Rights Center 1994) outlines a strategic approach to strengthening community-based forest management. Lynch and Talbot (1995) offer detailed steps for supporting community-based tenurial rights. The challenge for government is to find a way to identify existing community-based initiatives already struggling to build biodiversity conservation on traditional systems of knowledge and local tenurial institutions (c.f., Alcorn 1991; Alcorn 1995; Gadgil et al., 1993; Seymour 1994) in order for government to start joint management in the areas with the greatest potential for success.

Conclusion

What is an academic discussion in this book is literally a matter of life and death for people in Asia, Africa, and Latin America. Desperate people are taking the risk of losing their lives to stand up before the most powerful elements of their governments and local power structures in order to save forests upon which they depend for livelihood. Chico Mendez is famous as a martyr for defending local rights to Brazilian forests, but he was one of the tens of thousands who have been killed defending their home forests over the past several decades. In 1994, in Mindanao, Philippines, for example, indigenous peoples are fighting industrial plantation companies with arrows and other simple weapons, and many people have died in the struggle for rights to maintain their natural forests. Yet their deaths go unreported.

An Asian NGO leader once told me several years ago that the NGOs in his country were sorry to see international conservation experts coming into their country, because "they don't understand the causes of deforestation, they don't understand the dark side of the government, and as a consequence they fund things that actually undermine conservation." International conservationists who fuss that local people cannot be trusted to conserve forests are not seeing the forest for the trees. They are failing to understand livelihoods and politics in the countries where tropical forests are found. Peluso (1993) has echoed this analysis.

There is widespread evidence that indigenous communities continue to conserve forest patches and biodiversity in the face of many stresses, including escalating threats from outsiders who are extracting resources or settling on communities' lands, state expropriation of indigenous peoples' lands and resources, demographic changes, cultural change, failure to educate young people in traditional ecological knowledge, community institutions that are unable to interface effectively with outsiders, technological changes, and crop changes (Alcorn 1994; Lynch and Alcorn 1994).

But it is not necessary for a person or a community to have conservationist credentials in order to have property rights and the other basic human rights accorded citizens of any country. Conservationists should not forget this, lest they be accused of colonial approaches to conservation and undermine their own goals.

Communities have had, and continue to have, very good reasons to defend forests; and they have fought, and continue to fight, to defend their forests. Given the evidence from historical trends and present-day patterns, recognition of community tenure to forests is the single most effective policy reform option available for conservation of robust natural forests in most countries.

In Australia, Uganda, India, and several South Pacific nations, governments are recognizing that communities with traditional tenurial rights offer opportunities for partnerships and have begun seriously experimenting with joint protected area management. I urge open-minded biologists to consider the current poor performance of state forest and park departments, and the increasing conflicts with local people, and then consider the possibility that forests lying within state-owned national parks or other protected areas may best be protected through strengthening community ownership and partnerships with the state, instead of unrealistically expecting the state to be the sole protector (Alcorn 1993, 1994).

Acknowledgments

Portions of this paper are reprinted from another article (Alcorn 1995) with permission from the *Annals of the Missouri Botanical Garden.* I thank those who commented on earlier versions of this manuscript, particularly John Schelhas, Russell Greenberg, Judy Gradwohl, and Owen Lynch. The conclusions expressed in this paper are mine and should not be attributed to World Wildlife Fund or the U.S. Agency for International Development.

References

Alcorn, J. B. 1981. "Huastec noncrop resource management: Implications for prehistoric rainforest management." *Human Ecology* 9: 395–417.

Alcorn, J. B. 1983. "El te'lom Huasteco: Presente, pasado, y futuro de un sistema de silvicultura indígena." *Biótica* 8: 315–331.

Alcorn, J. B. 1984. *Huastec Mayan Ethnobotany.* Austin, TX: University of Texas Press.

Alcorn, J. B. 1989. "Process as resource: The traditional agricultural ideology of Bora and Huastec resource management and its implications for research." In *Resource Management in Amazonia: Indigenous and Folk Strategies,* edited by D. A. Posey and W. Balée, pp. 63–77. Bronx, NY: New York Botanical Garden.

Alcorn, J. B. 1990. "Indigenous agroforestry systems in the Latin American tropics." In *Agroecology and Small Farm Development,* edited by M. A. Altieri and S. B. Hecht, pp. 203–220. Boca Raton, FL: CRC Press.

Alcorn, J. B. 1991. "Ethics, economies, and conservation." In *Biodiversity: Culture, Conservation and Ecodevelopment,* edited by M. L. Oldfield and J. B. Alcorn, pp. 317–349. Boulder, CO: Westview Press.

Alcorn, J. B. 1993. "Indigenous peoples and conservation." *Conservation Biology* 7: 424–426.

Alcorn, J. B. 1994. "Noble savage or noble state?: Northern myths and southern realities in biodiversity conservation." *Etnoecológica* 2: 7–19.

Alcorn, J. B. 1995. "Economic botany, conservation, and development: What's the connection?" *Annals of the Missouri Botanical Garden* 82: 34–46.

Alcorn, J. B., and A. Molnar. 1996. "Deforestation and human–forest relationships: What can we learn from India?" In *Tropical Deforestation: The Human Dimension,* edited by L. Sponsel, T. Headland, and R. Bailey, pp. 91–121. New York: Columbia University Press.

Alcorn, J. B., and Toledo, V. 1995. "The role of tenurial shells in ecological sustainability: Property rights and natural resource management in Mexico." In *Property Rights in Social and Ecological Context: Case Studies adn Design Applications,* edited by S. Hanna and M. Munasinghe, pp. 123–140. Washington, DC: The World Bank.

Balée, W. 1994. *Footprints of the Forest.* New York: Columbia University Press.

Berkes, F., ed. 1989. *Common Property Resources.* London: Belhaven Press.

Berkes, F., D. Feeny, B. J. McCay, and J. M. Acheson. 1989. "The benefits of the commons" *Nature* 340: 91–93.

Bray, D. 1991. "The forests of Mexico: Moving from concessions to communities." *Grassroots Development* 15(3): 16–17.

Brown, M., and B. Wyckoff-Baird. 1992. *Designing Integrated Conservation and Development Projects.* Washington, DC: Biodiversity Support Program, World Wildlife Fund.

Conklin, H. C. 1954. *The Relation of Hanunoo Culture to the Plant World.* Ph.D. dissertation, Yale University. Ann Arbor: University Microfilms.

Crocombe, R. 1971. "An approach to the analysis of land tenure systems." In *Land Tenure in the Pacific,* edited by R. Crocombe, pp. 1–17. Melbourne: Oxford University Press.

Cunningham, A. B., and F. T. Mbenkum. 1993. *Medicinal Bark in International Trade: A Case Study of the Afromontane Tree* Prunus africana. Gland, Switzerland: Worldwide Fund for Nature.

Denniston, D. 1994. "Defending the land with maps." *World Watch* 7(1): 27–31.

DeWalt, B. R., and M. W. Rees. 1994. *The End of Agrarian Reform in Mexico: Past Lessons, Future Prospects.* San Diego: Center for U.S–Mexican Studies, University of California at San Diego.

Field, A. J. 1984. "Microeconomics, norms, and rationality." *Economic Development and Cultural Change* 32: 683–711.

Fox, J., ed. 1993. *Legal Frameworks in Forest Management in Asia.* Occasional Papers of the Program on Environment, Paper No. 16. Honolulu: East–West Center.

Gadgil, M., F. Berkes, and C. Folke. 1993. "Indigenous knowledge for biodiversity conservation." *Ambio* 22: 151–156.

Guha, R. 1983. "Forestry in British and post-British India." *Economic and Political Weekly,* October 29, 1983, pp. 1882–1896; and November 5–12, 1983, pp. 1940–1947.

Hardin, G. 1968. "The tragedy of the commons." *Science* 162: 1243–1248.

International Alliance of Indigenous-Tribal Peoples of the Tropical Forests. 1992. *Charter of the Indigenous-Tribal Peoples of the Tropical Forest.* Mimeo available from World Rainforest Movement, UK, and Cultural Survival, US.

Legal Rights and Natural Resources Center/Kasama sa Kalikasan. 1994. "Baguio declaration." NGO Policy Workshop on Strategies for Effectively Promoting Community-Based Management of Tropical Forest Resources. Manila: Legal Rights and Natural Resources Center (LRC)/Kasama sa Kalikasan (KSK).

Lewington, A. 1993. *Species in Danger, Medicinal Plants and Plant Extracts: A Review of Their Importation into Europe.* Cambridge, UK: TRAFFIC International and Worldwide Fund for Nature.

Lohmann, L. 1991. "Who defends biodiversity? Conservation Strategies and the Case of Thailand." *The Ecologist* 21(1): 5–13.

Lynch, O. J. 1990. *Whither the People? Demographic, Tenurial, and Agricultural Aspects of the Tropical Forestry Action Plan.* Washington, DC: World Resources Institute.

Lynch, O. J., and J. B. Alcorn. 1994. "Tenurial rights and community-based conservation." In *Natural Connections: Perspectives in Community-Based Conservation,* edited by D. Western, M. Wright, and S. Strum, pp. 373–392. Washington DC: Island Press.

Lynch, O. J., and K. Talbot. 1995. *Securing the Balance: National Laws and Community-Based Forest Management in Seven Asian and Pacific Countries.* Washington, DC: World Resources Institute

Malhotra, K. C., and M. Poffenberger, eds. 1989. *Forest Regeneration through Community Protection.* Calcutta: West Bengal Forest Department.

Messerschmidt, D. A., ed. 1993. *Common Forest Resource Management: Annotated Bibliography of Asia, Africa, and Latin America.* Rome: Food and Agriculture Organization of the United Nations (FAO).

Murray, G. 1989. Background paper for USAID/Costa Rica's FORESTA project. Unpublished.

Olofson, H. 1983. "Indigenous agroforestry systems." *Philippine Quarterly of Culture & Society* 11: 149–174.

Peluso, N. 1993. "Coercing conservation? The politics of state resource control." *Global Environmental Change* June 1993: 199–217.

Peters, C. M. 1994. *Sustainable Harvest of Non-Timber Plant Resources in Tropical Moist Forest: An Ecological Primer.* Washington, DC: Biodiversity Support Program, World Wildlife Fund.

Poole, P. 1995. *Indigenous Peoples, Mapping and Biodiversity Conservation: A Survey of Current Activities.* BSP Peoples and Forests Discussional Paper Series, No.1. Washington, DC: Biodiversity Support Program, World Wildlife Fund.

Porter, D., B. Allen, and G. Thompson. 1991. *Development in Practice: Paved with Good Intentions.* London: Routledge.

Raju, F., R. Vaghela, and M. S. Raju. 1993. *Development of People's Institutions for Management of Forests.* Ahmedabad, India: VIKSAT, Nehru Foundation for Development.

Scoones, I., M. Melnyk, and J. N. Pretty. 1992. *The Hidden Harvest: Wild Foods and Agricultural Systems, a Literature Review and Annotated Bibliography.* London: International Institute for Environment and Development (IIED).

Scott, J. C. 1976. *The Moral Economy of the Peasant.* New Haven, CT: Yale University Press.

Seymour, F. J. 1994. "Are successful community based conservation projects designed or discovered?" In *Natural Connections: Perspectives in Community-Based Conservation,* edited by D. Western, M. W. Wright, and S. Strum, pp. 472–496. Washington DC: Island Press.

Shepherd, G. 1992. *Managing Africa's Tropical Dry Forests.* London: Overseas Development Institute.

Toledo, V. 1992. "Biodiversidad y campesinado: La modernización en conflicto." *La Jornada del Campo* 9: 1–3.

Toledo, V. 1994. *La Ecología, Chiapas y el Artículo 27.* Mexico City: Ediciones Quinto Sol.

Ventocilla, J., H. Herrara, and V. Nuñez. 1995. *Plants and Animals in the Life of the Kuna.* Translated by E. King. Austin, TX: University of Texas Press.

Wali, A., and S. Davis. 1992. *Protecting Amerindian Lands: A Review of the World Bank Experience with Indigenous Land Regularization Programs in Lowland South America.* Latin American and the Caribbean Technical Department Regional Studies Program, Report No. 19. Washington, DC: World Bank.

Warner, K. 1991. *Shifting Cultivators: Local Technical Knowledge and Natural Resource Management in the Humid Tropics.* Rome: Food and Agriculture Organization of the United Nations (FAO).

Westoby, J. 1989. *Introduction to World Forestry.* Oxford, UK: Basil Blackwell.

Chapter 13

Land-Use Choice and Forest Patches in Costa Rica

John Schelhas

The conservation or destruction of forest patches in agricultural landscapes in the tropics ultimately depends on the cumulative effect of many individual decisions made by rural people over time. These decisions are not easily predicted, because they are made in response to complex relationships between production and consumption factors, policies and macrolevel factors, and culture and attitudes. The complexity of the factors determining rural people's land-use choices makes it difficult to draw broad generalizations about how to encourage forest patch conservation. A site-specific approach is needed that first understands existing patterns of land use in relation to the factors that influence them, and then builds forest patch conservation into these land-use systems in ways that will adapt to changing conditions.

This chapter will present a conceptual model for understanding how household land-use patterns are influenced by, and change in relation to, local, national, and global economic, cultural, and political factors. The model will be applied in discussing land-use patterns and forest patches in two sites in Costa Rica.

A Model of Land-Use Choice

The model of land-use choice in figure 13-1 integrates micro- and macrolevel factors to understand household economic and land-use strategies. Households are a fundamental unit for land-use decision making throughout the tropics; by focusing on households, the model is securely grounded in human actors and the effect that their actions

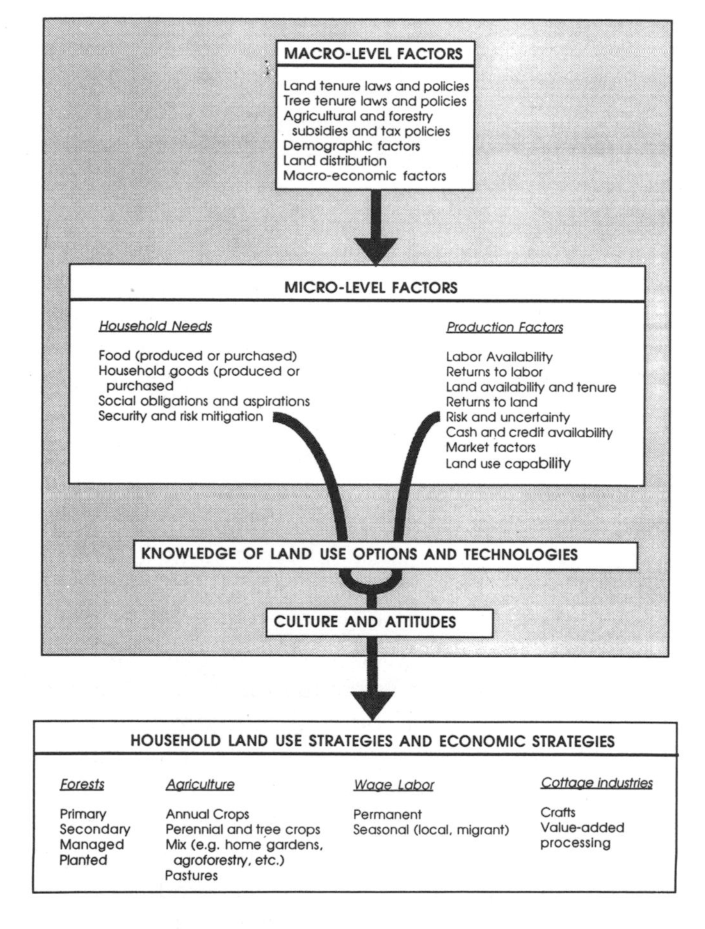

Figure 13-1

Conceptual model of household land-use strategies as combinations of different land uses, wage labor, and artisanal activities chosen in dynamic response to micro- and macrolevel factors, and influenced by cultural change and knowledge. (Modified from Schelhas and Greenberg 1993.)

have on the landscape. Households have multiple goals, which may include

1. maintaining or improving consumption of food and household goods;
2. providing hedges against short- and long-term risks and maintaining the capability to respond to family emergencies, illness, and accidents;
3. providing for future needs and desires, such as requirements for old age and educational opportunities and an inheritance for children; and
4. meeting social and political obligations (e.g., taxes, ritual expenses, status) (Chibnik 1990; Durrenberger and Tannenbaum 1992).

In order to meet these needs, households develop economic strategies from different land uses, wage labor alternatives, and artisanal activities (Chayanov 1966). Diversified economic strategies are common in the rural tropics, with households making use of different forest and nonforest zones, including primary and secondary forests, planting and livestock areas, forestry plantations, bodies of water, home gardens, secondary forests, and managed forests (Toledo et al. 1992), as well as engaging in wage labor (Collins 1987), commerce, and artisanal activities. Combinations of these different activities are selected by considering (1) household needs; (2) household resources, such as landholding size and tenure, land quality, household labor availability, and available cash and access to credit; and (3) production factors for different farm and on-farm options, including returns to land and labor, risk and uncertainty, and market factors.

Microlevel analysis of household decision making is fundamental, because it deals with the choices that rural inhabitants make and the landscape patterns that result. But household land-use choices must also be understood in relation to the broader context in which they occur, including macrolevel factors such as population density and growth, land distribution, and international commodity markets. Household land-use choices are also influenced by government through land tenure laws and policies, tree tenure laws and policies, and subsidies and taxes related to agriculture and forestry. Furthermore, land-use choices are influenced by culture and attitudes in a number of ways, including work preferences and consumption levels (Barlett 1980), food preferences (Persoon 1992), ritual and social obligations (Durrenberger and Tannenbaum 1992), and attitudes toward forests (Bengston 1994). Because these contextual factors influence household access to resources and create incentives and disincentives for different household land-use choices, it is important to understand the connections between macro and micro levels.

Forest Patches and Land-Use Choice

Forest patches play a variety of roles in household land-use and economic strategies, providing forest products for sale and subsistence, environmental services, and cultural and social values. The overall importance of forest patches to rural people and the specific benefits received from forest patches vary widely both from place to place and among different social groups at particular places. Many of these differences can be understood by applying the land-use choice model in figure 13-1. There are several key elements in this model that merit more detailed discussion in relation to forest patch use and conservation, including intensity of land use, policy contexts, demand for products and services, and cultural factors.

Intensity of Land Use

One key element running through the land-use decision-making model is intensity of land use. Boserup (1965, 1981) hypothesizes that agricultural intensification (increasing the yield per unit of land) is related to population density, with higher returns per unit of land required to support higher population densities. Boserup argues that intensification of land use is achieved at the cost of diminishing returns to labor. Therefore farmers only intensify when they are required to by increasing population density. Although Boserup focuses on *agricultural* intensification, she does discuss several examples of intensification related to forests, including conversion of forest to agricultural lands, elimination of forest fallows, and transformation of natural forests to tree plantations (Boserup 1965, 1981, 1990).

The Boserup model provides a useful framework for understanding land-use patterns and changes. Barlett (1982) used the Boserup model to analyze land-use choice and change in a Costa Rican village, finding that farmers with large landholdings were more likely to engage in cattle grazing, and farmers with small landholdings were more likely to engage in intensive cash cropping (Barlett 1982). In addition to being driven by population density, Barlett (1982) also found that intensification was driven by a desire to increase consumption levels of purchased goods. Other authors have added complexity to the Boserup model by showing that intensification choices also reflect household characteristics, risk, available technology, cash and credit availability, and market factors (Barlett 1982; Netting 1993; Schelhas, in press).

The Boserup model is particularly useful in understanding frontier land use. Expansion of previously geographically constrained groups onto new lands often results in disintensification of land use. When operating at lower population densities on the frontier, even intensive cultivators may turn to shifting cultivation because of its higher returns to labor (Netting 1969). On the frontier, a sparse population with access to abundant resources can live off the natural capital or "subsidy from nature." As a growing population exhausts previously abundant natural resources such as timber, virgin land, game, and freshwater supplies, other resources such as labor must be substituted (Boserup 1981: 5–6). Characteristic frontier land uses include hunting, long-fallow agriculture, and ranching; and although these land-use systems may exhaust natural resources, they are attractive to colonists because they provide high output per unit of labor (Boserup 1981: 135).

Frontier land use in the Neotropical lowlands of Central America and the Amazon has generally been consistent with the Boserup model, with natural capital being exhausted before land improvements are made. Low-intensity land uses such as pasture and shifting cultivation, which provide low returns to land but require low labor investments and yield high returns to labor, have generally dominated. In other parts of the world, frontier expansion has resulted in the replacement of natural forest with market-oriented agroforestry systems (Mary and Michon 1987). But in Latin America, although extensive forest land uses such as extraction of nontimber forest products and forestry may be appropriate for the low population densities of many Neotropical frontier areas, government policies for establishing land rights, speculative land markets, subsidies for cattle raising, and strong markets for beef and dairy products have instead encouraged conversion of forest to extensive agricultural and grazing lands.

Although frontier colonization in the Neotropics typically results in widespread conversion of tropical forests into pastures with low productivity, the destruction of forests is rarely complete and often not permanent. After an initial period of forest destruction during the unstable colonization period, forest and trees may return to some land. Trees may return because fields and pastures, established only to claim land, increase land values for sale, or capture government incentives, cannot be maintained in production (Bierregaard and Dale, chapter 10, this volume). In addition, as landholding objectives switch from increasing the exchange value of land to enhancing its productive value, trees and forestland uses are often integrated into farms to meet specific household and farm needs.

In predominantly agricultural landscapes, increases in forest cover may occur in association with disintensification caused by reduced population density or reduced labor availability. In Chiapas, Mexico, managed secondary forest increased (as did pasture) as landholders were drawn into employment in the oil industry during the oil boom (Collier et al., 1994). In Kenya, the presence of woodlots of black wattle (*Acacia mearnsii*), a tree cash crop used for tanning and charcoal, on small holdings was associated with household age, which decreases both labor availability and consumption needs (Dewees 1993). Disintensification can also occur concurrently with intensification. Reforestation of marginal lands (either due to poor soils or distance from a farmer's house) in Portugal was associated with increasing mechanization and irrigation, which were used to intensify agriculture on better lands (Bentley 1989).

The Boserup model of land-use intensification is helpful in understanding frontier land use; but because forest and extensive land use are not synonymous, care must be taken in using it to understand deforestation. Pasture and forest are both extensive land uses but have dramatically different biological conservation value. Biologically valuable forest land uses, ranging from natural forest management to shade-grown crops, differ in intensity—some more intensive than shifting agriculture or pasture. Increasing population density and overall intensity of land use are not always good predictors of biological diversity, although they are often related.

There are important roles for trees in even the most intensively managed landscapes. Cash cropping of trees was adopted by farmers in Haiti with average landholding sizes of 1.5 hectares (Murray 1986). Mortimore (1993) found dense scatters of multipurpose trees on almost all farms in a high-population-density area in Nigeria. Woodland and farm trees have remained a visible part of the landscape in the Machakos District, Kenya, under conditions of increasing population density and intensification of land use, with the density of fruit trees inversely related to landholding size (Tiffen et al., 1994). Home gardens with high plant diversity occupy 20 percent of the agricultural lands of Java, Indonesia, where population density in rural areas can reach 1,000 persons per square kilometer (Michon and Foresta 1992). In other areas of Indonesia, wild and extensive agroforestry systems are important for production and cover large land areas (Michon and Foresta 1992). In all of these examples, forest patches are associated with strong markets for forest products.

In general, more intensive land use can be expected to occur at greater population densities (or on small landholdings), on better lands, on more accessible lands, where opportunities for migration are limited, where few wage labor opportunities exist, and where cultivation and marketing of cash crops is possible. As intensification occurs, some forest will generally be converted to agricultural lands while remaining forest patches become more intensively managed. When forest patches are managed more intensively, there may be an overall trend toward ecosystem simplification (Foresta and Michon 1992), although intensification is often accompanied by diversification (Schelhas, in press). Very different types of forest patches can be expected, depending on the initial situation, the environmental factors, the policy environment, subsistence needs and market conditions, and the sociocultural context.

Macro-Level and Policy Contexts

The preceding section has touched on some of the ways that land-use choices are influenced by macrolevel factors and policies. For example, while population density is an important factor in land-use intensity, in many parts of the Neotropics inequities in land distribution mean that different landholders operate under different effective population densities. Such inequities can be so severe that large landholdings on good agricultural land are actually less intensively farmed than the many small landholdings on marginal lands (Durham 1979). Inequitable land distribution influences the pattern of land settlement and the extent of forest cover by driving small farmers to cultivate marginal lands that would otherwise be left in forest.

Policies that use conversion of forest to agricultural lands as proof that land is being used productively create an incentive for clearing land beyond that required by population growth or agricultural production needs. Land booms often occur during the opening of frontiers, when land values increase (at least temporarily) even as land is degraded (Hecht 1981). These speculative land markets often result in deforestation, since cleared, or "improved," land sells at a higher price than forested land. Forest and agricultural policies, related to timber harvesting, agricultural credit, food price, and subsidies for agrochemicals, alter the incentive structure for forest and agricultural land uses and can shift the ratio of forest to fields in the agricultural landscape. Tracing the impacts of such policies through microlevel choices to land-use patterns is difficult but extremely important, since even well-intended policies can create incentives for deforestation.

Demand for Products and Services

The presence and characteristics of forest patches is often closely related to the value of products and services they provide. Forest patches can provide a great diversity of timber and nontimber products, including plant foods, game, medicines, construction materials, fuelwood, fence posts, timber for sale, charcoal, fodder, soil amendments, and material for handicrafts and carpentry. These products are used both to meet household subsistence needs and to generate cash income. Forests and trees are also valued by farmers for the ecological services they provide, including maintaining soil fertility and watershed protection. Trees may also be incorporated into small niches in the farm landscape, such as living fences and hedgerows.

The importance of forest patches to farming systems and households can be highly variable, since trade-offs and substitutions are often possible. Some common trade-offs are those between subsistence and purchased products for household consumption, and between biological and purchased inputs. These trade-offs may be made in relation to cash and labor availability, or in relation to the returns to labor and land provided by different land uses. Several examples highlight the dynamic role of trees in household economies:

1. Planting or protecting trees is one way to address fuelwood needs; but if cash is available and land use-intensity high, gas or kerosene stoves may be chosen instead.
2. Trees can provide low-cost fertility inputs; but if cash is available, purchased fertilizers may provide greater returns to land and labor.
3. The usefulness of trees for erosion protection can be at least partially substituted for by soil conservation structures, mulches, and cultivation techniques.
4. Diversity in diets and nutritional benefits from forest products can also be met through purchased foods or exotic cultivars.

Forests in different places have different potential for production of timber and nontimber forest products. The extent to which this potential is tapped depends on markets and subsistence uses for forest products. People harvest forest products in different ways to meet their household needs and in response to local, national, and world markets, and they make choices between forest, agricultural, and off-farm options by comparing each option's investment requirements, returns, and risks. The sustainability of harvesting forest products can also vary considerably. Increasing market involvement can result in depletion of economically valuable plant species (Vasquez and Gentry 1989) or simplification of complex managed forests to favor higher production of economically valuable species (Michon and Mary 1994). On the other hand, demand for forest products can also lead to the continued presence of managed forests in even densely populated areas (Michon and Mary 1994; Veblen 1978). These forests may undergo significant floristic and faunal changes, but the biological impact of these changes must be analyzed on a case-by-case basis.

Cultural and Social Factors

While material factors have dominated in analyses of land-use choice, there are a number of ways in which culture influences land-use and natural resource–use choices. Barlett (1980) notes how culturally determined work preferences and consumption levels can influence intensification choices. Food preferences can have profound impacts on productive choices and forest patches. For example, Persoon (1992) describes how the belief that rice is a more socially acceptable food than sago in Indonesia is leading to a shift in cultivation patterns that is resulting in widespread forest destruction. Forest and wildlife can also have direct cultural value that leads to support for conservation, even when the perceived economic value of forest is low (Hartup 1994). The difficulty in developing predictive, materialist models of land-use choice highlights the importance of cultural factors, which can have wide-ranging influences.

The relationship between culture and land use is often complex. Van den Breemer (1992) describes how cultural taboos against rice cultivation on certain soils are ignored by Aouan society in the Ivory Coast when confronted with the practical benefits of rice cultivation. These benefits go beyond individual efforts to achieve higher incomes and are tied to collective decisions that defend the Aouan social order in the face of outside pressures. This one exception to their cultural beliefs, which permits widespread rice cultivation, is resulting in radical transformations of their environment via conversion of forests to agricultural lands (van den Breemer 1992). Discrepancies between environmental attitudes and behavior may be common (Hames 1991; Tuan 1968), and when choices must be made between maintaining the social or the ecological order, people may lean toward the former (Spooner 1987; van den Breemer 1992). Strong cultural ties between society and nature may support forest conservation, but people also adapt their cultural and social systems when faced with the exigencies of life in a changing social and economic environment. Culturally imbedded conservation values often break down during periods of cultural transition and do not work well across cultural boundaries (see, for example, Linares 1993). Yet the presence of forest in the landscape may differ greatly between two culture living side by side. Donovan (1994) notes the difference in forest cover between lands occupied by the Guaymi and other residents in southern Costa Rica. Although cultural patterns can adapt to tolerate very different levels of forest use and conversion, they can also be a key factor leading to forest patch conservation.

Forest Patches and Farming Systems

It is of fundamental importance to view landholder decisions about forest patches and resources as a part of household farming and economic systems. Forest land uses fit into household farming systems and economic strategies in several ways, including

1. requiring low labor and cash investment;
2. being appropriate for marginal lands (steep slopes or soils of low fertility);
3. providing multiple products to diversify production and reduce risk;
4. providing important subsistence or salable items;
5. providing amenity products or important minor products to meet specific household needs;
6. providing products with flexible harvesting times that can be used to meet emergency needs or taken advantage of during times when there is little agricultural or wage labor work; and
7. providing social and cultural benefits (Belsky 1993; Falconer and Arnold 1989; Schelhas, in press).

As changes in demographic, economic, social, cultural, and political conditions occur, the uses of forests patches and trees also change. Many changes that are taking place today are resulting in forest loss, but there is potential to develop forest patch management systems to meet the changing needs of landholders in the tropics.

Biodiversity Conservation and Rural People

It is also important to recognize that rural people and ecologists may look at the ecosystem from different perspectives. Spooner (1987) suggests that rural people may place a priority on avoiding disruptions to their social relationships even when ecosystem change is the result, while ecologists may be concerned about the survival of the system as a whole, based on their assumptions about the value of productivity and diversity. As a result, local people and ecologists may value ecosystems differently and may have different views about ecosystem changes.

While many forest-based land uses have biological conservation benefits, the protection of biological diversity is often not the critical value that leads landholders to choose these systems (Bulmer 1982; Michon and Foresta 1992). Because many of the benefits from biological conservation are diffuse and long term, they often accrue to society at large or to future generations. Similarly, many of the externalities of agriculture, such as soil erosion or pesticide use, have biological impacts off the farm or over the long term. Long-term and dispersed biological benefits and costs only play a significant role in landholder choice processes if social and cultural mechanisms are in place that provide incentives to individual landholders to make choices that meet broader societal needs, in addition to their household needs.

Examples from Costa Rica

To illustrate how land-use patterns reflect the social, economic, policy, and cultural context in which they occur, I will discuss land-use choice and forest patches in two cases in Costa Rica (see figure 13-2). The first will be a discussion of land-use patterns and change in Sarapiquí, a forested frontier area in the lowland tropics of northeastern Costa Rica.

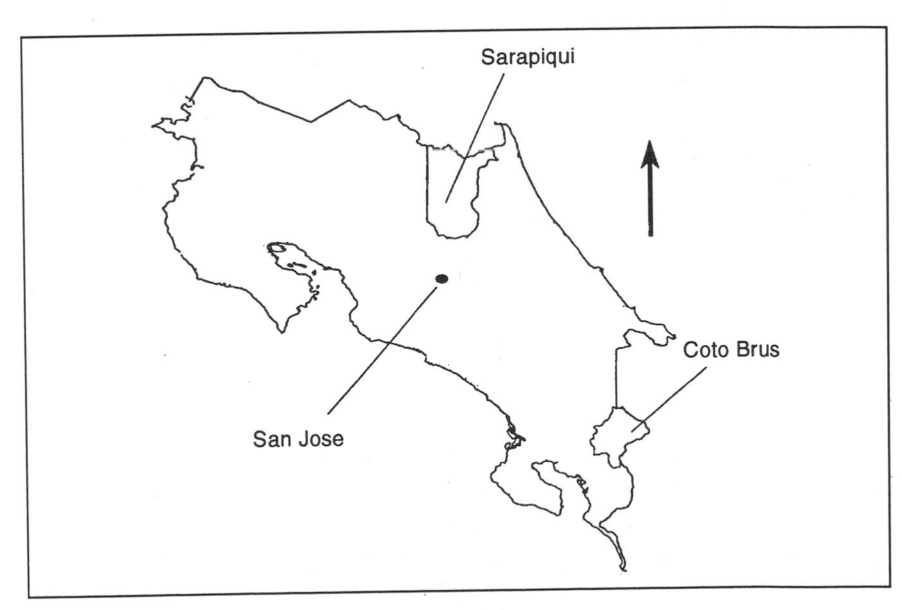

Figure 13-2
Coto Brus and Sarapiquí, Costa Rica.

The second will be a discussion of reforestation and tree management in Coto Brus, a midelevational site in southern Costa Rica with a slightly longer history of settlement.

Frontier Land Use in Sarapiquí

Recent land-use trends in Sarapiquí are characterized by widespread conversion of lowland tropical forests to cattle pasture. Between 1963 and 1983, forest cover in Sarapiquí declined from about 70 percent to 33 percent, while pasture increased from 24 percent to 57 percent and crops from 6 percent to 11 percent (Butterfield 1994a). During the same period, population increased from 4,856 to 18,909 in the 2103.9 square kilometer cantón (Schelhas, in press). The colonization process has been driven to some extent by overall population increase in the country, which had an annual growth rate of 2.91 percent for the 1980 to 1985 period (WRI 1994). But to a larger extent it has been driven by concentration of lands in the hands of a few in other areas of Costa Rica (Meehan and Whiteford 1985) and government policies that have been highly tolerant of squatting, especially on "underutilized" or "vacant" forested lands. Until recently, Costa Rican laws permitted squatters to acquire rights to any improvement they made on land after one year of uncontested occupancy, as well as the right to acquire title to land after ten years of occupancy. Early colonists in Sarapiquí cleared land well beyond their ability or interest to use it productively, solely to establish a claim. Sale of improvements was common, and because cleared land sold for twice the price of forested land (Oviedo Solano 1988), the deforestation process was hastened. The result was widespread clearing of forests, in which many trees were burned or allowed to rot, with only the most valuable timber trees utilized or left standing for later harvest.

The broader context in which colonization took place also influenced the rate of deforestation. Individuals from many social strata were drawn to the frontier in search of land and economic opportunity. Successful farm establishment, made difficult for everyone by the lowland tropical environment, was particularly hard for poor and previously landless farmers who lacked cash for investment and had little farm management experience. Annual cropping provided poor returns and was risky due to the poor soils, lack of a dry season, and bird pests, making cattle raising the most attractive land use. Entry into cattle was difficult and slow for poorer landholders because of high capital costs for fencing, pasture establishment, and animal purchases, and debts were often incurred for food and materials during the early years of colonization.

Wealthier people who came to the frontier in search of new economic opportunities commonly purchased the small farms that had been carved out of the forest. Poor colonists, often in debt to the local store or money lender, tended to sell off their farms to these willing buyers. Wealthier landholders aggregated many smaller farms into larger landholdings, generally managed as cattle ranches because pasture demonstrated productive use of the land and discouraged squatter invasions while requiring relatively low labor and cash inputs. Cattle ranching was also encouraged by a strong U.S. market for imported beef and subsidized credit for cattle ranching, part of Inter-American Development Bank and World Bank export diversification programs for Costa Rica. Cattle ranching has been closely tied to land speculation, since significant increases in the value of land can be realized by improving access to and titling land (both of which are more readily accomplished by larger and wealthier landholders than by smaller and poorer landholders). Large cattle ranches often have very low returns to land; but this is in large part because they are managed with the objective of claiming and holding land, not to produce cattle, and investments in land improvements are seldom made (Schelhas 1994). The original colonists, after selling their farms, often have gone on to colonize other forested areas. The combination of the efforts of squatters to carve farms out of forests and the gradual aggregation of these smaller farms into large cattle ranches for land speculation has led to a very extensive system of land use characterized by rapid deforestation and low productivity of land. The expansion of this pattern over all forest lands in the region has been stopped largely only by the presence of national parks.

Government policies, population growth, land and wealth distribution, and international markets have played an important role in shaping the landscape in Sarapiquí, which has become dominated by cattle pasture. But microeconomic trends have also favored cattle pasture. With the exception of banana plantations and a few small citrus operations along major paved roads, nearly all the farms greater than five hectares in Sarapiquí raise cattle on some of their land. Cattle raising has many features that make it economically attractive to both large and small holders, including low risk, low labor investment requirements, and high returns to labor (Barlett 1982). Financing for cattle is more available than for agriculture, both from banks, which brand the cattle themselves as collateral for loans rather than requiring land as collateral, and from other farmers, who lend cattle through share arrangements. Cattle herds also serve as savings accounts for farmers, because animals can be easily sold in economic hard times. Markets for beef and dairy products have remained relatively strong and stable, and live animals and cheese can be easily transported from remote regions without perishability problems. While large landholders generally produce beef, small landholders have dual-purpose operations producing both animals for sale and cheese for the domestic market. Dual-purpose operations are particularly attractive to small landholders because cheese provides weekly income. Small landholders have been able to intensify their cattle production, increasing their stocking rates and returns to land, even on the infertile, lowland tropical soils of Sarapiquí by intensively managing pastures and animal rotations (Schelhas 1994).

Other research in Costa Rica has found that cattle raising is more common on large landholdings, where the low returns to land are acceptable, while farmers with smaller landholdings are more likely to engage in intensive cultivation of basic grains and annual cash crops (Barlett 1982). In Sarapiquí, climatic and cultural factors combine to limit the cultivation of annual crops and instead promote the intensification of cattle raising. Rice and beans are the key foods in the Costa Rican diet, but they are not easily cultivated in Sarapiquí. Rice cultivation is difficult due to poor soils, bird pests, and frequent spoilage of crops caused by rain during harvests; and rice is generally only planted by very extensive methods as a step in the conversion of forest to pasture when much of the labor is already being performed. Beans are generally only cultivated with extensive slash-mulch methods because rainfall at harvest time often spoils the crop. Extensive bean cultivation, which requires very little labor investment but produces low returns to lands (table 13-1), is more common among larger landholders (Schelhas, in press). Smaller farmers receive better returns to land from intensive cattle production than from extensive bean production (which require a minimum one year fallow), and intensification of beans is not attractive due to the high risks of crop loss caused by climatic factors (Schelhas, in press). Root crops are better adapted to the lowland tropical environment than grains, and they are cultivated on a small scale. But the cultural preference for rice and beans in the diet and the environmental factors hindering their cultivation ultimately combine to favor a sort of subsistence cattle production system, in which farmers use their weekly income from cheese sales to purchase rice and beans imported from outside the region.

Because returns to labor from cattle raising diminish with intensification (table 13-1), farmers with smaller landholdings either choose off-farm employment or diversify their land use by adding intensive, perennial cash crops appropriate for the lowland tropical environment, such

Table 13-1

Comparison of annual investments and returns for different land-use options in Sarapiquí.

	Cattle/ cheese (< 25 ha)	Cattle/ cheese (> 25 ha)	Cattle/ beef (> 25 ha)	Black pepper	Beans	Wage labor
Labor investment (jornales[a]/ha)	26.3	13.7	6.7	333.3	24.2	NA
Cash investment ($/ha)	23.1	31.0	65.3	299.0	23.9	NA
Returns to land ($/ha)	156.7	74.6	41.2	3951.0	128.5	NA
Returns to labor ($/jornal[a])	4.4	10.4	7.1	11.9	6.1	3.78
Risk	Low	Low	Low	High	High	Low
Hectares in land use (mean)	13.8	81.3	187.5	0.4	0.5	NA

[a]1 jornal = 6 hours.
Source: Schelhas, in press.

as black pepper and ornamental plants (Schelhas, in press). These crops offer high returns but also entail high risk due to fluctuations in international commodity prices. As a result, many farmers continue to graze cattle, because it is a low-risk land use capable of producing steady returns, while converting a small plot of land to more risky cash crops in an effort to increase their overall cash income. This diversification strategy strikes a balance between farmers' needs to earn income and reduce risk.

For landholders with smaller landholdings (less than 14 hectares), off-farm employment provides better returns than any intensive land use (table 13-1). Wage labor in Costa Rica generally pays a reasonable minimum wage, overtime, health benefits, vacation, and an annual bonus. A number of smaller plantations producing nontraditional export crops such as ornamental plants, citrus, and black pepper are found within many communities, providing nearby opportunities for wage labor. The immediate returns and short-term security of wage labor draw many people with smaller landholdings into the wage labor market, although land is often retained with hopes of eventually making it productive or as an investment.

Farmers do not convert all their lands to crops and pastures. Landholders leave forest along streams and on very steep slopes for watershed protection, and most retain woodlots to meet future subsistence needs for fuelwood and timber or for future sale (recognizing that prices are rising with increased national scarcity of timber). Valuable timber trees are often left standing in pastures when they are initially cleared, but over time these are harvested for sale or subsistence use and they rarely regenerate. When timber is sold, it is often in response to a cash need, such as the need to invest in farm improvements or to meet emergency expenses, and woodlots serve as savings accounts, gradually accumulating value while requiring little maintenance and having flexible harvest times. These types of forest patch use and maintenance are characteristic of both small and large landholders, since both benefit from diversification, although the absolute size of the forest patches is greater on large landholdings.

Large landholders in the region are experimenting with selective harvesting through natural forest management systems, but it appears that returns to land will be lower than in even the most extensive cattle ranching (Schelhas, in press). Timber management should be attractive to many large landholders, although leaving a large percentage of a farm in forest can make it a target for a squatter invasion. Small landholders express interest in reforestation, especially with fast-growing, high-value timber species that have the potential to provide high returns to land. Although promising species for reforestation are being discovered (Butterfield 1994b), the high initial investment in tree planting and weeding, the 15- to 25-year wait for returns, and the uncertainty of returns from an untested land use combine to limit the attractiveness of reforestation to small landholders (Schelhas 1994).

There are several ecotourism operations in Sarapiquí that protect large forest patches. Such lands can be legally declared as forest reserves, which means that the government will aid the landholder in evicting squatters. Most ecotourism operations are owned by foreigners, but they do provide employment for some local people. In Sarapiquí, ecotoursism protects several large forest tracts but does not employ enough people to alter the larger patterns of conversion of forest to pasture (Schelhas 1991).

The landscape in Sarapiquí, a result of farmers' tendency to diversify their land use, is a mosaic of pasture, annual and perennial crops, and forest. Forest currently plays a small but important role in farming systems, providing subsistence products and watershed benefits, and serving as a store of value. These forest patches can be expected to persist in the landscape, although under current conditions they seem destined to occupy only a small percentage of the land. If forests are to play a more important role in household land-use systems, the benefits they provide to landholders must be increased. There is a particular need to provide benefits that are competitive with cattle raising, which is the dominant land use in the region. Based on existing land use patterns and forest uses, the alternatives with the most potential are (1) increasing the value of forest management through the addition of valuable understory plants or harvestable animals, and (2) the use of forests as a source of low returns and store of value to balance high-risk involvement in cash cropping (partially replacing cattle raising, which currently fills this role) (Schelhas 1994).

The example of frontier land use in Sarapiquí conforms to the Boserup model, where, under conditions of low population density, extensive land use relying on nature's capital dominates. Although the rate of forest destruction is exaggerated by policy incentives, cattle remains the economically most attractive land use in Sarapiquí, driving widespread conversion of forest to pasture. However, in response to the recent closing of the frontier in southwestern Sarapiquí in 1982, landholders have begun to intensify and diversify land use. Many landholders indicate that they would like to increase the forest component on their lands. However, there are short-term economic obstacles to reforestation. Natural regeneration takes many years to produce secondary forests of only moderate value, and tree planting entails high initial costs and relatively long waits for returns.

Reforestation in Coto Brus

During my research in Sarapiquí in 1989, few landholders had acted on the interest they expressed in reforestation. As a result, I was not able to gather any data that could lead to a better understanding of the extent and type of reforestation that might occur following the initial clearing associated with the opening of the frontier. In 1994 I began research on reforestation in Coto Brus, Costa Rica, to address these questions.[1] The cantón of Coto Brus, in southern Costa Rica (see figure 13-

[1]Additional research on forest patches is being carried out by three Cornell graduate students: Terry Jantzi—environmental attitudes and forest patches, Tom Thacher—participation in reforestation incentive programs, and Kim O'Conner—riparian land use.

2), was initially colonized about the same time as Sarapiquí, although its population experienced more rapid growth in the 1960s and 1970s leveling off in the early 1990s (Manger 1992). Coto Brus differs from Sarapiquí in a number of other ways, including (1) greater isolation from population centers and markets for agricultural crops, (2) less plantation agriculture, and (3) environmental factors such as higher elevation, more fertile soils, and the presence of a distinct dry season. Coffee production is the most important economic activity in Coto Brus, although cattle pasture has occupied more land since the early 1970s (Manger 1992). Coto Brus entered an economic slump in the early 1990s brought about by a decline in world coffee prices. No agricultural crop has emerged to replace coffee, and population is believed to have decreased in recent years due to outmigration. The current land-use pattern in Coto Brus is a mosaic of pasture, coffee, and forest patches.

Many small farmers in Coto Brus indicate an interest in maintaining and increasing the amount of forest on their land. These changing attitudes appear to partially reflect changing conditions and attitudes at the national level, including (1) near complete loss of large areas of continuous forest outside national parks, (2) government concern over availability of sufficient timber to meet national needs in the future, and (3) a strong conservation message disseminated in the Costa Rican media over the past five years. But despite the widespread attention that forest conservation has received, there has not been a dramatic increase in forest cover and little is known about how individual landholders are responding to the shift from forest abundance to forest scarcity.

Interviews and farm mapping with eight community leaders associated with the Coto Brus Agroecology Foundation in 1994 provide some preliminary indications of some of the ways that farmers maintain forest patches under their own initiative. Streamside corridors and other watershed protection forests are one of the principal types of forest patches valued by landholders.[2] Farmers widely report that leaving trees along streams and around springs helps to maintain water flows, and many are widening the existing streamside corridors through natural regeneration and tree planting. There is little opportunity cost for

[2]The forestry law prohibits agricultural use or removal of vegetation from land (1) within a 200-meter radius of springs on slopes and 100 meters for springs on flat lands, (2) 10 meters on both sides of rivers on flat lands and 50 horizontal meters in steep topography, and (3) 100 meters from the shores of natural lakes. These regulations are only minimally enforced.

streamside corridors in Coto Brus, since stream banks are often very steep and riparian areas are not used to grow any particular crop. The only consideration countering the perceived watershed benefits of streamside corridors is the need to allow cattle access to the streams, which is generally done by cutting trails though the understory.

Living fences, primarily using *Gliricidia sepium* and *Dyphysa robinoides,* are very common in Coto Brus and, in pasture areas, often form a low, narrow, dense canopy. Sun-grown coffee has dominated Coto Brus in recent decades, although as a result of the decline in coffee prices in the early 1990s, many farmers have begun to integrate nitrogen-fixing trees, principally *Erythrina sp.,* into their coffee plots. A dense shade canopy leads to disease problems during the rainy season in Coto Brus; therefore, a low open canopy is maintained by pollarding and continued regular thinning of branches. Pruned leaves and branches accumulate on the ground, reducing erosion and providing nutrient inputs. Shade-grown coffee, while not as productive as open-grown coffee, requires fewer purchased fertilizers and pesticides and enables farmers to continue to profit from coffee even when prices are low (Budowski 1993).

Forest patches are also maintained to protect and produce timber, and their size is more or less proportional to farm size. Large farms (over 100 hectares) often have from one-third to one-half of their land area in forest. At the extreme end, the 10,000-hectare Hacienda Alturas de Coton is primarily in forest (92 percent forested in 1984), although armed guards have been required to keep squatters off the property (Edelman and Seligson 1994). The presence of valuable timber trees is often a key criterion in the decision of which forested areas to leave uncleared, although many farmers express a concurrent interest in providing wildlife habitat based more on nature appreciation than economics. Reforestation is common, and many of the large landholders in the cantón are reforesting land both with and without government incentives.

On small farms in Coto Brus, timber is generally grown and protected in small pockets. Small forest patches with valuable timber are protected, as are areas where natural regeneration of valuable timber trees occurs. Tree planting is often undertaken on a small scale (less than one hectare) independent of any outside incentives, generally in the hopes of producing marketable timber in the future; but large-scale reforestation without incentives is not common due to the high initial investment required and the long wait for returns.

In response to widespread overall deforestation in Costa Rica and projected national shortages of timber, the Costa Rican government has implemented two reforestation incentive programs (Butterfield 1994b). One program, for landholders with title, covers the full planting and maintenance costs for the initial five years of reforestation and requires no repayment of funds (Butterfield 1994b). The second, for landholders without title, covers 70 percent of the costs for three years through a revolving fund (Butterfield 1994b). By covering or defraying the initial planting and maintenance costs for reforestation, these programs address one of the principal obstacles to reforestation—high initial costs—but they do not address the problem of the long wait for returns.

From 1988 to 1993, 141 farmers participated in the two reforestation incentive programs, reforesting a total of 380 hectares (see table 13-2). From 1988 through 1990, virtually all reforestation was with two exotic species, Caribbean pine (*Pinus caribaea*) and eucalyptus (*Eucalyptus deglupta*). In recent years, poor growth and marketability of these two species has led to a near complete shift to native species, with the most popular being *Terminalia amazonia*, a fast-growing, versatile, economically valuable native hardwood.

A 50 percent sample of participants in the reforestation incentive programs in Limoncito, one of the four districts of Coto Brus, was surveyed in January 1994 to collect some preliminary information on what types of farmers were participating in the program and the ways in which they were reforesting. The average landholding size of participants in Limoncito was 36.3 hectares, slightly larger than the average of 21.6 hectares for the district as a whole (DGEC 1987), which is espe-

Table 13-2
Reforestation in Coto Brus under government incentive programs.

Year	Hectares	No. new reforesters	No. repeat foresters	Total reforesters
1988	9	5	0	5
1989	18	11	1	12
1990	24	17	1	18
1991	172	47	5	52
1992	71	19	3	22
1993	136	42	5	47
Total	430	141		

Source: DGF, CAC-CotoBrus, and Sabanillas Development Association records

cially significant since the reforestation programs are targeted at small and medium-sized landholdings. Of the 22 participating landholders surveyed, 9 percent were large absentee landholders with full-time businesses elsewhere, 23 percent were landholders with full-time off-farm employment, and 50 percent were landholders over 55 years of age. Farms belonging to older landholders were often already informally divided and allocated to children; and in all but two cases, the older farmer received some financial support from a family member engaged in off-farm employment. Only 18 percent of the participants in the reforestation program in Limoncito were active, full-time farmers.

These results suggest that, even with the substantial incentives provided by the Costa Rican government, the people who are choosing to reforest are those who are not relying on their land for their income. Participants in the reforestation program are making subsidized investments in underutilized land that will increase in value over the long term. The results also highlight the different land-use choices made by households in the same geographic area and the dynamic relationship between forest patches and agricultural land that are the result of household and farm characteristics.

Landholders in Coto Brus are showing increasing interest in forest patches, including riparian corridors, woodlots, reforestation, agroforestry, and living fences. This change can be attributed to a combination of factors. Population has leveled off, or even decreased, as people seek economic opportunity elsewhere in Costa Rica. Peasant farming is gradually declining and wage labor is increasing in both Coto Brus and the country as a whole. Pasture in Coto Brus is underutilized, in part due to distance from markets. Landholders increasingly are viewing timber production as a land use with long-term economic potential. Due to poor markets for cattle products and improving markets for timber, timberland and pastureland now sell for roughly the same price, reducing the incentive to convert forest to pasture to increase land value. And landholders are becoming more aware of the environmental services that trees and forest provide, such as fertility maintenance and watershed protection, and are maintaining and incorporating trees and forest into their farms for these reasons.

Conclusions

Land-use patterns and forest patches in Sarapiquí and Coto Brus, Costa Rica, are a product of the social, economic, cultural, and policy context in which they occur. The rapid deforestation that is associated with the opening of frontiers in the tropics is destructive to biological diversity and wastes valuable natural resources. Although frontier conditions favor extensive land use reliant on natural rather than human capital, widespread deforestation is not inevitable. Frontier deforestation in Costa Rica has been exaggerated by social inequities and government policies. Concentration of land into large landholdings leads to squatter invasions by the landless; and the government's approach of intervening in the process only to mediate squatter disputes, with a key issue being who is making productive use of the land, leads both large and small landholders to convert forestlands to extensive agriculture and pasture. Land reform or tax policies that promote intensive land use in areas most appropriate for agriculture could greatly reduce forest destruction at the national level.

Forest patches have a place in intensive and extensive agricultural landscapes. Both large and small landholders diversify their land uses and economic strategies, and forest patches can play an important role in these strategies. Larger tracts of forest can be managed under natural forest management, as plantations on larger individual landholdings, or where communities hold land in common. Forest patch types more appropriate for intensively managed landscapes, such as high-value woodlots, tree crops, shade-grown crops, and living fences, can be integrated into smaller farms.

Changing macroeconomic trends, including increasing wage labor and expansion of intensive export crops, create conditions favorable to reforestation. Tree planting can be encouraged by programs that make it easier for the farmer to bear the high initial costs, although many farmers are reforesting without them. The cost, in cash or labor, of reforestation is high, and there is a need for the development of low labor strategies to enrich natural regenerated forest and to hasten and increase economic returns. While it is unlikely that forestland uses will widely replace pastures in the near future, forest patches and silvipastoral strategies are increasing.

Finally, while it is true that interest in long-term and noneconomic forest values often requires some minimum level of development, interviews with landholders in Sarapiquí and Coto Brus suggest that relatively poor rural people value forests for their aesthetic value and importance to their heritage. Farmers often talk about wildlife and forests with respect and affection, and many forest conservation and reforestation decisions appear to be only partially motivated by economics. With this in mind, forest patch conservation in Costa Rica may be best enhanced by a three-fold approach that includes (1) eliminating policy incentives for forest destruction and creating incentives supporting forest management; (2) improving technical assistance for forest land uses that meet household needs, particularly those which compete with cattle or complement intensive cash crop production; and (3) environmental education programs that strengthen existing forest values and stimulate new appreciation of forests.

Acknowledgments

Partial funding for field research has been provided by a Tinker Foundation Field Research grant, a Research Fellowship from the Jessie Smith Noyes Foundation and the Organization for Tropical Studies, and the Cornell Program in Ecological and Social Science Challenges of Conservation (National Science Foundation grant # BIR-9113293). I would like to thank Marcos Roble, Hernan Villalobos, Venicio Aguilar, and Gerardo Vargas for their assistance with research in Coto Brus. My research and writing on agricultural intensification in relation to forest conservation benfited greatly from discussions with and encouragement from Bob Netting. I would like to thank the following people for commenting on an earlier draft of this manuscript: Russ Greenberg, Janis Alcorn, Tom Thacher, Kim O'Conner, and Terry Jantzi.

References

Barlett, P. F. 1980. "Adaptive strategies in peasant agricultural production." *Annual Review of Anthropology* 9: 545–573.

Barlett, P. F. 1982. *Agricultural Choice and Change: Decision Making in a Costa Rican Community.* New Brunswick, NJ: Rutgers University Press.

Belsky, J. M. 1993. "Household food security, farm trees, and agroforestry: A comparative study in Indonesia and the Philippines." *Human Organization* 52(2): 130–141.

Bengston, D. N. 1994. "Changing forest values and ecosystem management." *Society and Natural Resources* 7(6): 515–533.

Bentley, J. W. 1989. "Bread forests and new fields: The ecology of reforestation and forest clearing among small-woodland owners in Portugal." *Journal of Forest History,* October 1989.

Boserup, E. 1965. *The Conditions of Agricultural Growth: The Economics of Agrarian Change under Population Pressure.* New York: Aldine.

Boserup, E. 1981. *Population and Technological Change: A Study of Long-Term Trends.* Chicago: University of Chicago Press.

Boserup, E. 1990. *Economic and Demographic Relationships in Development.* Baltimore: Johns Hopkins University Press.

Budowski, G. 1993. "The scope and potential of agroforestry in Central America." *Agroforestry Systems* 23: 121–131.

Bulmer, R. N. H. 1982. "Traditional conservation practices in Papua New Guinea." In *Traditional Conservation in Papua New Guinea: Implications for Today,* edited by L. Morauta, J. Pernetta, and W. Heaney, pp. 59–77. Monograph 16. Boroko, Papua New Guinea: Institute of Applied Social and Economic Research.

Butterfield, R. P. 1994a. "The regional context: Land colonization and conservation in Sarapiquí." In *La Selva: Ecology and Natural History of a Neotropical Rain Forest,* edited by L. A. McDade, K. Bawa, H. A. Hespenheide, and G. S. Hartshorn, pp. 299–306. Chicago: University of Chicago Press.

Butterfield, R. P. 1994b. "Forestry in Costa Rica: Status, research priorities, and the role of La Selva Biological Station." In *La Selva: Ecology and Natural History of a Neotropical Rain Forest,* edited by L. A. McDade, K. Bawa, H. A. Hespenheide, and G. S. Hartshorn, pp. 317–328. Chicago: University of Chicago Press.

Chayanov, A. V. 1966. *The Theory of Peasant Economy* (1925), edited by D. Thorner, B. Kerblay, and R. E. F. Smith. Homewood, IL: Irwin.

Chibnik, M. 1990. "Double-edged risks and uncertainties: Choices about rice loans in the Peruvian Amazon." In *Risk and Uncertainty in Tribal and Peasant Economies,* edited by Elizabeth Cashdan, pp. 279–302. Boulder, CO: Westview Press.

Collier, G. A., D. C. Mountjoy, and R. B. Nigh. 1994. "Peasant agriculture and global change." *BioScience* 44(6): 398–407.

Collins, J. L. 1987. "Labor scarcity and ecological change." In *Lands at Risk in the Third World: Local-Level Perspectives,* edited by P. D. Little, M. M Horowitz, and A. E. Nyerges, pp. 19–37. Boulder, CO: Westview Press.

Dewees, P. A. 1993. *Trees. Land, and Labor.* World Bank Environment Paper No. 4. Washington, DC: The World Bank.

Dirección General de Estadisticas y Censos (DGEC). 1987. *Censo Agropecuario 1984.* San Jose, Costa Rica: DGEC.

Donovan, R. 1994. "BOSCOSA: Forest conservation and management through local institutions (Costa Rica)." In *Natural Connections: Perspectives in Community-Based Conservation,* edited by D. Western, R. M. Wright, and S. Strum, pp. 215–233. Washington, DC: Island Press.

Durham, W. H. 1979. *Scarcity and Survival in Central America: Ecological Origins of the Soccer War.* Stanford, CA: Stanford University Press.

Durrenberger, E. P., and N. Tannenbaum. 1992. "Household economy, political economy, and ideology: Peasants and the state in Southeast Asia." *American Anthropologist* 94: 74–89.

Edelman, M., and M. A. Seligson. 1994. "Land inequality: A comparison of census data and property records in twentieth-century southern Costa Rica." *Hispanic American Historical Review* 74(3): 445–491.

Falconer, J., and J. E. M. Arnold. 1989. *Household Food Security and Forestry: An Analysis of Socio-Economic Issues.* Rome: FAO.

Foresta, H. de, and G. Michon. 1992. "Complex agroforestry systems and conservation of biological diversity (II)." In *Harmony with Nature, Proceedings of the International Conference on Tropical Biological Diversity,* pp. 488–497. Kuala Lumpur, Malaysia: Malayan Nature Society.

Hames, R. 1991. "Wildlife conservation in tribal societies." In *Biodiversity: Culture, Conservation, and Ecodevelopment,* edited by M. L. Oldfield and J. B. Alcorn, pp. 172–199. Boulder, CO: Westview Press.

Hartup, B. K. 1994. "Community conservation in Belize: Demography, resource use, and attitudes of participating landowners." *Biological Conservation* 69: 235–241.

Hecht, S. 1981. "Deforestation in the Amazon Basin: Magnitude, dynamics, and soil resource effects." *Studies in Third World Societies* 13: 61–108.

Linares, O. F. 1993. "Palm oil versus palm wine: Symbolic and economic dimensions." In *Tropical Forests, People and Food: Biocultural Interactions and Applications to Development,* edited by C. M. Hladik, A. Hladik, O. F. Linares, H. Pagezy, A. Semple, and M. Hadley, pp. 595–606. Paris: UNESCO and Parthenon Publishing Group.

Manger, W. F. 1992. "Colonization on the southern frontier of Costa Rica: A historical cultural landscape." MS thesis, Memphis State University.

Mary, F., and G. Michon. 1987. "When agroforests drive back natural forests: A socio-economic analysis of a rice-agroforest system in Sumatra." *Agroforestry Systems* 5: 27–55.

Meehan, P. M., and M. B. Whiteford. 1985. "Expansion of commercial cattle production and its effects on stratification and migration: A Costa Rican case." In *Social Impact Analysis and Development Planning in the Third World,* edited by W. Derman and S. Whiteford, pp. 178–195. Boulder, CO: Westview Press.

Michon, G., and H. de Foresta. 1992. "Complex agroforestry systems and the conservation of biological diversity: Agroforests in Indonesia, the link between two worlds." In *Harmony with Nature, Proceedings of the International Conference on Tropical Biological Diversity,* pp. 457–473. Kuala Lumpur, Malaysia: Malayan Nature Society.

Michon, G., and F. Mary. 1994. "Conversion of traditional village gardens and new economic strategies of rural households in the area of Bogor, Indonesia." *Agroforestry Systems* 25(1): 31–58.

Mortimore, M. 1993. "Northern Nigeria: Land transformation under agricultural intensification." In *Population and Land Use in Developing Countries,* edited by C. L. Jolly and B. B. Torrey, pp. 42–69. Washington, DC: National Academy Press.

Murray, G. F. 1986. "Seeing the forest while planting the trees: An anthropological approach to agroforestry in rural Haiti." In *Politics, Projects, and People: Institutional Development in Haiti,* edited by D. W. Brinkerhoff and J. C. Garcia Zamor, pp. 193–226. New York: Praeger.

Netting, R. McC. 1969. "Ecosystems in process: A comparative study of change in two west African societies." *National Museums of Canada Bulletin* 230, pp. 102–112.

Netting, R. McC. 1993. *Smallholders, Householders: Farm Families and the Ecology of Intensive, Sustainable Agriculture.* Stanford: Stanford University Press.

Oviedo Solano, M. A. 1988. Annex 1.b. In *Wildlands Consolidation and Development in the Cordillera Volcanica Central, Costa Rica,* edited by W. Araya, R. Butterfield, F. Morales, S. Poisson, and C. Schnell. San José, Costa Rica: Organization for Tropical Studies.

Persoon, G. 1992. "From sago to rice: Changes in cultivation in Siberut, Indonesia." In *Bush Base, Forest Farm: Culture, Environment, and Development,* edited by E. Croll and D. Parkin, pp. 187–199. London: Routledge.

Schelhas, J. 1991. "A methodology for assessment of external issues facing national parks, with an application in Costa Rica." *Environmental Conservation* 18(4): 323–330.

Schelhas, J. 1994. "Building sustainable land use on existing practices: Smallholder land use mosaics in tropical lowland Costa Rica." *Society and Natural Resources* 7(1): 67–84.

Schelhas, J. In press. "Land use choice and change: Intensification and diversification in the lowland tropics of Costa Rica." *Human Organization.*

Schelhas, John and Russell Greenberg. 1993. "Forest Patches in the Tropical Landscape and the Conservation of Migratory Birds." Migratory Bird Conservation Policy Paper No. 1. Washington, DC: Smithsonian Migratory Bird Center.

Spooner, B. 1987. "Insiders and outsiders in Baluchistan: Western and indigenous perspectives on ecology and development." In *Lands at Risk in the Third World: Local-Level Perspectives,* edited by P. D. Little, M. M. Horowitz, and A. E. Nyerges, pp. 58–68. Boulder, CO: Westview Press.

Tiffen, M., M. Mortimore, and F. Gichuki. 1994. *More People, Less Erosion: Environmental Recovery in Kenya.* Chichester, UK: Wiley.

Toledo, V. M., A. I. Batis, R. Becerra, E. Martínez, and C. H. Ramos. 1992. "Products from the tropical rain forests of Mexico: An ethnoecological approach." In *Sustainable Harvesting and Marketing of Rain Forest Products,* edited by M. Plotkin and L. Famolare, pp. 99–109. Washington, DC: Island Press and Conservation International.

Tuan, Y. 1968. "Discrepancies between environmental attitude and behaviour: Examples from Europe and China." *Canadian Geographer* 12(3): 176–191.

van den Breemer, J. P. M. 1992. "Ideas and usage: Environment in Aouan society, Ivory Coast." In *Bush Base, Forest Farm: Culture, Environment, and Development,* edited by E. Croll and D. Parkin, pp. 97–109. London: Routledge.

Vasquez, R., and A. H. Gentry. 1989. "Use and misuse of forest harvested fruits in the Iquitos area." *Conservation Biology* 2(4): 350–361.

Veblen, T. T. 1978. "Forest preservation in the western highlands of Guatemala." *Geographical Review* 68(4): 417–434.

WRI/UNEP/UNDP. 1994. *World Resources 1994–95.* New York: Oxford University Press.

Chapter 14

Reading Colonist Landscapes: Social Interpretations of Tropical Forest Patches in an Amazonian Agricultural Frontier

John O. Browder

Landscapes are our unwitting cultural autobiographies.
—Peirce F. Lewis (1979)

Introduction

To a large degree, tropical forests are human social spaces. "Very little tropical forest is pristine in the sense of never having been inhabited by man" (Webb 1982). Indeed, growing evidence suggests that tropical forests are largely anthropogenic structures.

> *[W]hat today looks like a vast area of pristine tropical forest is actually a mosaic of forest patches that have "recovered" from small-scale agriculture and forest clearance over many years in the past. Most tropical forests have been submitted to numerous cycles of management and abandonment by human societies since remote times* (Webb 1982).

Prominent archaeologists and ethnoecologists are discovering ways to read tropical forest landscapes as the cultural biographies of ancient and contemporary indigenous peoples (e.g. Balée 1989; Gomez-Pompa et al., 1987; Posey 1985; Roosevelt 1990). In this chapter I discuss how we might read one important feature of the landscapes created by contemporary pioneer farmers in the Brazilian Amazon—their forest remnants or patches—through the use of satellite image interpretation.

Study Site

The Brazilian Amazon state of Rondônia encompasses an area of 243,000 square kilometers, roughly the size of the former West Germany or the U.S. state of Montana (figure 14-1). Physiographically characteristic of the southern Amazon fringe, the dominant form of vegetation is classified as open tropical moist forest or transition forest, extending over approximately 75 percent of the state (World Bank 1981).

In the early 1970s, surveys indicated that between one-third and one-half of Rondônia's surface area was suitable for commercial crop cultivation (EMBRAPA 1975; FJP 1975). To relieve social pressures arising from agricultural modernization and the displacement of small family farmers and sharecroppers in south and southeastern Brazil, the government opened the Rondônian frontier to landless pioneers (Carnasciali et al. 1987; Mueller 1980; Romeiro 1987). From 1970 to 1990, Rondônia's population soared from just over 100,000 to an estimated 1.5 million (IBGE 1975; author's estimates). Not surprisingly, the principal environmental impact of this rapid settlement has been the alarming conflagration of the state's tropical forests.

Estimates of the percentage of Rondônia's natural forest area converted to other uses rose from 0.3 percent in 1975 to an estimated 24 percent by 1990 (Mahar 1989). In a process replicated throughout lowland Amazonia, colonists typically begin by clearing and burning small areas of forest and planting annual crops and coffee or cacao. After a period of five to ten years of nutrient mining, they shift to cattle production as the predominant land use.

Not surprisingly, given these patterns, the colonist's farm is typically a fragmented landscape and an ideal laboratory in which to examine natural forest patches. Numerous researchers are interested in explaining colonists' decisions to clear forest, but rarely do they ask about the factors that influence colonists' decisions to retain forest patches. What

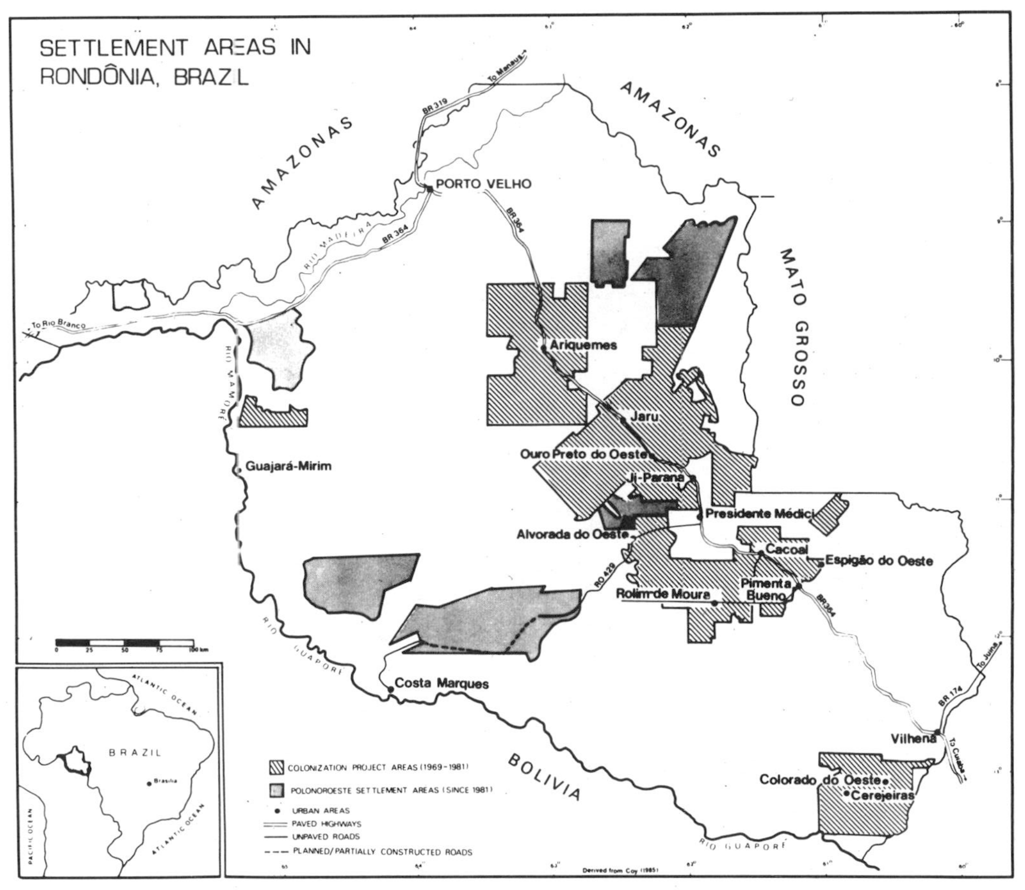

Figure 14-1
Map of Rondônia.

complementary uses do forest patches serve in an agronomic production profile that is increasingly dominated by cattle and pasture expansion? What social factors (i.e., social traditions, demographic forces, and public policies) influence colonist land-use decisions regarding forest fragments?

In considering these questions in relation to pioneer frontier landscapes in Amazonia, I believe it is important to state two points that are typically under-emphasized in discussions of tropical forests.

1. Forest patches are not just remnants of larger natural ecosystems. They are also, quite importantly, social spaces. Their existence, size, shape, and, in some cases, biological composition are socially determined by forces operating at two general levels of society. One is the local level, that of the household and village, where variations in individual agroecological knowledge, microeconomic characteristics, and local demographic factors influence decisions concerning land use, in general, and use of forest patches, in particular. The other is the larger macroeconomic and political level encompassing national governments and the transnational forces that drive the global economy. Public policies emanate from a matrix of national and global interests and create incentives to use forest lands in certain ways (Browder 1988). In discussing forest patches and their functions, it is important not to lose sight of their essential social nature.

2. There is a tendency to view the process that produces forest remnants as only destructive, with irreversible degenerative effects on forests. Given this view, forest remnants can only diminish in size. However, one of the most rapidly expanding landscapes, after pasture, is secondary forest. Forests not only shrink and disappear, they also reappear and regenerate. Secondary-forest patches, however, are quite different in biological composition from primary-forest patches, and successional processes are highly varied and complex (Brown and Lugo 1990; Uhl, et al., 1981).

Colonist Uses of Forest Patches

Data from a 1992 survey of 240 colonist households in three different settlement areas in Rondônia indicate that most colonists do not permanently use forest patches on their properties for commercial ends as do some indigenous peoples, such as the Huastec Maya of Mexico, who commonly plant coffee in their forest groves called *te'lom* (Alcorn

1989).[1] However, colonists do use and depend on forest patches for noncommercial household goods and services. In 1994, for example, only 25.8 percent of the farmers studied exchanged timber cut from their forest patches for money, goods, or services. Even fewer, 11.7 percent, commercialized any nontimber forest products originating from their forest patches. A much higher proportion (60.8 percent) of colonist households in 1991 extracted wood from their forest patches for fuel and building materials, while 70.0 percent collected various nontimber forest products for household consumption. In both sets of cases, however, considerable differences in local practices were observed, indicating either variations in spatial distributions of forest types or socioeconomic differentiation among agrarian communities, or both (table 14-1).

Reasons given by colonists for not using forest patches for either commercial or subsistence purposes included colonists' lack of knowledge about the existence of anything useful to extract, low price, lack of time, and perceptions of the forest patch as being too dangerous and inhospitable a place to visit. Despite the implementation of recent state government environmental protection legislation, surprisingly few (2.2 percent) of those respondents not extracting anything in 1991 cited excessive government restrictions.

The Process of Forest Fragmentation in the Agricultural Landscape

The colonist's farm in Rondônia is a restless landscape. At the ground-level perspective, an observer might be tempted to conclude that farmers just seem to roll back the forest over time in a relentless pursuit of fresh nutrients for agricultural production. However, the view from an orbital platform in space reveals a more complex reality. Forest fragmentation proceeds in fits and starts, is spatially discontinuous, and, in the aggregate, produces a tapestry of diverse vegetational formations.

[1]These settlement areas, located in the municipios of Rolim de Moura, Mirante de Serra, and Alto Paraiso, all largely in tropical forest cover, represent three major soil types and have distinctive social histories, broadly reflecting the ecological and social diversity found in Rondônia. All three settlement areas are located approximately 60 kilometers from the interregional highway (BR 364) and were settled in the same time period (1978–81). The household surveys were undertaken in July 1992.

Table 14-1

Forest Product Extraction and Commercialization from Colonist Forest Fragments, Rondônia, Brazil, 1991.

	RM	OP	AP	Mean
Farmers marketing				
Timber products	6.5%	25.5%	40.2%	25.8%
Non-timber forest products	4.9	10.3	18.3	11.7
Farmers extracting				
Wood products	26.2	62.9	84.1	60.8
Nonwood forest products	36.1	69.1	96.3	70.0
Reasons for not extracting				
Nothing to extract	36.2	32.1	11.4	28.1
Low price, weak market	10.6	15.1	42.8	20.7
Lack time	4.2	17.0	17.1	12.6
Forest too dangerous	14.9	3.8	8.6	8.8
Government restrictions	2.1	1.9	2.8	2.2

RM: Rolim de Moura Municipio ($N=$ 61 households); OP: Ouro Preto D'Oeste Municipio ($N=$ 97); AP: Alto Paraiso Municipio ($N=$ 82); Mean: Weighted mean.

Source: 1992 household survey of 240 colonist farmers.

Over time, forest fragments often grow in size and complexity as abandoned crop fields regenerate into secondary forest. For example, a detail of two recent (1985 and 1991) SPOT satellite scenes of the Rolim de Moura colonization project (figure 14-2) illustrates the point that forest fragments can expand in size through natural regeneration quite rapidly if left alone. What factors influence colonist's decision either to spare certain primary-forest areas or to allow forest clearings to regenerate over time? Four cases illustrate the diversity of social factors that must be considered in reading colonist landscapes.

Interiorization of Forest Clearings

The process of forest fragmentation typically begins at the roadside and proceeds as a roughly linear and progressive pushing-back of the forest within the rectangular property boundaries of each farm. (Most farms in Rondônia were initially 100 hectares in size, dimensioned at 500 by 2,000 meters.) Basic microeconomic common sense dictates that farmers first concentrate production (i.e., clear forests and plant crops) in

Figure 14-2

Forest regeneration (Rolim de Moura, Rondônia): abandoned pasture in 1986 (*left*) and 1991 (*right*).
Source: SPOT, 1986, 1991.

the areas closest to the road providing access to their farms and progressively clear outward from the road over time (figure 14-3).

There are numerous instances, however, when this idealized transitional sequence (road to nonforest to forest) is interrupted by interior clearings deep inside the forest remnants. Such clearings expedite the process of forest fragmentation and begin to give the colonist landscape the prototypical appearance of a mosaic of forest patches. Three reasons for this "economically irrational" clearing behavior may be given:

1. In some cases these interior forest clearings occur on patches of relatively fertile soils. Farmers that received government credits to plant cacao or rubber are often required by creditor agencies to plant the subsidized perennials on their best soils available. Farmers not receiving such credits, but knowing how to recognize different soil types, might be inclined to clear forest and plant their most important cash crop (e.g., coffee) on the best soils as well.

2. Many farmers, suspicious of neighbors and government extension agents, simply wish to conceal some of their cropping activities in their forest remnants.

3. In Rondônia's hastily demarcated settlement areas, where property lines exist only on paper, farmers have established a customary practice of marking their interior property boundaries by clearing for-

Figure 14-3
Overview of two abutting farm lots (in block) showing cleared areas and remnant forests between access roads. The landscape overall exhibits many instances of small interior forest clearings, many made to establish property boundary markers for neighbors, Rolim de Moura, Rondônia.
Source: SPOT 1991.

est and sometimes planting a perennial crop or pasture. Farmers who clear an area of forest along the entire width of their interior property line effectively create an isolated forest island or patch on their farm, although these boundary-line openings are often most susceptible to secondary successional regrowth.

Stream Buffer Patches

Given the rectilinear configuration of land use commonly found on Rondônian farms, it is sometimes surprising that forest patches are so irregular in shape and size. One explanation of this irregularity is that some forest patches follow biophysical or topographic contours, rather than those dictated by economic efficiency. For example, Brazilian law requires that rural property owners leave intact all natural vegetation spanning 100 meters on either side of streams as a buffer against soil erosion. The law, when obeyed, promotes soil conservation, inhibits erosion, especially from heavily degraded pastures, and may provide a microrefugia for various forms of wildlife, a prospect that has yet to be systematically investigated in Brazilian Amazonia. Although compliance with this law is variable, many farmers do leave some area of natural forest as stream buffer patches on their properties, creating a physiographically oriented fragmentation pattern on the colonist landscape (figure 14-4).

Large Forest Reserves and Tiny Resource Island Patches

Public policy plays an important role in structuring colonist landscapes, and an understanding of land-use laws is critical to reading these landscapes. One such law requires all rural property owners to maintain 50 percent of their property in natural forest cover. While enforcement has been spotty and voluntary compliance irregular, the law has provided some impetus for colonists to retain large patches or a patchwork of small forest fragments.

Another Brazilian policy fosters the creation of tiny resource island patches. One such law prohibits the cutting of Brazil nut trees, consid-

Figure 14-4
Stream buffer patches, Rolim de Moura, Rondônia.
Source: SPOT 1991.

ered to be a valuable economic resource. Farmers are more likely to comply with this law if the Brazil nut trees are within close eyesight of a public road (figure 14-5). Such minuscule forest fragments provide no significant environmental services for either watershed protection or wildlife habitat. In many cases, they no longer produce nuts and have been barren since the surrounding forest was removed, thereby eliminating the habitat of the pollinating agent of these trees. Such remnants only exist to satisfy the language, but not the intent, of a well-meaning Brazilian law.

Amenity and Legacy Patches

Colonists also freely choose to retain natural forest patches regardless of legal constraints or inducements for purposes of amenity and legacy. Many farmers value the quality-of-life benefits provided by forest patches: shade, cool air, freshwater, song bird habitat, occasional source

Figure 14-5

Resource island patch (Rolim de Moura, Rondônia) comprised of two isolated Brazil-nut trees whose interlocking crowns, spanning about 60 square meters, reflected enough light to be detected in this SPOT, multispectral image.
Source: SPOT 1991.

of edible wild plants, and periodic game hunting area. It is not uncommon to find farmers protecting small forested areas just for such purposes (figure 14-6).

While most colonists do not commercialize extractive products of their forest patches, many understand that the future value of their farm, or rather their children's farm, depends on the availability of "unspent land," or forest land. The common desire among family farmers to leave a material legacy to their children is an important motive cited by colonists for retaining forest patches.

While both public policies and more subjective values play a role in determining colonist retention and management of natural forest patches, household demographics are also potentially important factors. Many colonists who arrived with small children several years ago are now entering the "empty nest" phase of the family life cycle. With less remunerated family labor, it is more difficult for aging first-generation colonists to clear primary forests and keep existing clearings productive. Some forest patches may exist, or come into being, because household

Figure 14-6

One-hectare natural forest amenity patch located next to colonist dwelling, Rolim de Moura, Rondônia.
Source: SPOT 1991.

consumption needs and forest-clearing capabilities have declined over time. The relationship between household composition and reliance on extractive activities of natural forest patches would be an interesting subject for future research.

Conclusions

Colonist landscapes in the tropical forest lands of Rondônia are fragmented spaces, strewn with forest patches of various shapes and sizes. This seeming hodgepodge of diverse vegetation results from several forces shaping colonist society. As such, the colonist landscape is very much an assemblage of social spaces. Forest clearing establishes property lines, often ambiguous in the adjoining back ends of colonists' lots. Various public policies restrict colonist clearing behavior, such as required stream buffers, forest reserves, and resource island patches. Cultural values favor colonist maintenance of forest patches for amenity, recreation, and legacy. Reading colonist landscapes necessarily entails understanding the different forces that shape colonist decision making at the household level. Each farm lot is a roadmap showing the accumulated effects of land-use decisions made by individual colonist households over time.

In conclusion, five principles apply to reading colonist landscapes:

1. Forest patches are dynamic, human ecological spaces both degenerate and regenerate over time as a result of interacting social and biophysical processes. Forest patches are social spaces, living artifacts of the prevailing social order conjoined with both biophysical processes and the demographic and socioeconomic characteristics of agricultural households and local farming communities. Understanding the multilevel social context in which fragmentation and regeneration occur must precede and inform policies intending to improve forest fragment management.

2. The process of fragmentation is spatially uneven. While the general configuration and rate of fragmentation may appear to be similar within a given area, closer examination shows it is a fractured process, reflecting the diverse biological composition of different forested areas and social differentiation within the agrarian population.

3. Colonist relationships to forest patches are affected by immediate socioeconomic conditions, household labor capacity, market prices

for extractive versus agricultural products, regulatory prices, and individual value preferences toward environmental amenity and the perceived welfare of future generations.

4. Rondônia's colonists typically do not manage their existing forest fragments for commercial purposes. Most, however, do extract forest products and gain subsistence benefits from natural forest groves. Some colonists also appreciate the ecological functions of the forest as a source of clean water and wildlife habitat, and they value the amenities provided by the forest for hunting, fishing, and recreation. Increasingly, colonists are planting commercial tree seedlings on abandoned clearings to promote economic regeneration of degraded forest areas. For instance, Matricardi and Abdala (1993) estimate that about 1,300 farmers in Rondônia (representing approximately 2.5 percent of all the farms in the state) have planted seedlings of among 10 different native commercial timber species.

5. The promotion of natural forest remnant conservation and management by Amazonian colonists must begin by recognizing that colonists are adaptive forest farmers whose resource-use patterns and lifestyles coevolve with the forest environment around them. As colonists learn about forest species and ecosystem dynamics, they successfully manipulate the forest in ways that diversify the sources of value—for cash income as well as for subsistence needs—that might be derived from tropical forest patches.

References

Alcorn, J. B. 1989. "An economic analysis of Huastec Mayan forest management." In *Fragile Lands of Latin America: Strategies for Sustainable Development,* edited by J. O. Browder, pp. 182–208. Boulder, CO: Westview Press.

Balée, W. 1989. "The culture of Amazonian forests." In *Resource Management in Amazonia: Indigenous and Folk Strategies,* edited by D. A. Posey and W. Balée. *Advances in Economic Botany* vol. 7, pp. 1–22.

Browder, J. O. 1988. "Public policy and deforestation in the Brazilian Amazon." In *Public Policy and the Misuse of Forest Resources,* edited by R. Repetto and M. Gillis, pp. 247–298. Cambridge, UK: Cambridge University Press.

Brown, S., and A. E. Lugo. 1990. "Tropical secondary forests." *Journal of Tropical Ecology* 6: 1–32.

Carnasciali, C. H. 1987. "Consequencias sociais das transformaçoes tecnológicas na agricultura do Paraná." In *Os Impactos Sociais da Modernização Agricóla,* edited by G. Martine and R. C. Garcia, pp. 125–167. São Paulo: Hucitec.

EMBRAPA (Empresa Brasileira de Pesquisa Agropecuaria). 1975. "Mapa esquemático dos solos das regioes norte, meio-norte e centro-oeste do Brasil." *Boletim Técnica,* 17. Rio de Janeiro: Centro de Pesquisas Pedológicas, EMBRAPA.

FJP (Fundação João Pinheiro). 1975. "Levantamento de reconhecimento de solos, da aptidão agropastoril, da formaçoes vegetais e do uso da terra em ârea do território federal de Rondônia." Report prepared for the Superintendencia do Desenvolvimento do Centro-Oeste. Belo Horizonte: FJP.

Gomez-Pompa, A., J. Salvador Flores, and V. Sosa. 1987. "The 'pet kot': A man-made tropical forest of the Maya." *Interciencia* 12: 10–15.

IBGE (Instituto Brasileiro de Geografia e Estatística). 1975 *Anuário Estatistico do Brasil.* Rio de Janeiro: IBGE.

Lewis, P. F. 1979. "Axioms for reading the landscape." In *The Interpretation of Ordinary Landscapes,* edited by D. W. Meinig, pp. 11–32. Oxford, UK: Oxford University Press.

Mahar, D. 1989. *Government Policies and Deforestation in Brazil's Amazon Region.* Washington, DC: The World Wildlife Fund and Conservation Foundation.

Matricardi, E., and Abdala Wilson. 1993. "Mogno em Rondônia." Unpublished report commissioned by John O. Browder and the Natural Resources Defense Council and funded by the United States Fish and Wildlife Service.

Mueller, C. C. 1980. "Recent frontier expansion in Brazil: The case of Rondônia." In *Land, People and Planning in Contemporary Amazonia,* edited by F. Barbira-Scazzocchio, pp. 141–153. Cambridge, UK: Cambridge University Press.

Posey, D. A. 1985. "Native and indigenous guidelines for new Amazonian development strategies: Understanding biological diversity through ethnoecology." In *Change in the Amazon Basin: Man's Impact on Forests and Rivers,* edited by John I. Hemming, pp. 156–181. Manchester, UK: Manchester University Press.

Romeiro, A. R. 1987. "Alternative developments in Brazil." In *The Green Revolution Revisited: Critique and Alternatives,* edited by B. Glaeser, pp. 79–110. London and Boston: Allen and Unwin.

Roosevelt, A. 1990. "The historical perspective on resource use in tropical Latin America." In *Economic Catalysts to Ecological Change.* Working Paper, Tropical Conservation and Development Program. Gainesville, FL: University of Florida.

Uhl, C., K. Clark, H. Clark, and P. Murphy. 1981. "Early plant succession after cutting and burning in the upper Rio Negro region of the Amazon Basin." *Journal of Ecology* 69: 631–649.

Webb, L. J. 1982. "The human face in forest management." In *Socioeconomic Effects and Constraints in Tropical Forest Management,* edited by E. G. Hallsworth, pp. 143–157. Chichester, UK: Wiley.

World Bank. 1981. *Brazil: Integrated Development of the Northwest Frontier.* Mission Report. Washington, DC: World Bank.

Chapter 15

Sacred Groves in Africa: Forest Patches in Transition

Aiah Randolph Lebbie and Mark Schoonmaker Freudenberger

Introduction

Sacred forests, "sacred groves," or "traditional forest reserves" are unique forest patches that have survived due to strong cultural forces in many parts of Africa. Rural communities often set aside and restrict access to wooded areas representing ecological spaces that are distinctly different from the surrounding agricultural or pastoral landscapes. Some sacred forests are biologically diverse vestiges of original forests left unaltered by human interference. Ranging in size from less than one hectare to several thousand hectares, these traditional forest reserves form a patchwork of biotic islands with a high potential for the conservation of remnant biological communities.

The rapidly growing literature on sacred areas in Asia, the Americas, and Africa describes the varied cultural and ecological functions of sacred forests. Until recently, these forest patches have been viewed as cultural curiosities. However, a closer look at these cases shows that sacred forests are distinct common property regimes maintained by strong institutional authorities and reinforced by a wide range of sanctions that limit excessive extraction of forest products and wildlife. While these sacred groves are valuable, indigenous natural reserves protected by deep cultural and historical traditions, they also provide practical benefits to local populations such as river-catchment protection, windbreaks, shade, and useful forest products such as firewood, fruits, and medicinal plants. Sacred forests around the world share the common feature of existing independent of government laws and regulations. Governments rarely take initiatives to work with local populations to conserve and expand these islands of ecological diversity.

Sacred forests, which have been extensively studied in Asia, generally are clusters of forest vegetation that honor a deity, provide sanctuary for spirits, remind present generations of ancestors, or protect a sanctified place from exploitation. Groves of trees are treated as sacred by virtue of their location, cultural meaning, and history. Forests are also protected because the collection of forest products sustains temples and other religious institutions (Chandrakanth et al., 1990; Chandrakanth and Romm 1991; Daniels et al., 1993; Gadgil and Vartak 1976; Mansberger 1987; Messerschmidt 1985; Sponsel and Natadecha-Sponsel 1993; Wachtel 1985).

Sacred forests are also found in North America. The Navajo set aside places in forests such as mountaintops, sweat lodges, burial grounds, old homesites, lightning-struck trees, and areas reserved for the collection of medicinal plants (Eibender and Wood 1991). Most of these areas cannot be logged without destroying their spiritual value. Among the Pueblo Indians, the Jemez Mountains of New Mexico are sacred, though the area is subject to many competing land uses (Gibson 1987). Similar sacred areas exist among other Native American groups (Kosek 1993).

This chapter contributes to the growing literature on sacred forests by presenting a case study of these unique forest formations found in the Moyamba District, Sierra Leone. It concludes with an assessment of the range of options policy makers should consider when seeking to protect these biologically diverse areas.

Sacred Forests and Forest Remnants in Africa

Sacred forests have long been studied by colonial administrators, academic researchers, and development workers. These observers note that traditional village institutions regulate access to and use of the flora and fauna in biologically rich areas for a wide variety of cultural reasons (Castro 1990; Dorm-Adzobu et al., 1991; Ellenburg et al., 1988; Freudenberger 1993; Gerdén and Mtallo 1990; Lahuec 1980; Richard 1980; Shepard 1992; Trincaz 1980). Powerful individuals in rural communities are often vested with special authority to punish the unauthorized use of traditional forest reserves due to their association with deities, spirits, or other supernatural bodies. Restrictions also govern the extraction of forest products (though rarely ban total use), depending on the kinds of sacred groves involved.

Relatively few detailed studies have been conducted on sacred forests in Africa. Several recent exploratory studies indicate that sacred forests may be quite prevalent in contemporary rural Africa, even in communities influenced by Islam and Christianity (Dorm-Adzobu et al., 1991; Freudenberger 1993; Gerdén and Mtallo 1990). In the Babati District of central Tanzania, 46 traditional forest reserves were identified, totaling roughly 288 hectares and ranging in size from 1/2 to 100 hectares. Some forests are used as gathering places during male circumcision and traditional instruction of women; other forested areas are set aside as cemetery grounds. Certain protected areas are believed to bring rain. In each of these areas severe fines are imposed by the local residents for breaking the rules that limit access to the forests (Gerdén and Mtallo 1990). Similar studies in The Gambia (Freudenberger 1993) and Ghana (Dorm-Adzobu et al., 1991) indicate that sacred groves are an integral part of the rural landscape and remain important repositories of cultural and ecological value.

European explorers first visiting coastal Sierra Leone in the early 1500s noted that "secret societies" held animist rites within forested areas near village settlements (Little 1948a). During the colonial period, administrators and researchers drew attention to the groups that practiced religious rituals in sacred forests, often located around springs, streams, and on mountaintops (Gaisseau 1954; Lem 1948). Anthropologists studying "rain-shrine communities" among the Tongo of present-day Zambia noted that shrines were venerated for the powers they held over rainfall and harvests and that "the immediate area around a shrine is sacrosanct, and no one may cut wood, dig roots, or burn the bush in that area save on prescribed occasions" (Colson 1951).

The former president of Kenya, Jomo Kenyatta, studied the history of sacred forests among the Kikuyu (Kenyatta 1962). Both Kenyatta and subsequent researchers describe in considerable detail sacred groves in such places as the Kirinyaga district near Mount Kenya (Castro 1988, 1990). During the 1930s, over 200 sacred groves were recorded in the district, often located on hilltops and along ridges, and ranging in size from one-twentieth of a hectare to one hectare. Kikuyu generation sets managed these sites, now mostly converted to state forest plantations (Castro 1988, 1990).

Rural communities often use sacred forests to foster shared beliefs and group solidarity. Initiation rites and other rituals often take place in sacred forests for extended periods of time in order to teach cultural traditions and practices. Youth are taught proper behavior, including relations with nature. As one study noted, "cultural taboos are an example of this, where restrictions on the utilization of certain plants, animals, or areas can prevent the depletion of natural resources important for the continued livelihood of the population" (Gerdén and Mtallo 1990).

Sacred forests are areas where the community has established a covenant with deities or other sacred entities to refrain from certain uses of the environment. Sacred forests are usually controlled by a traditional authority, which may be a fetish priest in charge of a god of the grove, the chief of a village, or the members of specific groups such as "secret societies" (Little 1965). Within these traditional organizations, specialists are often delegated as conservators or guardians who possess considerable governance powers in rural communities of the coastal rainforest belt of West Africa (Little 1948a; Freudenberger 1993; Nti-amoa-Baidu 1990). While these organizations and their functions have been somewhat transformed by modernization, they remain responsible for the provision of various social services and local administrative functions.

Village-level institutions responsible for the conservation of sacred forests were viewed with great suspicion by colonial authorities since these associations also played a role in organizing resistance to colonial rule (Bromley 1992; Little 1965, 1967). Contemporary African governments continue to distrust the activities of these traditional village-level organizations because they represent a challenge, if not a threat, to the authority of state institutions. Indeed, in many coastal West African states sacred forests are used by secessionist or rebel groups as hiding places and warfare training grounds.

Sacred groves of present-day Africa are threatened by changing cultural mores and practices. Religious conversion to Islam and Christianity, the waning power of the elder gerentocracy supportive of sacred groves, and continued pressures to extract resources combine to undermine the viability of sacred groves. In some areas of Africa, such as on the slopes of Mount Kenya, sacred groves are but relics of a past era

(Castro 1990). In marked contrast to the situation in many other African countries, the cultural significance of the sacred groves in Kenya has been eroded by the weakening of the generation-set system, the popularity of new religious beliefs and practices, the disappearance of communal celebrations and workshops, and the increasing privatization of land tenure (Castro 1990). The conversion from animist practices to Christianity and Islam, in particular, reduces the ideological and popular support for the sacred forests.

A Case Study of Sacred Forests in Moyamba District of Sierra Leone

Sierra Leone is one of the smallest countries in West Africa with a land area of 71,620 square kilometers, and a population of slightly less than five million people. During the early nineteenth century the country was largely forested, but logging for export and subsistence agriculture have reduced the area of mature forests to just under 5 percent (3,600 square kilometers) of the land area. Most remaining forests are now protected in forest reserves, but uncontrolled logging, with little or no subsequent forest management, results in further encroachment by farmers, making the future of such forests very bleak.

This case study explores the cultural mechanisms that have conserved sacred forests in the Moyamba District of Sierra Leone. The study was conducted in early 1994, during a conservation assessment of sacred groves in 6 of 14 chiefdoms in the Moyamba District. The Moyamba District is approximately 6,899 square kilometers in size and is located in southwestern Sierra Leone (figure 15-1). Like much of Sierra Leone, the vegetation in the district has been greatly modified as a result of swidden agriculture, further exacerbated by declining fallow periods. With the exception of riparian forests and three national forest reserves, the sacred forests are the only remaining fragments of a once luxuriant rainforest. These sacred forests provide a unique opportunity to explore issues surrounding the sociocultural values of sacred groves in Sierra Leone, primarily because numerous traditional village organizations still exist that protect these biologically diverse fragments.

Typical Flora and Fauna of Moyamba District Sacred Groves

The sacred forests of the Moyamba District are islands of ecological diversity (Lebbie and Guries, forthcoming). Within an ecosystem other-

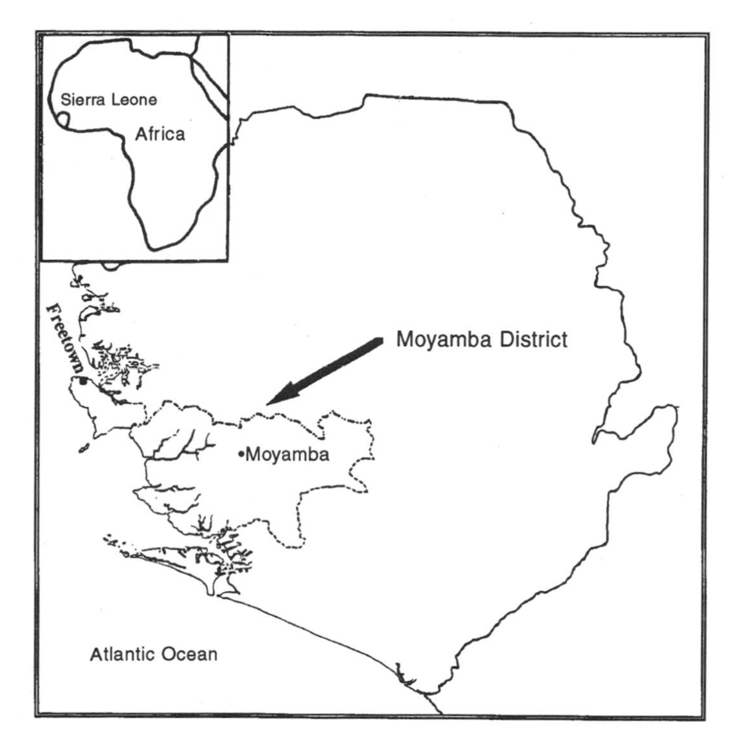

Figure 15-1
Map of Sierra Leone showing the location of the Moyamba District.

wise much disturbed by human interventions, these forest fragments are unmistakable, standing out as forest islands very distinct from the denuded, agro-pastoral landscape. Trees in the forests measure up to 200 centimeters dbh (diameter at breast height) and exceed 30 meters in height, creating closed canopies. The species composition in some of these forests is reminiscent of the Upper Guinean rainforest and includes *Piptadeniastrum africanum, Brachystegia leonensis, Pentaclethra macrophylla, Klainedoxa gabonensis, Pycnanthus angolensis, Parinari excelsa, Enantia polycarpa, Acioa scabrifolia,* and *Diospyros spp.* Lianas, though few in number, can reach 30 cm dbh, an indication of the maturity of these forests.

In some of the sacred forests, the largest trees are sometimes laden with epiphytes, as their bark provides suitable habitat for the growth of these plants. Both snags and dead fallen logs are adorned with fungi, which help in recycling the nutrients locked within these dead trees. The deep silence in the forest is broken only by the occasional sizzling of insects and the piping of hornbills and other birds. At night, the nocturnal animals take control with the keening of the dwarf galago (*Galago demidovii*) emanating from these forests. The beautiful stripes of the African giant snails are unmistakable, and numerous millipedes are present. Tree frogs are often seen, leaping only when dislodged by a snake. Three species of monkeys, *Cercopithecus campbelii, Cercopithecus petaurista,* and *Colobus polykomos,* have been noted in some sacred groves. As with similar forest patches in other parts of the world, the sacred forests act as refuges for wildlife as well as being repositories of genetic material for yet unforeseen uses.

Traditional Village Organizations and Natural Resource Management

The main tribal group in the Moyamba District is the Kpaa Mende, locally known as Mende (Little 1967). The district is subdivided into 14 chiefdoms, each independently administered via a hierarchical organization of "paramount chiefs," section chiefs, and numerous chiefs, each of whom represents a village. At the core of the village social structure is a set of institutions actively involved in the governance of the community (figure 15-2). The most important of these traditional village-level organizations are called *poro, wunde,* and *sande.* These and several other similar, though more specialized, traditional institutions (*njayei, humui, kpikili, gbangbani,* and *hunting*) make collective decisions regarding the social, economic, and religious life of rural communities. Numerous parallel organizations also contribute to village political and economic life. Various village development committees, which channel exterior aid to clinics, schools, churches, and mosques, coexist with animist traditions. A hierarchy of state institutions links the village to the broader national sphere. The village chief is an intermediary between state institutions and the local governance organizations of Mende society.

Colonial travelers, administrators, and anthropologists noted during their initial contacts with Sierra Leonean rural communities the presence of traditional village organizations called *poro* (Alldridge 1910; Brown 1937; Little 1948a, 1965, 1967; Wallis 1905). The *poro* are groups of circumcised men adhering to the beliefs and practices established by the organization. *Poro* members have a strong sense of identity and cohesiveness due in part to sharing circumcision rituals and the associated training in the traditions and proper comportment of the culture. Mende men of all social categories are members of this association,

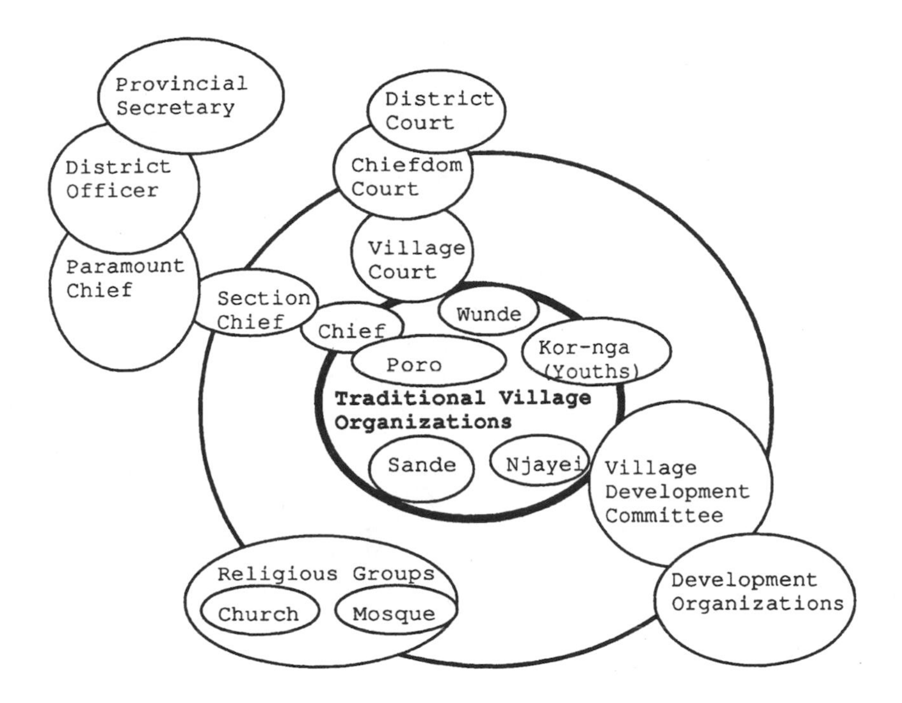

Village Organization	Head	Sacred Grove Conservator
Poro	Sowa	Yamba
Wunde	Lahwa	Markamee
Sande	Sowie	Sowie

Figure 15-2
Idealized organizational schema of villages in the Moyamba District.

provided that they have taken part in the secret rituals. Sacred forests are an integral part of the *poro* culture, because that is where deities and spirits of the community reside and where circumcision takes place.

Women belong to similar groups called *sande,* also organized in association with forests. The aim of the *sande* is to educate young girls for an accepted pattern of life. They are taught to be hard working and modest in their behavior toward elderly people. Non-*sande* and non-*poro* members do not have access to the secrets of the initiated or to the sacred forests, and for this reason westerners considered them "secret societies."

The authority of the *poro* and many complementary traditional village organizations rests in their ability to use a powerful force called *hale* (medicine) (Little 1948b). *Hale* is any object that possesses latent powers of good and evil. All Mende refer to traditional village organizations such as *poro, sande,* and others as *hale. Hale* is used by *poro* and *sande* groups to enforce the rules set down by the association of initiated members. The breaking of sanctions results, in the worst cases, in the ostracization of the malefactor from the village. Usually the severity of punishments is gradually meted out to encourage modification of behavior rather than outright exclusion.

The power of *hale,* used to enforce proper behavior toward natural resources, is currently enforced in much the same way as observed by colonial travelers:

> *A sign of the* poro *placed upon a farm, or tree, is sufficient to deter any would-be thieves from helping themselves from the products of either. Not so very long ago the writer of this article, being on the march and being tired and thirsty, halted in the vicinity of a tree laden with ripe oranges. A boy was sent to gather some of the fruit, but, upon going up to the tree, he returned without touching any of the oranges. He explained that he dare not do so, as there was* poro *on the tree, and added that if anyone picked or ate a single orange from that tree, the medicine* (hale) *would catch him, and he would die. And although everyone stood looking at that delicious fruit with longing eyes, not a man would touch it until the medicine man came and removed the fetish.* (Wallis 1905)

The colonial administration was threatened by the *poro* society because of its ability to influence collective behavior. When the British took over the administration of the Sierra Leone protectorate in the late nineteenth century, the British government passed an ordinance, referred to as the Poro Ordinance, declaring *poro* associations illegal and forbidding the placing of the *poro* symbols on fishing grounds, palm trees, and other resources. The colonial administration claimed that the placement of such signs on these resources diminished trade and, hence, export earnings (Little 1965).

The legitimacy of the *poro* was never destroyed. Three decades ago, one anthropologist noted that "both the harvesting of palm fruit and fishing were regulated by *poro,* since the placing of the society's sign prohibited the use of the plantations or fishing grounds concerned until it was removed" (Little 1965). Similar restrictions are still prevalent today. Traditional village organizations such as the *poro* and *wunde* continue to institute sanctions to ensure the sustainable management of community resources. *Hale* is currently used in the Moyamba District to prevent the excessive exploitation of fisheries in lakes and streams. For example, in the Kowa chiefdom, fishing in four streams is regulated by the *poro* and the *wunde* societies. A ban is placed on the harvesting of fish for three to six months by two men assigned by each society. Totem plants, such as *Afromomum spp.* or *Selaginella myosurus,* are placed near the fishing areas to indicate that restrictions on fishing are in force. The seasonal ban on fishing is lifted by the two authorities following annual sacrifices in the sacred prayer forests to the ancestral spirits. Overfishing is thus avoided, or at least some regulation is imposed to control the seasonal harvest.

Sacred Forests in Moyamba District

Remnant forests so evident on the Moyamba District landscape are often the sacred forests of the *poro, sande, wunde,* and similar traditional village organizations. Sacred forests are excellent examples of common property regimes built on the cultural precepts and authority of traditional Mende institutions (Bromley 1992). These forests are the collective property of the community of initiates and no single person controls use of the resources. Every chiefdom with sacred forests (*kaima lorgboi*) possesses its own rules and administrative structures, which ensure that the groves are protected from overexploitation by members of the community.

In a survey of 6 out of 14 chiefdoms in the Moyamba District, a total of 392 sacred groves were located in association with 235 villages. (Because of civil unrest at the time the survey was being conducted, a complete survey of all 14 chiefdoms in the region was not possible.) On average, most villages have two sacred groves; but in a few instances, up to five sacred groves were identified. The different types of sacred groves encountered in this study area can be classified based on the associated village organizations and their functions (table 15-1). Over half of the sacred forests (52 percent) are *sande lorgboi* (women's sacred forests), more than 38 percent are *poi lorgboi* (men's poro forest), approximately 4 percent are *wunde lorgboi* (men's wunde forest), and over 2 percent are *hemi* (sacred prayer forest). The other sacred forests, which include *ndogbonyamui* (fetal burial forest), *kabandae lorgboi* (legendary or mythical forest), and *humui lorgboi, kpikili lorgboi, gbangbani,* and *hunting* (all traditional village organizations), account for just under 4 percent of the study area. In total, the traditional village organizations' sacred forests make up over 97 percent of the sacred forests in the 6 chiefdoms, and the community sacred forests account for just

Table 15-1

Sacred groves in 6 of 14 chiefdoms in the Moyamba District of Sierra Leone.

	Traditional Village Organizations' Sacred Forests									Community Sacred Forests	
Chiefdom	Poi lorgboi	Wunde lorgboi	Sande lorgboi	Ndogbo-nyamui	Njayei lorgboi	Humui lorgboi	Kpikili lorgboi	Hunting (non-Mende)	Gbangbani (non-Mende)	Hemi	Kabandae lorgboi
Kori	55	1	96	1	1		1	1		4	
Kamajei	12	5	16		2	1					
Fakunya	25	3	26								
Kaiyamba	23	3	27		1			1	1		
Kowa	15	1	22	1	2					4	2
Dasse	20	1	17							1	
Total	150	14	204	2	6	1	1	2	1	9	2
Percentage	38.3	3.6	52.0	0.5	1.5	0.3	0.3	0.5	0.3	2.3	0.5

under 3 percent. We have classified these latter forests as community sacred forests, because access to them is not governed by an individual's membership in any of the traditional village organizations. (Note that the *ndogbonyamui (bolé)* is a sacred forest associated with the *sande* society.)

Poi Lorgboi: Men's Poro Forest

The *poro* (the anglicized literal usage of *poi* in Mende) are widely respected political powers and arbiters of culture in Mende societies (Little 1949). Each *poro* association has a forest that is used as a meeting place for its members to make important decisions on matters pertaining to the welfare of the entire village or chiefdom. However, several villages may share one *poi lorgboi.* In addition, some of these groves also serve as burial sites for the group's members, especially those in the higher echelon of the *poro* society. For example, the *sowa* (head of initiation rites) and the *yamba* (custodian of the grove) would be buried in these forests upon their death. The custodian enters the forest on a regular basis to check for any signs of human interference.

Most of the *poro* forests examined in the Moyamba District ranged in size from two to seven hectares. Reports of an 83-hectare *poro* forest in the district could not be confirmed because of civil unrest at the time the study was being conducted.

Every initiated member of the *poro* association is expected to uphold the rules governing the sacred groves, because protecting the forest is the affair of the entire membership. Non-*poro* members in the community are expected to respect the *poro* forests, and illegal entry can result in heavy fines. Such fines currently consist of the payment of a live goat, a bushel of clean rice, a five-gallon tin of palm oil, a fowl, and "cold water" (this includes palm wine, bamboo wine, and whisky). In certain situations, trial by jury can lead to the offender being ostracized from the community, particularly when the offender proves to be obstinate.

Hunting and the collection of forest products in the *poro* forests are forbidden even to members, except during the initiation rite of new members when sticks, palm fronds, and vines can be cut to make a temporary dwelling house for the initiates. Some minimal harvesting of medicinal plants that have become depleted in the surrounding vegetation is allowed, but only when the *poro* is meeting in the forest and the head of initiation rites and the custodian of the forest have granted authorization. Trapping of rodents by initiates during the two- to four-week-long initiation ceremony is often allowed in order to teach trapping skills that the boys will need when they become farmers.

Intervals between the initiation ceremonies can range between three to seven years. Two *poro* forests were discovered at Kamato and Waiima, both in the Kori chiefdom, in which initiation ceremonies were last held in 1964 and 1971, respectively. The level of protection accorded these groves, and the extent to which the groves' resources are utilized, indicate that cultural and spiritual values take precedence over the material benefits that could be derived from the groves.

Wunde Lorgboi: Men's Wunde Forest

The *wunde* is a traditional village organization that exists only in the Moyamba District. Members of this society are expected to be *poro* initiates before seeking membership in the association. The *wunde* is the most powerful traditional society and a highly respected village organization in the Moyamba District. Among their numerous sociocultural functions is warfare training (Addison 1936; Little 1967).

In comparison to the *poro* forests, the *wunde* forests are few; though like the *poro* forests they are well protected from encroachment. Fines are imposed on trespassers, including members that break the *wunde* laws. Similar fines imposed by the *poro* are also employed by the *wunde* society, except that the fines are twice as great and exclude the giving of a fowl. The heavy fines imposed are intended not only to stop the offender from repeating his actions, but also to inform the larger community as a whole that such offenses are not tolerated. In addition, the wunde association is actively involved in the management of other natural resources, such as the regulation of freshwater fishing.

Sande Lorgboi: Women's Sacred Forests

The *sande* is a women's traditional village organization, and among its numerous functions is the training of the *bogbeni* (young girls newly initiated into the *sande* society) in appropriate behavior toward their future husbands and the wider community. Instruction in sexual matters, household duties, and child care are imparted as well to the new initiates.

The sacred forests of the *sande,* generally between one and two hectares in size, are more prevalent than sacred forests of the other traditional village organizations (table 15-1). These forests are reported to be valuable storehouses for many of the medicinal plants used by the *sande* association. Some of the medicinal plants, as revealed by the society's herbalists, thrive only in the mature *sande* forests (Lebbie and Guries 1995). No hunting or trapping is allowed in these forests.

Access to the *sande* forests is open only to initiated women; illegal entry can result in monetary fines. In some cases, if the culprit cannot be readily identified, fetish ceremonies are performed in the society's sacred house or in the *sande* forests. These rites are believed to induce swollen stomach and hernia of the scrotum, particularly in the guilty man. These afflictions represent the use of psychological deterrents to ward off potential violators, and as one may imagine, they have been remarkably effective.

Ndogbonyamui (Bolé): Fetal Burial Forests

Women use the *ndogbonyamui (bolé)* to perform postmortem caesarian surgery following death of a pregnant woman to remove the fetus and ascertain the cause of death before burial. Only women who have been initiated into the *sande* society are allowed to enter these forests. The flora and fauna contained in the fetal burial forests are considered abominable, and on no account should they be harvested. Illegal entrance and eating of any fruit or animal, particularly by men, is thought to result in impotence, a swollen abdomen, and often death. The fines imposed for breaching these rules also serve as the penance to cleanse the individual from succumbing to the fatal consequences of the curse imposed. Some men are allowed to enter these forests, but only in instances when the women cannot muster the physical strength to dig the graves to bury the fetus and the mother. Before the men leave, certain male-cleansing ceremonies must be performed. In these forest reserves, sociocultural values have taken absolute precedence over material factors in determining the use of the fetal burial forests.

Hemi: Sacred Prayer Forests

There are two categories of sacred prayer forests: one used by the entire community and the other used separately by men and women. In prayer forests libations are poured and requests made to ancestral spirits for guidance, prosperity, and the general well-being of the entire village or chiefdom community. Although these forests are separated by sex, people are allowed to enter them as long as they do not harvest anything. Hunting and trapping are not allowed, and it is assumed that if an individual decides to hunt or trap in these forests, the spirits of the dead will ensure that the attempts are unsuccessful. Though the rules are not strictly enforced, the community's reliance on moral suasion to ensure conformity by every community member serves as an incentive to protect these forests from degradation.

Kabandae Lorgboi: Legendary or Mythical Forests

Other forest patches often protected on the Moyamba District landscape include the legendary (mythical) forests, or *kabandae lorgboi.* A legendary or mythical forest may be created following the occurrence of some strange or unusual event. In other cases, the presence of a strange object such as a huge tree, rock, or cave provides the rationale for the surrounding forest to be declared a mythical forest (Harris 1954). Some of these forests are located on mountaintops and serve to protect the community's main source of water supply. To understand better the rationale behind their creation, we shall cite two key examples of legendary (mythical) forests, tracing the role of animist beliefs in their creation and the maintenance of the biological resources within them.

KORTIBUIYA MYTHICAL FOREST. The mythical forest known as Kortibuiya became revered when villagers claimed that a white-skinned person from a nearby forest disappeared into it. Activities such as the hunting and trapping of venerated animals were forbidden, and vegetation could not be cut, leading to the protection of a spring in the forest. Later, the community saw fire emerging out of a huge *Ceiba pentandra* tree in the forest. They interpreted the fire as an indication that the devil or a genie dwelling in the mythical forest had abandoned it. The forest was subsequently reduced to less than one hectare in size as a result of frequent wild fires and expanding farmlands.

The continued survival of such sites depends primarily on the reoccurrence of the strange events that led to their declaration in the first place. When such an event is not repeated, the destruction of such forests is likely, as is indeed evident from the Kortibuiya case. Moreover, the presence of elderly people in the community supportive of animist beliefs is essential for reminding members of the need to revere such sites.

BANDA HILL MYTHICAL FOREST. The people of Njama, a village in the Kowa chiefdom, have continued to protect a 20-hectare forest known as Banda Hill, believed to be the place where the ancestors (*Banda-nga*) of the Njama village descended. During the time when the ancestors occupied the hill, they were contantly at war with the surrounding villages. In order to protect themselves, they dug a huge trench around the hill where they remained unscathed. During a temporary truce in the region, they moved to another hill, called Tungia which became a resting place. When peace was restored in the entire region, ancestors moved down to present-day Njama, and Banda and Tungia Hills became revered sites. Farming, hunting, trapping, and palm wine tapping were forbidden on the upper reaches of Banda Hill; however, the people were allowed to farm on the lower slopes. If farming were practiced on the upper reaches of the hill, death was believed to follow shortly thereafter. It was also believed that when a death was imminent in the village, or in other surrounding villages, the ancestors could be heard singing and dancing on the slopes of Banda Hill.

In the second half of this century, a man by the name of Mattia Maluwa cultivated rice on the upper reaches of the Banda Hill against the wishes of the community elders. He set fire to the brush to clear his field, and villagers claim that he killed 200 monkeys in the fire. One night before he could begin harvesting his rice field, he dreamed that the 200 monkeys he had killed in the fire were the souls of the ancestral spirits. In the dream, it was made known to him that he would not live to repeat his actions or harvest his rice. He explained the dream to the entire community, and before harvesting could begin Mattia Maluwa died. Since his death, the Njama villagers have become more wary of farming on Banda Hill. Vegetation on the hill has regenerated, though it is still in the early stages of secondary-forest succession.

The Sacred Forests of Other Traditional Village Organizations

The other traditional village organizations, such as *njayei, humui, kpikili, hunting,* and *gbangbani,* are also organized around forests (table 15-1). Together they account for just over 2 percent of the sacred forests in the six chiefdoms of the Moyamba District. The roles of these village organizations are more specialized than the roles of others previously discussed. For example, the *gbangbani* is a men's village organization of the Limba tribe of Sierra Leone, and its purposes are social and moral, with no political function. The *hunting* society is also a non-Mende organization of the Krio tribe, and it provides both social and some governance functions for a motley collection of people. Both the *humui* and the *kpikili* societies have closer ties, but the *kpikili* is considered to be an advanced grade of the *humui* society. While the rules of the *humui* society determine restrictions of intercourse, marriage, and sexual conduct, the *kpikili* society specializes in medical as well as ritual washing or cleansing. The *njayei* society, on the other hand, is concerned with the cure of insanity and other forms of mental conditions, as well as the propagation of agricultural fertility.

Survival of Sacred Forests in the Moyamba District

Sacred forests have been created for many reasons in the Moyamba District. Their survival hinges on strong cultural values to the local communities, coupled with the authority of traditional village institutions like the *poro* and *wunde,* among others, to regulate access and use. Several of the *poro* and sacred prayer forests examined turned out to be former settlement sites. Before the villages' ancestors could settle in an area, they would make a sacrifice of *hale,* referred to as *kpakpa,* which would be buried around the future settlement site. The sacrifice was to protect the village and its members from any harm and to assure prosperity to them and their subsequent generations. The destruction or dislocation of the *kpakpa* would result in its latent and negative nature becoming active, bringing doom upon the entire community. Many of the community members claimed that the exact locations of these *kpakpa* sacrificial areas in the sacred forests are not known; and rather than risk any evil consequences, the local communities and the traditional village institutions such as the *poro* have decided to protect these sacred groves forever. In addition, many rural communities maintain considerable respect for the dead. Since some of these forests have also been used as burial sites for some members of the traditional village organizations, it is often considered taboo to disturb them. For these reasons, some of the sacred forests have received maximum protection.

Threats to Sacred Groves in Sierra Leone

The long-term viability of sacred groves is generally quite secure in the Moyamba District. Endogenous threats are unlikely due to the strength of indigenous social institutions (*poro* and *wunde*) regulating the use of natural resources. But we did observe threats to the sacred forests from external commercial interests, such as the mining of precious minerals.

The discovery of diamonds in the Kono District of Sierra Leone led to mining operations within sacred groves and the subsequent clearing of the forests (Conteh 1979). Government-sanctioned mining operations were permitted in the sacred groves in the belief that fair compensation would be paid to ensure the relocation of the desecrated groves. Though many of the groves were relocated, the survival of these new forests is in question. Many communities failed to identify with the new groves because they lacked the cultural and spiritual significance of the original sacred areas. In other regions of the Moyamba District, large-scale mining of bauxite and rutile have led to the destruction of many sacred forests and other sacred sites. Mining companies did not provide compensation for the destruction or relocation of these forests, since most of the affected communities could not prevent them from destroying their sacred forests and other sacred sites. The literature on sacred forests in other parts of Africa also suggests that local communities may not necessarily possess the authority to defend their sacred forests from extractive commercial activities (Castro 1990; Conteh 1979).

In the Moyamba District, expanding agricultural activities and declining fallow periods have left many of the sacred forests isolated in a sea of grassland. Fires from nearby farms easily traverse the grassland, consuming portions of the forest vegetation. When fires are repeated on an annual basis, the size of groves is reduced following the invasion of fire-tolerant *Imperata* grass. In one community, the invasion is so severe that most of the agricultural lands have been abandoned, leaving a small and threatened sacred forest.

Biological processes occurring within forest fragments are influenced by their size, degree of isolation, and various edge effects (Alverson et al. 1994). Sacred groves are not immune from such threats. Our field observations of the sacred groves in the Moyamba District show that their small size predisposes them to severe windthrows during rainstorms. Repeated heavy windstorms knock down large trees and leave some sacred groves looking like secondary forests even though a few of the larger girth trees remain. One possible solution is to increase the size of some of the more isolated and smaller groves to ensure that they are not adversely affected by such threats.

The disappearance of traditional animist beliefs, which are necessary to the survival of sacred forests, is linked to ascendance of new age groups who are disrespectful of the older generations' belief systems. Certainly the decline in traditional norms associated with sacred forests on the slopes of Mount Kenya (Castro 1990) is a sobering reminder of the rapidity with which changes in social systems can lead to significant ecological perturbations. Although a decline in respect for the spirits of sacred forests associated with the *poro, wunde, sande,* and other groups is not yet common in the Moyamba District, mythical forests are likely candidates for future sacred grove destruction. In one village, some of the young boys noted that since they had not experienced some of the mysteries surrounding a particular legendary forest, they were at liberty to harvest resources from the forest. These infractions contribute to the breakdown of sanctions and belief systems needed to protect such forests. *Poi lorgboi, wunde lorgboi,* and *sande lorgboi* are probably not threatened to the same degree as mythical forests because the spiritual forces inhabiting the sacred areas are still highly venerated and feared within Mende culture. Moreover, the authority of the traditional institutions is still strong and respected by all. Nonetheless, the protection of these forests is a pressing concern in Sierra Leone as well as in other parts of Africa.

Policy Implications

Sacred groves are indigenous common property resources that merit further study by both the academic community and development practitioners. The case of the sacred forests of the Moyamba District exemplifies the functions and uses of these unique forest patches, and their conservation could be a cornerstone of environmental strategies in Africa. In some parts of Africa sacred forests survive with little interference from external agents, yet in other situations national and international interventions may be needed to protect these unique ecological resources. Two policy options merit consideration.

State Management of Sacred Forests

National government policy makers conclude in some instances that rural institutions are incapable of managing sacred forests. This assumes that traditional organizations are neither able nor willing to exclude external interests from excessive exploitation of the flora and fauna of a sacred area. In addition, the state may be convinced that the community itself cannot police its own resources. During the colonial era, French foresters recommended that the customary tenure arrangements regulating use and access to sacred forests be abolished and that control over these forests be delegated to professional foresters supported by a corpus of legal codes and regulations.

The French forester Aubréville strongly recommended in the 1930s that sacred forests, or *bois fêtiches* in West Africa, be systematically set aside as state reserves (Bergeret 1993). During the late colonial period sacred forests of the Casamance in southern Senegal were gazetted as forest reserves (Sambou, pers. comm.), a policy that, to this day, remains in force in some African countries. For instance, following a decision by the Gambian village of Pirang to clear-cut a sacred forest in order to replace it with a mango plantation, the foresters of the Gambian–German Forestry Project gazetted the 64-hectare forest (Ellenberg et al., 1988). As in most other African countries, various provisions of the Gambian forest laws permit the transfer of forest management control to the state. National governments can acquire title to sacred forests simply by declaring them state lands under articles allowing for expropriation in the public interest. Communities living contiguously to these gazetted forests are usually granted various use rights to forest resources so long as these extractive activities do not destroy the ecological integrity of the protected space. Once the forest is placed under state control, government foresters attempt to design and implement forest management plans.

The conversion of sacred forests to state reserves obliges the state to assume the high administrative costs of forest management. Forestry services in Africa generally lack the financial means to manage effectively gazetted state reserves, be they national game parks or state forests (Brandon and Wells 1992; Freudenberger 1992). Structural adjustment programs in some countries result in slashed budgets of forestry departments, which in turn severely constrain the ability of technical services to take on the task of managing small and dispersed forest patches. To compound further the difficulties in managing gazetted forests, sufficient technical knowledge may simply not be available to institute viable forest management plans.

Despite these difficulties, government appropriation of sacred forests should not be excluded as a policy option, for as the case of the Pirang forest park in The Gambia illustrates, sacred forests face threats from many fronts. In the Moyamba District of Sierra Leone, we suspect that "legendary" and "mythical" sacred forests are the most vulnerable to conversion to other uses. Unless they are replaced by new sacred forests created by the rural populations themselves, the breakdown of community sanctions may lead to their disappearance. In situations where particularly unique flora and fauna are located in threatened sacred forests, decisive government intervention may be needed to conserve the forested areas. Perhaps a more serious concern is that the state might expropriate the sacred forests and fail to enforce its claim to management, while the local community no longer considers itself responsible for management. This transition from a common property regime (*res communes*) to a state property regime may in fact be nothing more than an open-access regime (*res nullius*) (Bromley 1992).

Comanagement of Sacred Forests between Local Communities and the State

Development theory and practice in many African countries currently favor the transfer of resource management powers from the state to local resource-user groups (Bromley 1992). New and mutually beneficial partnerships are being forged between the state and local populations to facilitate comanagement of natural resources (Painter 1991; Thomson and Coulibaly 1994). Some rural communities are gradually acquiring the legal power to control the use of their surrounding natural resources. In many interesting experiments in this new, decentralized, "microterritorial" (*gestion du terroir*) approach, the state and local communities negotiate rules and regulations governing the use of natural resources. Enforcement provisions are crafted by both government officials and local communities to ensure compliance with locally defined resource-use agreements. In effect, both the state and the local communities are engaged in a process of defining new tenure arrangements.

Effective comanagement partnerships between the state and traditional village institutions are the product of dialogue, consensus building, and negotiation. The state and nongovernmental organizations bring together various stakeholders to negotiate the future uses of a particular sacred forest. Forums can be held at the district level by forestry professionals, nongovernmental development organizations, and research institutes to discuss the ecological and cultural status of sacred forests, as well as policy measures to enfranchise village-level organizations to manage their own sacred forests.

Sacred forests should not only be protected, but measures should be taken to increase their size. The institutions (such as *poro, sande,* and *wunde*) responsible for the overseeing of sacred forests in such places as the Moyamba District of Sierra Leone demonstrate the possibilities of expanding sacred forests.

In the Moyamba District village of Konda, two sacred groves are located adjacent to each other. They are separated only by a four-hectare cultivated piece of land that was once considered a sacred forest but has been cleared to prevent rebel forces from hiding there. During a forest inventory conducted in preparation for this study, medicinal plants (which herbalists had long thought extinct) were discovered in the sacred prayer forest. The herbalists responded favorably to our proposal that the two forests be once again joined together in order to provide a more favorable habitat for the growth of the rare medicinal plants. The *poro* members, including the village chief, convened a village meeting to discuss the matter. After some debate over forest size and possible ill effects should the expanded forest begin to harbor wild animals, the four-hectare plot was declared a sacred forest and sanctions were instituted to ensure its preservation. If the newly reserved land had belonged to an individual family, the *poro* society would have paid compensation for the loss of productive farmland. This initiative warrants monitoring, for it demonstrates how communities can devise and enforce rules to improve the management of forest resources.

In these types of cases, government could sanction the authority of traditional village organizations, such as the *poro* and *sande,* to manage sacred forests. This would entail little more than the articulation and dissemination of a national policy on sacred forests. A sacred forest-policy declaration might state that traditional institutions possess the rights and responsibilities to enforce restrictions on the use of sacred forests and that local courts may punish infractions. This approach would incur relatively few financial costs, since government technical agencies might do little more than monitor the ecological status of sacred forests while district-level court systems might handle appeals to the rulings of local courts.

The conservation of sacred forests hinges on the negotiation of new partnerships between the state and local communities. Secular and sacred values can be merged together to assure the future of sacred forest patches. Secular African states tend to discount the spiritual and cultural contents of indigenous resource management practices, yet these religious codes and practices have conserved biologically diverse sacred forests despite profound changes in the ecologies and economies of Africa. African states may find it necessary to intervene to protect sacred forests threatened by extractive activities and commercial interests. Rather than impose a uniform policy of placing sacred forests under state control, governments should seek a more subtle response that respects the capacity of local communities to manage for themselves these areas. By listening respectfully to the people who depend on sacred forests for their spiritual and cultural well-being, governments will hear innovative policy proposals.

Acknowledgments

The authors wish to thank with appreciation NYZS/The Wildlife Conservation Society for a research grant provided to Aiah Lebbie to conduct field work on the sacred groves in the Moyamba District of Sierra Leone. We also thank the people, the paramount chiefs, and the chiefs of the six chiefdoms in the Moyamba District for their cooperation and support. Special thanks go to Professor Raymond P. Guries for direction and encouragement during the field work and comments on the manuscript. The efforts of paramount chief Thomas B. Gbappi IV, Abibu Johnson, Dr. Duncan McCollin, Professor Timothy C. Moermond, Dr. John Schelhas, Professor Abu Sesay, Mohamed Imam Bakarr, and Rhonda H. Kranz are gratefully acknowledged. We also thank Professor Leslie E. Sponsel and Kevin Bohrer for their invaluable comments and suggestions on the manuscript.

References

Addison, W. 1936. "The Wunde society: Protectorate of Sierra Leone, British West Africa." *Man* 26: 207–209.

Alldridge, T. J. 1910. *A Transformed Colony*. London: Seeley and Co., Limited.

Alverson, W. S., W. Kuhlman, and D. Waller. 1994. *Wild Forests Conservation Biology and Public Policy.* Washington, DC: Island Press.

Bergeret, A. 1993. "Discours et politiques forestières coloniales en Afrique et Madagascar." In *Colonisations et Environnement,* edited by J. Pouchepadass, pp. 23–48. Paris: Société Françias d'Histoire d'Outre-mer.

Brandon, K., and M. Wells. 1992. "Planning for people and parks: Design dilemmas." *World Development* 20: 557–570.

Bromley, D. W. 1992. "Property rights as authority systems: The role of rules in resource management." In *Emerging Issues in Forest Policy,* edited by P. Nemetz, pp. 453–470. Vancouver: UBC Press.

Brown, G. W. 1937. "The poro in modern business: A preliminary report of field work." *Man* 3: 8–9.

Castro, A. H. 1988. "Facing Kirinyaga: A socio-economic history of resource use and forestry intervention in southern Mount Kenya." Ph.D. dissertation, University of California, Santa Barbara.

Castro, P. 1990. "Sacred groves and social change in Kirinyaga, Kenya." In *Social Change and Applied Anthropology: Essays in Honor of David W. Brokensha,* edited by M. S. Chaiken and A. K. Fleuret, pp. 277–289. Boulder, CO: Westview Press.

Chandrakanth, M. G., and J. Romm. 1991. "Sacred forests, secular forest policies and people's actions." *Natural Resources Journal.* 31(4): 741–756.

Chandrakanth, M. G., J. K. Gilless, V. Gowramma, and M. G. Nagaraja. 1990. "Temple forests in India's forest development." *Agroforestry Systems* 11(3): 199–211.

Colson, E. 1951. "The plateau tonga of northern Rhodesia." In *Seven Tribes of British Central Africa,* edited by E. Colson and M. Gluckman, pp. 94–163. Oxford, UK: Oxford University Press.

Conteh, J. S. 1979. "Diamond mining and Kono religious institutions: A study in social change." Ph.D. dissertation, Indiana University, Bloomington.

Daniels, R. J. R., M. D. S. Chandran, and M. Gadgil. 1993. "A strategy for conserving the biodiversity of the Uttara Kannada District of South India." *Environmental Conservation* 20(2): 131–138.

Dorm-Adzobu, C., O. Ampadu-Agyei, and P. G. Veit. 1991. *Religious Beliefs and Environmental Protection: The Malshegu Sacred Grove in Northern Ghana.* World Resources Institute, Center for International Development and Environment. From the Ground Up, Case Study No. 4.

Einbender, L., and D. B. Wood. 1991. "Social forestry in the Navajo Nation." *Journal of Forestry* 89(1): 12–18.

Ellenberg, H., A. Galat-Luong, H. J. von Maydell, M. Muhlenberg, K. F. Panzer, R. Schmidt. 1988. *Pirang: Ecological Investigations in a Forest Island in The Gambia.* Hamburg: Siftung Walderhaltung in Afrika, Hamburg.

Freudenberger, M. S. 1992. "The great gum gamble: A planning perspective on environmental change in northern Senegal." Ph.D. dissertation. University of California, Los Angeles.

Freudenberger, M. S. (ed.). 1993. "Institutions and natural resource management in The Gambia: A case study of the Foni Jarrol District." Madison, WI: LTC Research Paper no. 114.

Gadgil, M., and V. D. Vartak. 1976. "The sacred groves of western Ghats in India." *Economic Botany* 20: 152–160.

Gerdén, C. A., and S. Mtallo. 1990. *Traditional Forest Reserves in Babati District, Tanzania: A Study in Human Ecology.* Uppsala, Sweden. Swedish University of Agricultural Sciences, International Rural Development Centre, Working Paper 128.

Gibson, D. 1987. "Cutting through a sacred forest." *Sierra* 72(1): 135–136.

Gaisseau, P. 1954. *Sacred Forest: Magic and Secret Rites in French Guinea.* New York: Knopf.

Harris, W. T. 1954. "Ceremonies and stories connected with trees, rivers, and hills in the protectorate of Sierra Leone." *Sierra Leone Studies (New Series)* 2: 91–97.

Kenyatta, J. 1962. *Facing Mount Kenya.* New York: Vintage.

Kosek, J. 1993. "Ethics, economics, and ecosystems." *Cultural Survival Quarterly* Spring: 19–23.

Lahuec, J. 1980. "Le parc d'un village Mossi (Zaongho) du traditionnel au moderne." *Cahiers ORSTOM. Série Science Humaines* 17(3–4): 151–154.

Lebbie, A. R., and R. P. Guries. 1995. "Ethnobotanical value and the conservation of sacred groves of the Kpaa Mende in the Moyamba District, Sierra Leone." *Economic Botany* 49(3): 294–308.

Lem, F. H. 1948. "Le culte des arbres et des génies protecteurs du sol au Soudan français." *Bulletin de l'Institut Français d'Afrique Noire* 10: 538–559.

Little, K. L. 1948a. "The poro as an arbiter of culture." *African Studies* 7(1): 1–15.

Little, K. L. 1948b. "The function of medicine in Mende society." *Man* 48(142): 127–130.

Little, K. L. 1949. "The role of the secret society in cultural specialization." *American Anthropologist* 51(2): 119–212.

Little, K. L. 1965. "The political function of the poro: Part I." *Africa* 35(4): 349–365.

Little, K. L. 1967. *The Mende of Sierra Leone.* Oxford, UK: Alden and Mowbray Ltd.

Mansberger, J. R. 1987. "In search of the tree spirit: Evolution of the sacred tree (*Ficus religiosa*)." M.A. thesis. University of Hawaii.

Messerschmidt, D. A. 1985. "People and resources in Nepal: Customary resource management systems of the upper Kali Gandaki." In *Proceedings of the Conference on Common Property Resource Management.* Prepared by the National Research Council, pp. 455–480. Washington, DC: National Academy Press, April 21–26.

Ntiamoa-Baidu, Y. 1990. "Coastal wetlands conservation: The Save the Seashore Birds project." In *Living with Wildlife: Wildlife Resource Management with Local Participation in Africa,* edited by A. Kiss, pp. 91–95. Washington, DC: World Bank.

Painter, T. 1991. "Approaches to improving natural resource use for agriculture in Sahelian West Africa: A sociological analysis of the "aménagement/gestion des terrois villageois" approach and its implications for non-government organizations." New York: CARE International. Technical Report No. 3.

Richard, P. 1980. "Proto-arboriculture, reboisement, arboriculture paysanne des *Savanes septentrionales* de Côte d'Ivoire." *Cahiers ORSTOM.* Série Science Humaine 17(3–4): 257–264.

Shepard, G. 1992. *Managing Africa's Tropical Dry Forest: A Review of Indigenous Methods.* London: Overseas Development Institute.

Sponsel, L. E., and P. Natadecha-Sponsel. 1993. "The potential contribution of Buddhism in Developing an Environmental Ethic for the Conservation of Biodiversity." In *Ethics, Religion and Biodiversity: Relations between Conservation and Cultural Values,* edited by L. S. Hamilton and H. F. Takeuchi, pp. 75–97. Cambridge, UK: White Horse Press.

Thomson, J., and C. Coulibaly. 1994. "Decentralization in the Sahel: Regional synthesis." Club du Sahel/CILSS.

Trincaz, P. 1980. "L'Importance de l'arbre dans l'imaginaire de Cheikh Mamadou Sane: Du rêve religieux à la réalité du village thérapeutique dans la forêt casamançaise." *Cahiers ORSTOM.* Série Science Humaine 17(3–4):309–310.

Wachtel, P. S. 1985. "Asia's sacred groves." *International Wildlife* (March/April): 24–27.

Wallis, B. 1905. "The poro of the Mende." *Journal of the African Society* 4(14): 183–189.

Part IV

Management

Chapter 16

Managing Forest Remnants and Forest Gardens in Peru and Indonesia

Miguel Pinedo-Vasquez and Christine Padoch

A recent volume reviewing the situation of tropical forest peoples begins with the categorical statement, *"Il n'y a pas de forêt vierge"* (Commission Européenne 1993). While the assertion that no undisturbed forests exist on the planet is controversial, the ubiquity and importance of management of forests by village people are indisputable. Traditional tropical forest management technologies and the roles managed forests play in village economies have only recently been appreciated by those studying forests. With a few notable exceptions (Gordon 1969; Conklin 1957), researchers in the past rarely focused on these important and widespread resource management activities. Recently, however, traditional forestry and agroforestry have attracted attention. Reports from the Neotropics (Alcorn 1984; Balée 1994; Denevan and Padoch 1988; Gómez-Pompa et al., 1987; Hecht 1982; Posey 1984, 1985), as well as from Southeast Asia (Fox 1977; Guo and Padoch, in press; Kunstadter et al., 1978; Michon 1985; Torquebiau 1984; Wiersum 1982), have confirmed the diversity and wide prevalence of these technologies.

In this chapter we will discuss examples of forest management by village folk from opposite sides of the globe and from widely divergent societies and forests. The examples serve to highlight the diversity of techniques traditional peoples employ in managing trees, and the efficacy and complexity of their technologies. The forests or forest gardens that are formed and maintained by traditional managers in our two examples differ widely in their origins, in the species they harbor, and in the products they offer. They coincide, however, in many features of their management and in how management activities affect species diversity. The most remarkable similarity between the different systems is their concentration on techniques of managing individual plants, rather than all members of a species or type, as an indiscriminate group. This strategy results in maintenance or even promotion of high species diversity. Forest managers also traditionally attempt to promote more than one product or function in their woodlands, and the benefits gained tend to be diverse and changing. In these aspects the forests tended by traditional managers may be particularly valuable and contrast most significantly with those managed in conventional ways by professional foresters.

Managing the Flooded Forests of Amazonia

We conducted a three-year-long study to see how deliberate management of one area of floodplain forest by traditional Amazonian methods affected its structure and floristic composition, as well as its yield of economic products. Here we will present only a brief summary of the results; this study is described in greater detail and data on the results of forest management are presented in their entirety elsewhere (Pinedo-Vasquez 1995). The floodplains of the Amazon River and its major whitewater tributaries, the *varzea,* occupy a mere 1–2 percent of the total area of the Amazon Basin. The importance of the *varzea* in Amazonian history, however, belies its small size. Its fertile soils and bounteous aquatic resources, its accessibility and relatively concentrated human populations have distinguished the *varzea* from the generally sparsely settled uplands. For centuries the *varzea* has sustained a broad range of resource extraction and management, agriculture, and settlement (Roosevelt 1987; Maroni 1988).

The yearly floods that bring fertile sediments and make *varzea* lands economically desirable also make them difficult to exploit by modern methods. Techniques for effectively exploiting *varzea* lands and resources were known, however, to their aboriginal Amerindian settlers, as well as to the inheritors and transformers of those traditions: the *ribereño* (or riberinho) villagers who continue to make a living there.

Studies conducted on the forest management practices of traditional populations suggest that Amazonians use an array of management techniques to increase yields of a variety of valuable forest products (Anderson et al., 1985; Balée & Gély 1989; Padoch and de Jong 1990). The use of such techniques has produced changes in the species composition and structure of varzea forests in several regions of Amazonia (Anderson et al., 1985).

Our three-year field study focused on the tropical moist forests of the Napo–Amazon *varzea* floodplain in lowland Peru. This floodplain lies at the confluence of the Napo and Amazon Rivers. Situated between three and four degrees south latitude and at a mere 109 meters above sea level, the climate of the site is warm and humid.

Amazonians have long used the Napo–Amazon floodplain for hunting, fishing, farming, and collecting various forest products. According to the 1990 National census, the study area of approximately 7,820 hectares is now inhabited by approximately 10,503 people, who populate eight villages, including the district capital of Francisco de Orellana. This area of rather dense settlement is, however, but a narrow strip bordering a huge expanse of uplands with a mere 0.7 person per square kilometer. Ribereños, who are largely descendants of several indigenous groups, are by far the most numerous component of the population.

From the time of the Spanish and Portuguese colonization to the late 1960s, the extraction of *varzea* resources such as manatees (*Trichechus inunguis*), turtles (especially charapa [*Podocnemis expansa*]), fish (especially paiche [*Gigas americana*]), and timber (especially lupuna [*Ceiba petandra*]) was the most important economic activity of local people living within the Napo–Amazon floodplain (San Roman 1975). However, from the mid-1970s to the present day, a mix of commercial and subsistence agriculture has become the dominant land use in this and other *varzea* floodplains of the lowland Peruvian Amazon. Farming is very diverse, complex, and dynamic (Padoch and de Jong 1992). Napo-Amazon villagers produce some crops such as rice in near monocultures, others like manioc in diverse swiddens, and still others such as fruits in complex agroforests. Using distinctive sets of agricultural practices and crops, ribereños take advantage of the ecological diversity, river fluctuations, and changing economic opportunities that characterize life in the *varzea.*

Several types of *varzea* forests are under different degrees of management along the Napo–Amazon *varzea* floodplain. Capinurales—forests dominated by the tree known locally as capinuri (*Maquira coriaceae*)—are some of the most economically and ecologically important forests in this area. Residents of the Napo–Amazon floodplain extract several products from capinurales for both domestic use and occasional sale in regional markets. These include wood for plywood and resin from the capinuri. Edible fruits from several leguminous shimbillo species (*Inga* spp.), charichuelo species (*Rheedia* spp.), and others are eaten, occasionally sold in local and regional markets, and highly prized as attractants of game animals and fish that are often abundant in these zones. Latex of oje (*Ficus* spp.) trees is widely used as an antihelminthic, and a variety of other construction materials, firewood, and medicinal plants are also harvested for consumption and sale.

Capinural forests mainly occupy low natural levees along the Napo–Amazon *varzea* floodplain. Individual capinurales vary from 50 to 75 hectares in size, covering between 20 and 30 percent of the Napo–Amazon floodplain. Much of this zone is periodically farmed, and seedlings of capinuri tend to become established quickly in farm fields, as are juveniles in fallows. In mature capinurales individuals of this species are found in all forest layers. Such adaptability is a great advantage in the management of capinural forests. Although seedlings, juveniles, and adults respond well to natural and human disturbances, individuals are mainly established in dense stands after a long process of management and the application of a complex of management techniques.

Management Techniques Used by Ribereños in Capinural Forests

The management of capinurales can be divided into three stages, with the precise number and kinds of techniques used varying from one stage to the other. People begin managing seedlings of capinuri and other valuable species in their agricultural fields. During this first stage two principal management techniques are used: *huactapeo* and *jaloneo*. *Huactapeo* is a technique that consists of three operations: selective weeding, cleaning (including the removal of the roots and stems of species that regenerate by sprouting), and controlled burning (including the elimination of regeneration materials such as the roots and stems of sprouting species by burning). The second, *jaloneo*, is a thinning technique that includes the selection of healthy, well-formed seedlings and the uprooting of seedlings with defects. By uprooting seedlings, managers reduce the probability of their resprouting, which would likely take place if offending seedlings were merely cut.

The management of seedlings lasts throughout the agricultural phase, until yields of bananas decline and the fields become fallows (*purmas*). During this second stage local people manage the juveniles of capinuri and other valuable species by applying three principal techniques: *huahuancheo, raleo,*and *mocheado. Huahuancheo* is mainly a liberation technique by which juveniles of competitor species that are not considered valuable are removed from the fallow. This technique involves selection and marking of seedlings slated for removal as well as the choice of an area from which the largest juveniles of competitor species will be removed. In this selected and marked area, juveniles of competitor species are removed. Care is taken to choose appropriate techniques and tools to reduce subsequent damage in the forests. Cutting (including the removal of all marked individuals by felling), girdling, or burning the bottom of the tree follows. By removing these individuals, space and light are made available for growth in height and diameter of stems of capinuri, capirona, and other valuable species.

Another thinning technique that is used is known as *raleo.* This operation is employed to remove defective, diseased individuals or those damaged by pests. This technique is similar to *huahuancheo* although the object of the operation is different. It includes selection of juveniles of capinuri and other valuable species whose stems have defects or signs of infection and pest damage. This operation is mainly conducted by people who are specialized in recognizing such signs. They mark the individuals to be eliminated by removing a small section of bark from the trunk. The selected and marked individuals are felled using axes or machetes. Damage produced by the fall of juveniles is reduced by removing vines before cutting. Again, the cut trees are harvested and used as construction materials or firewood.

The third technique used, *mocheado,* is mainly a pruning operation that involves the removal of branches of selected individuals of umbrella-shaped shrubs such as tamara (*Crataeva benthamii*). Pruning of shrubs is done to increase light availability for the development of seedlings and juveniles of valuable species. This includes three distinct operations: selection, including the identification of individuals of shrub species that are suppressing seedlings and juveniles of valuable species; marking of individuals to be pruned; and cutting and removal of low and high branches of the marked individuals of shrubs. Cutting the lower branches increases growing space and cutting the highest branches increases light, promoting the growth of juveniles of valuable species.

These three management techniques are applied in swidden fallows every three to five years. During this second management stage the majority of juveniles of capinuri and other economic species tend to form dense stands, and gradually the fallow becomes an established capinural forest.

The third stage begins when the fallow has become a mature capinural forest. Two management techniques were found to be the most commonly used in the Napo–Amazon *varzea* during this stage: *anillado* and *desangrado*. *Anillado* involves the killing of selected stems of competitor species using girdling and fire. The technique comprises five activities: selection, marking, removal of understory vegetation (principally vines), girdling, and burning. The last involves burning a small area of the trunk from which the bark was removed until the sapwood is affected. The *anillado* technique usually causes the tree to die rapidly and avoids resprouting from the roots or stem. Amazonians use it to kill stems of large species such as *Ficus insipida* that are difficult to control because they tend to sprout easily from their roots.

Desangrado, a girdling technique, is one of the most commonly used in the Napo–Amazon *varzea.* Employing the *desangrado* technique, villagers remove small stems of competitor species, individuals of stranglers, and woody climbing vines such as garabato (*Stigmaphyllon* spp.). The *desangrado* technique involves two operations: (1) selection, in which all individuals of a vine and other species that are climbing or strangling the trunk and/or covering the canopy of valuable species are selected for removal, and (2) girdling, which involves the removal of bark, cambium, and sapwood in a ring extending around the selected individual at the bottom of its trunk. From the ring fissure, sap, resins, and water are lost. The abundance of resin or sap attracts ants, termites, and other insects that not only extract the sap or resin but also damage any new sprouts. The infestation of insects thus controls the sprouting of vines and helps kill them.

These two management techniques are applied on an average of once every six to eight years. By killing vines and selected dominant trees local people maintain relatively dense stands of capinuri.

Effects of Traditional Management on Capinural Forests

The suite of traditional management techniques outlined above is employed to increase the value of capinural forests to local villagers. The changes their application produce in the forests of the Napo–Amazon floodplain were determined by comparing plots that had been managed as capinurales for varying periods with control areas that had not been managed.

Management was found neither to increase nor decrease the number of trees (greater than 10 centimeters diameter at breast height) per area found in capinural forests. The average number of stems in an unmanaged site (317 per hectare) was well within the range (305–341 per hectare) found in the managed sites. The application of these techniques does, however, result in an increase in the commercial volume per hectare of timber in capinurales. The mean commercial volume of managed capinurales was 81 cubic meters per hectare for areas that had been managed as mature capinural forests for 8 years, 89 cubic meters per hectare for those managed for 16 years, and 85 cubic meters per hectare for those managed for 24 years. All these values were significantly higher than the estimated mean of 54 cubic meters per hectare for the unmanaged capinural.

The application of traditional management techniques also rather surprisingly resulted in a statistically significant increase in the number of species. The mean numbers of species in managed sites were 32 per hectare (8 years), 30 per hectare (16 years), and 36 per hectare (24 years), while in the unmanaged sites it was 22 per hectare. The increase in the number of species in managed capinurales was observed to be due to the ability of several pioneer species to remain in the gaps that were opened by the removal of some individuals from the canopy. Such openings were not present in the unmanaged sites; as a result pioneer species were not found in capinural forests that had not been managed. Some of the pioneer species that are found in managed plots but absent from mature, unmanaged capinural forests—such as *Carica* spp.—are economically important. It should be noted that while the unmanaged sites were located in mature capinural forests, neither these nor any of the managed sites could be classed as old-growth forests. The Napo–Amazon *varzea* is a highly dynamic floodplain, and disturbance, whether by humans or by the rivers' unpredictable waters, is a constant.

Managing Forest Gardens in Kalimantan

Over the last five years we have also been studying the resource management practices, including the manipulation of forests and forest gardens, of the Tara'n Dayaks of West Kalimantan (Borneo) in Indonesia. Despite their vastly different environments, remarkably unlike regional population densities, and disparate histories, the Tara'n and the Amazonian ribereños manage forests in some similar ways and with some similar effects.

Many descriptions of the resource-use practices of Borneo's farmers mention tree, principally rubber, cultivation as an important economic activity (e.g., Dove 1988; Freeman 1970; Padoch 1982). A few notable studies have also described complex agroforestry and forest management systems (Chin 1985; Morris 1953; Weinstock 1983). The Tara'n live north of the Kapuas River in the Balai subdistrict of the Sanggau Regency. The population of the subdistrict in 1989 was 21,206, the great majority of them Tara'n Dayaks. The subdistrict comprises an area of 396 square kilometers, and as a whole has a surprisingly high population density (54 people per square kilometer) for an interior, rural region in Kalimantan. The territory of Tae, a village that includes five hamlets with their fields and forests, is our principal study site. It covers a total area of almost 16 square kilometers, and supports about 88 people per square kilometer.

Mature forests in the Balai subdistrict tend to be small fragments and largely restricted to hilltops. Much of the area, however, is wooded. Patches of younger and older vegetation on hillsides attest to the continuing practice of swidden farming. Swidden cultivation is, however, waning, while intensive wet-rice farming is assuming ever greater importance. As Tara'n farmers rely less on swiddens, and extend their permanent fields, the hillsides are converted increasingly to managed forests or forest gardens. Over the last five years fewer than half of Tae households have farmed any swiddens at all. Of those who did very few cleared more than a hectare. The Tara'n of Tae and neighboring villages combine subsistence production of rice, small livestock, vegetables, and fruits with production for market. Sale of natural rubber is the principal source of income for most households throughout much of the year, with seasonal marketing of fruits very important in some years.

The Tara'n of Balai make and manipulate many kinds of managed forests. The various types can be distinguished by their different origins, management techniques and priorities, and rights of access and inheritance (Padoch 1993; Padoch and Peters 1993.) Perhaps the most interesting are those known as *mawa'n* in Tara'n or *tembawang* in the Indonesian language.

Mawa'n originate as the highly diverse fruit gardens that commonly surround villages, traditional longhouses, or rice farm or work huts. Upon building a house in a new site, or making a work hut in a new swidden, wet-rice field, or rubber garden, Tara'n surround the dwelling with a diversity of fruit trees, shrubs, and herbs. The Tara'n assert that many of the fruit trees found around their houses were not deliberately planted but resulted from their discarding fruit seeds with the household refuse or from their spitting the seeds while eating fruit in the doorway. Seedlings of desirable species, whether deliberately planted or fortuitously discarded, are often protected. Herds of free-roaming pigs and chickens that share villages with most Dayaks are often a problem around principal dwelling sites. Particularly vulnerable and valuable seedlings are frequently shielded from animals by small bamboo fences until they are large enough to survive.

In their earliest stages, therefore, future *mawa'n* are house gardens, tended intensively and, like house gardens around the world, used for a great variety of purposes (Fernandes and Nair 1986; Niñez 1984). Such areas are not designated *mawa'n* by the Tara'n until the house is moved or the field hut abandoned. Tara'n Dayaks report that on average they change house sites every 20 to 25 years, often moving only a few hundred meters away. Swidden huts are normally abandoned after two to three years' use. Productive fruit trees surrounding the houses, however, are rarely cut. A long history of moves and plantings of gardens have formed a landscape in Balai dotted with *mawa'n* that mark where dwellings were located many decades or even a century or so ago. Some individual, mature *mawa'n* are 10 hectares or more in size and, in total they cover hundreds of hectares in Tae village territory.

In their manner of creation, therefore, *mawa'n* differ greatly from Amazonian capinurales. Many of the fruit and other trees that form an early *mawa'n* are planted. As they mature, however, *mawa'n* come to more closely resemble forests. While planting of valuable trees continues in many *mawa'n* for decades, the Tara'n also not only tolerate but actively encourage the establishment of spontaneous vegetation. All desirable plants, whether planted or not, are protected and managed in more mature *mawa'n*.

Management techniques used in *mawa'n* vary widely, but the suite of activities invariably includes cleaning of undergrowth, including selective weeding, and harvesting of products, including the making of paths and, usually, continued planting of desirable species.

The timing and intensity of clearing vary with the species that make up the plot. *Mawa'n* rich in fruits such as durian (*Durio zibethinus*) and illipe nut (*Shorea* spp.) are usually slash weeded once or twice a year, coinciding with fruit harvests. Areas under trees, often extensive in size, are cleared where ripe fruits are apt to fall. Clearing around illipe nut trees may be especially intensive, with controlled burning of understory vegetation sometimes used. These weedings, although often done to enhance ease of exploitation rather than specifically to remove undesirable species and promote useful ones, nonetheless have that effect. Slash weeding is done quickly but is selective. Judgments are continually made as to which seedlings, vines, and herbs to spare and which to remove; those individuals that obstruct the growth of useful trees, rattans, bamboos, or bushes are slashed. Very casual management throughout the year, such as the removal of a vine or other competing plant from the vicinity of an economic plant when and if it is noticed, may also play an important role in this and many other traditional management systems.

Some *mawa'n* managers plant rubber, among the fruit trees. Rubber, which came into these regions in the late 1930s or early 1940s, has become a very important cash crop. Tara'n plant most of their rubber in managed swidden–allow agroforests known as *tanah adat* (Padoch and Peters 1993), but many *mawa'n* also feature interplanted rubber trees. Where rubber is tapped, cleaning around each tree is done quite frequently to maintain easy access. Again such weeding is not done expressly to manipulate the species composition or structure of the plot, but it finally has that effect.

Mawa'n are by definition rich in fruits. Most include a variety of the familiar domesticated fruits of Southeast Asia, such as rambutan, langsat, mangosteen, and always durian. But *mawa'n* are also obviously managed for diversity. One transect through a 10-meter by 200-meter *mawa'n* revealed 44 species among 224 individual trees or stands of bamboo five centimeters dbh or more. While this floristic diversity does not equal levels reported from some mixed dipterocarp forests of the region (Ashton 1984; Kartawinata et al 1981), it surpasses that of other natural forests of Borneo and is certainly high for a heavily managed garden in a high population density area. Since unmanaged forests are extremely rare in this region, no direct comparisons between managed and unmanaged forests in Tara'n territory were done.

Of the 44 species found in the 0.2-hectare transect, 30 had edible fruits or shoots. Many were semidomesticated or wild relatives of the well-known market species; especially common were relatives of mango and rambutan. Many spontaneously occurring fruits, construction materials, fiber sources, and other plants are allowed to remain or are encouraged in *mawa'n.* It is often impossible to ascertain whether any particular individual tree was actually planted or not. Tara'n forest managers claim to plant, at least occasionally, 74 different fruit species, and they identify more than 100 species of fruit in the managed forests of Tae. Many fruit trees have multiple uses, including as construction timbers and as medicine.

Animals fare less well than fruits in *mawa'n.* Tara'n acknowledge that game animals and larger birds are rarely found in their managed woodlands these days. Formal research on the effects of management and human use of *mawa'n* on animals, birds, insects, or other fauna has yet to be done in Tae.

Mawa'n are important as sources of many subsistence products, especially fruits; for many villagers, they are also important sources of cash. Much, if not most, tengkawang or illipe nuts (*Shorea spp.*), an important "forest product" export of West Kalimantan, are harvested from managed *mawa'n.* Apart from tengkawang, for the last 10 years Tae villagers have been marketing durian fruits in quantity. While the production of durian varies greatly from year to year, in particularly good years, sales of the fruit can provide the principal cash income for some Tae families. At the height of the 30-day-long durian season of 1991, three of the five hamlets of Tae sent a daily total of 10,000 fruits to market. The price of Tae durians is also generally rising with improved access in the last few years to markets in West Kalimantan and the Malaysian state of Sarawak. Those families who tap rubber trees in *mawa'n* also get most of their cash income from the forest-gardens. The effects of increased market activity on *mawa'n* management and access are complex and are discussed in detail elsewhere (Peluso and Padoch, in press).

The Lessons of Traditional Forest Management

The ribereños of the Napo–Amazon *varzea* floodplain and the Dayaks of Tae make a living in very different social and natural environments. Tara'n live in uplands, surrounded by neighbors and with no room to expand; ribereños reside in floodplains next to nearly vacant terra firme lands. The two groups also manage forests in different ways for different products. *Mawa'n* are largely planted, while capinurales are not; capinurales are largely managed for timber, while *mawa'n* are valuable for their fruit. There are, however, striking similarities in the way these disparate forests are managed and the results that are obtained.

Management in both places tends to be long term, often beginning in an agricultural stage, with techniques evolving and changing in type and intensity with the stage of forest or forest-garden development. Among the technical aspects that tie traditional management in Amazonia and Borneo and distinguish it from conventional forest management, perhaps the most striking and most important is the effect on plant species diversity. The majority of conventional forest management techniques are designed to reduce the number of species per area in order to be able to increase the commercial volume per area (Graaf 1986; Jonkers 1987). Because of such effects, conventional management systems for timber production, such as the CELOS silvicultural system for natural regeneration in the tropics (Graaf 1986) have been criticized as ecologically unsustainable (Prance 1979). In contrast, many traditional forest management techniques are designed to produce changes in the structure and maintain or even increase the original floristic diversity in order to augment the yield of timber and other forest products (Pinedo-Vasquez et al., 1990). While conventional management techniques are based on species selection and exclusion, traditional management techniques tend to select individuals that are limiting the development of their desired trees, rather than all individuals of particular species. The use of such techniques has managed not only to obtain high yields of timber or fruits per area but also to reduce the rotation times for timber, roundwood, fruits, and other forest product extraction, while maintaining high species diversity.

The forest patches we have described are all managed for more than one economically valuable product. They supply a variety of goods consumed at home and a selection of commodities that are traded locally and regionally. Some products from these managed woodlands, particularly rubber in Borneo and plywood in Amazonia, make their way to international markets. It is not the final destination of the goods that determines how management of the forest is carried out, but rather who makes the management decisions. When the price of one commodity rises sharply on local or far-off markets, ribereño or Dayak managers may indeed respond by intensifying their management and production of the desired product. A fall in price will surely lead to less time and attention spent on the species. These vagaries of the market, however, tend not to lead to single-species, single-purpose management. When done by villagers forest manipulation tends to be conservative and multipurpose.

The results of our studies suggest that some traditional management techniques might indeed be useful in designing and implementing large, medium or small programs of forest management at the industrial level.

More detailed experimental studies are required, however, to measure the ecological sustainability and the economic viability of the practices by ribereños, Tara'n Dayaks, and other traditional peoples.

Acknowledgments

We would like to thank many members of the ribereño community of Portugal, Peru, and the Tara'n Dayak community of Tae, Indonesia, for their help, counsel, and patience. We would also like to thank Adi Susanto for his expert field advice and Andrea Quong for her valuable aid in editing earlier drafts.

References

Alcorn, J. B. 1984. *Huastec Maya Ethnobotany.* Austin, TX: University of Texas Press.

Anderson, A., A. Gély, J. Strudwick, G. Sobel, and M. Pinto 1985. "Uma sistema agroflorestal na *varzea* do estuario Amazonico." *Acta Amazonica, Suplement* 15: 195–224.

Ashton, P. A. 1984. "Biosystematics of tropical woody plants: A problem of rare species." In *Plant Biosystematics,* edited by W. F. Grant, pp. 497–516. New York: Academic Press.

Balée, W. 1994. *Footprints of the Forest: Ka'apor Ethnobotany—The Historical Ecology of Plant Utilization by an Amazonian People.* New York: Columbia University Press.

Balée, W., and A. Gély 1989. "Managed forest succession in Amazonia: The Kayapo case." In *Resource Management in Amazonia: Indigenous and Folk Strategies.* Advances in Economic Botany 7, edited by D. Posey and W. Balée, pp. 129–158. New York: New York Botanical Garden.

Chin, S. C. 1985. "Agriculture and resource utilization in a lowland rainforest Kenyah community." *Sarawak Museum Journal Special Monograph No. 4.* Kuching, Malaysia: Sawarak Museum.

Commission Européenne 1993. *Situation des Populations Indigènes des Forêts Denses Humides.* Brussels: CECA-CEE-CEEA.

Conklin, H. C. 1957. *Hanunoo Agriculture.* Rome: Food and Agriculture Organization (FAO).

Denevan, W. M., and C. Padoch, eds. 1988. *Swidden-Fallow Agroforestry in the Peruvian Amazon.* Advances in Economic Botany 5. New York: New York Botanical Garden.

Dove, M. 1988. *Swidden Agriculture in Indonesia.* New York: Mouton.

Fernandes, E. C. M., and P. K. R. Nair 1986. "An evaluation of the structure and function of tropical homegardens." *Agricultural Systems* 21: 279–310.

Fox, J. J. 1977. *Harvest of the Palm: Ecological Change in Eastern Indonesia.* Cambridge, MA: Harvard University Press.

Freeman, D. 1970. *Report on the Iban.* London: Athlone Press.

Gomez-Pompa, A., J. S. Flores, and V. Sosa. 1987. "The 'Pet-Kot': A man-made tropical forest of the Maya." *Interciencia* 12: 10–15.

Gordon, B. L. 1969. "Anthropogeography and rainforest ecology in Bocas de Toro Province, Panama." Office of Naval Research Report. Berkeley: Department of Geography, University of California.

Graaf, N. R. de. 1986. *A Silvicultural System for Natural Regeneration of Tropical Rainforest in Suriname.* Wageningen, the Netherlands: Agricultural University.

Guo, H., and C. Padoch. In press. "Patterns and management of agroforestry systems in Yunnan: An approach to upland rural development." *Global Environmental Change.*

Hecht, S. B. 1982. "Agroforestry in the Amazon Basin: Practice, theory and limits of a promising land use." In *Amazonia: Agriculture and Land Use Research,* edited by S. B. Hecht, pp. 331–371. Cali, Colombia: CIAT.

Jonkers, W. B. J. 1987. *Vegetation Structure, Logging Damage and Silviculture in a Tropical Rainforest in Suriname.* Wageningen, the Netherlands: Agricultural University.

Kartawinata, K., R. Abdulhadi, and T. Partomihardjo 1981. "Composition and structure of a lowland dipterocarp forest at Wanariset, East Kalimantan." *Malaysian Forester* 44: 397–406.

Kunstadter, P., E. C. Chapman, and S. Sabhasri, eds. 1978. *Farmers in the Forest: Economic Development and Marginal Agriculture in Northern Thailand.* Honolulu: University Press of Hawaii.

Maroni, F. 1988. "Noticias auténticas de famoso rio Maranon." *Serie Monumenta Amazonica* B4. Iquitos, Peru: CETA.

Michon, G. 1985. "De l'homme de la foret au paysan de l'arbre: Agroforesteries indonesiennes." Ph.D. thesis, University of Montpellier, France.

Morris, H. S. 1953. *Report on a Melanau Sago Producing Community in Sarawak.* London: Her Majesty's Stationery Office.

Niñez, V. 1984. "Household gardens: Theoretical considerations on an old survival strategy." *Food Systems Research Series No. 1.* Lima: International Potato Center.

Padoch, C. 1982. *Migration and Its Alternatives among the Iban of Sarawak.* Koninklijk Instituut voor Taal-, Land-, en Volkenkunde (Royal Institute of

Anthropology and Linguistics). Leiden, the Netherlands: Martinus Nijhoff.

Padoch, C. 1993. "Managing forest fragments and forest gardens in Kalimantan." In *Forest Remnants in the Tropical Landscape: Benefits and Policy Applications,* edited by J. K. Doyle and J. Schelhas, pp. 45–47. Proceedings of the Symposium presented by the Smithsonian Migratory Bird Center. Washington, DC: Smithsonian Institution.

Padoch, C., and W. de Jong. 1990. "The impact of forest products trade on an Amazonian place and population." In *New Directions in the Study of Plants and People.* Advances in Economic Botany 8, edited by G. T. Prance and M. J. Balick, pp. 151–158.

Padoch, C., and W. de Jong. 1992. "Diversity, variation, and change in ribereño agriculture." In *Conservation of Neotropical Forests: Working from Traditional Resource Use,* edited by K. Redford and C. Padoch, pp. 158–174. New York: Columbia University Press.

Padoch, C., and C. Peters 1993. "Managed forest gardens in West Kalimantan, Indonesia." In *Perspectives on Biodiversity: Case Studies of Genetic Resource Conservation and Development,* edited by C. S. Potter et al., pp. 167–176. Washington, DC: AAAS.

Peluso, N., and C. Padoch. In press. "Changing resource rights in managed forests of West Kalimantan." In *Borneo in Transition: People, Forests, Conservation, and Development,* edited by C. Padoch and N. L. Peluso. Kuala Lumpur, Malaysia and Oxford, UK: Oxford University Press.

Pinedo-Vasquez, M. 1995. "Human impact on varzea ecosystems in the Napo-Amazon, Peru." Unpublished doctoral thesis, Yale School of Forestry and Environmental Studies. Pinedo-Vasquez, M., D. Zarin, P. Jipp, and J. Chota-Inuma 1990. "Use-values of tree species in a communal forest reserve in Northeast Peru." *Conservation Biology* 4: 405–416.

Posey, D. A. 1984. "A preliminary report on diversified management of tropical forest by the Kayapo Indians of the Brazilian Amazon." In *Ethnobotany in the Neotropics.* Advances in Economic Botany 1, edited by G. T. Prance and J. Kallunki, pp. 112–126.

Posey, D. A. 1985. "Indigenous management of tropical forest ecosystems: The case of the Kayapo Indians of the Brazilian Amazon." *Agroforestry Systems* 3: 139–158.

Prance, G. T. 1979. "Notes on the vegetation of Amazonia. III. The terminology of Amazonia forest types subject to inundation." *Brittonia* 31: 26–38.

Roosevelt, A. 1987. "Chiefdoms in the Amazon and Orinoco." In *Chiefdoms in the Americas,* edited by R. Drennan and C. Uribe, pp. 153–185. Lanham, MD: University Press of America.

San Roman, J. 1975. *Perfiles Historicos de la Amazonía Peruana.* Lima: Ediciones Paulinas.

Torquebiau, E. 1984. "Man-made dipterocarp forest in Sumatra." *Agroforestry Systems* 2: 103–127.

Weinstock, J. A. 1983. "Rattan: Ecological balance in a Borneo rainforest swidden." *Economic Botany* 37(1): 58–68.

Wiersum, K. F. 1982. "Tree gardening and taungya on Java: Examples of agroforestry techniques in the humid tropics." *Agroforestry Systems* 1: 53–70

Chapter 17

Timber Management of Forest Patches in Guatemala

Scott A. Stanley and Steven P. Gretzinger

Introduction

A researcher studying native flora and fauna in tropical forest patches may be compared to a historian examining the architectural design of a burning Victorian house. In both cases, immediate action is necessary to ensure that a valuable resource does not disappear.

For Central America, the house is burning quickly; most of its broadleaf primary forests have been reduced to scattered fragments. Given political and socioeconomic trends in the region, a realistic method for conserving remaining patches is to make them economically productive through natural forest management.

Forest patches have been successfully managed by traditional societies in the tropics (Gómez-Pompa and Kaus 1990; Nations and Nigh 1980). Although many patches today have been degraded by logging, fire, or agriculture, rural dwellers remain dependent on them for various products. Ironically, despite the fact that such patches may possess substantial timber and nontimber resources, they are often considered of little monetary value. Also apparently ignored are the ways in which sustainable use of patches by local communities could help reduce impacts on critical protected areas (Terborgh 1992). Forest patches usually form part of a neglected backdrop rather than a primary management target.

In 1990, the Production from Natural Forests Project was initiated at CATIE (Tropical Agronomical Research and Higher Education Center) with USAID funding to show that natural forest management in Central America is both technically feasible and financially attractive. Since its inception, the project—in conjunction with rural dwellers, industry, and governmental and nongovernmental agencies in Guatemala, Honduras, Nicaragua, and Costa Rica—has developed and implemented field inventory techniques, management plans, silvicultural treatments, low-impact harvesting methods, and data analysis to promote the conservation of remaining broadleaf forests.

This chapter focuses on timber production from forest patches in the Bio-Itzá Reserve of San José, located in the buffer zone of the Maya Biosphere Reserve, Petén, Guatemala. We begin with a discussion of the Petén and its problems and present natural forest management as a partial solution. We then look at the general ecology of commercial tree species in the region and present a field-sampling technique uniquely suited for tropical patches. Finally, we develop recommendations for sustainable timber productions based on a financial analysis and a silvicultural index model applied to the Bio-Itzá case.

Setting

In 1990, the Guatemalan government established the 1.5-million-hectare Maya Biosphere Reserve (MBR) in that country's northernmost region of the Petén. Occupying roughly 10 percent of Guatemala's territory and harboring most of its broadleaf forests (PAFG 1992), the MBR borders preserves in Mexico and Belize, and forms part of the largest contiguous protected area in Central America. Unfortunately, traditional control measures have largely been ineffective in protecting the MBR's resources. This problem is particularly acute in the buffer zone which is now mainly comprised of scattered forest fragments.

The Guatemalan government recently initiated a conservation strategy based on the leasing of long-term forest rights to established communities within the reserve's multiple-use and buffer zones. In 1994, San Miguel la Palotada and the Bethel Cooperative became the first villages to receive such rights. In response to successful first-year operations, other communities are interested in pursuing a similar path.

Located in the buffer zone of the MBR, the 3,600-hectare Bio-Itzá Reserve is an appropriate site for community-based forest management (figure 17-1). Established in 1991 by indigenous Maya Itzá inhabitants of San José, the reserve's primary function is to protect the forest from degradation by immigrants, and support the traditional forest-based culture. To achieve these goals, the reserve promotes sustainable resource extraction by local residents, encourages tourism, and offers opportunities for scientific research. In 1992, community leaders solicited CATIE assistance to execute an inventory and prepare a management plan for the reserve.

The Bio-Itzá Reserve belongs to the moist, hot, subtropical life zone where annual rainfall varies between 1,160 and 2,000 millimeters, the dry season extends to five months, periodic droughts are common, and temperatures fluctuate between 22 and 27°C (Guillen et al., 1993). Although the reserve seldom exceeds 275 meters above sea level, a unique combination of human activities, soil types, and topography has fostered a diverse array of marshland, bottom and upland hardwood, and secondary successional forest communities characteristic of the Petén (Gretzinger 1994; Lundell 1937).

Portions of the Bio-Itzá Reserve were selectively logged in 1979 for mahogany (*Swietenia macrophylla*) and spanish cedar (*Cedrela odorata*). In 1981, some of the forest was subjected to wildfires. In 1989, logging was repeated for the above species as well as santa maria (*Calophyllum brasiliense*) and canxan (*Terminalia amazonia*). Felling sites, log decks, skid trails, roads, and burned areas are still evident. Although the forest is currently exploited for nontimber products (Guillen et al. 1993), no timber management is currently being conducted.

Problem

The Petén's vast forests, sparse population, and vague land tenure policies attract immigrants, resulting in a 12-fold population increase since 1964 to over 300,000 people (Schwartz 1990). This influx is dominated by farmers unfamiliar with lowland tropical forests and reliant upon slash-and-burn agriculture. An estimated 1.5 percent of Petén forest is converted annually to other uses with the loss of five million cubic meters of wood (Leonard 1987; Schwartz 1990). Ninety-eight percent of this loss attributed to agriculture and fire (PAFG 1992).

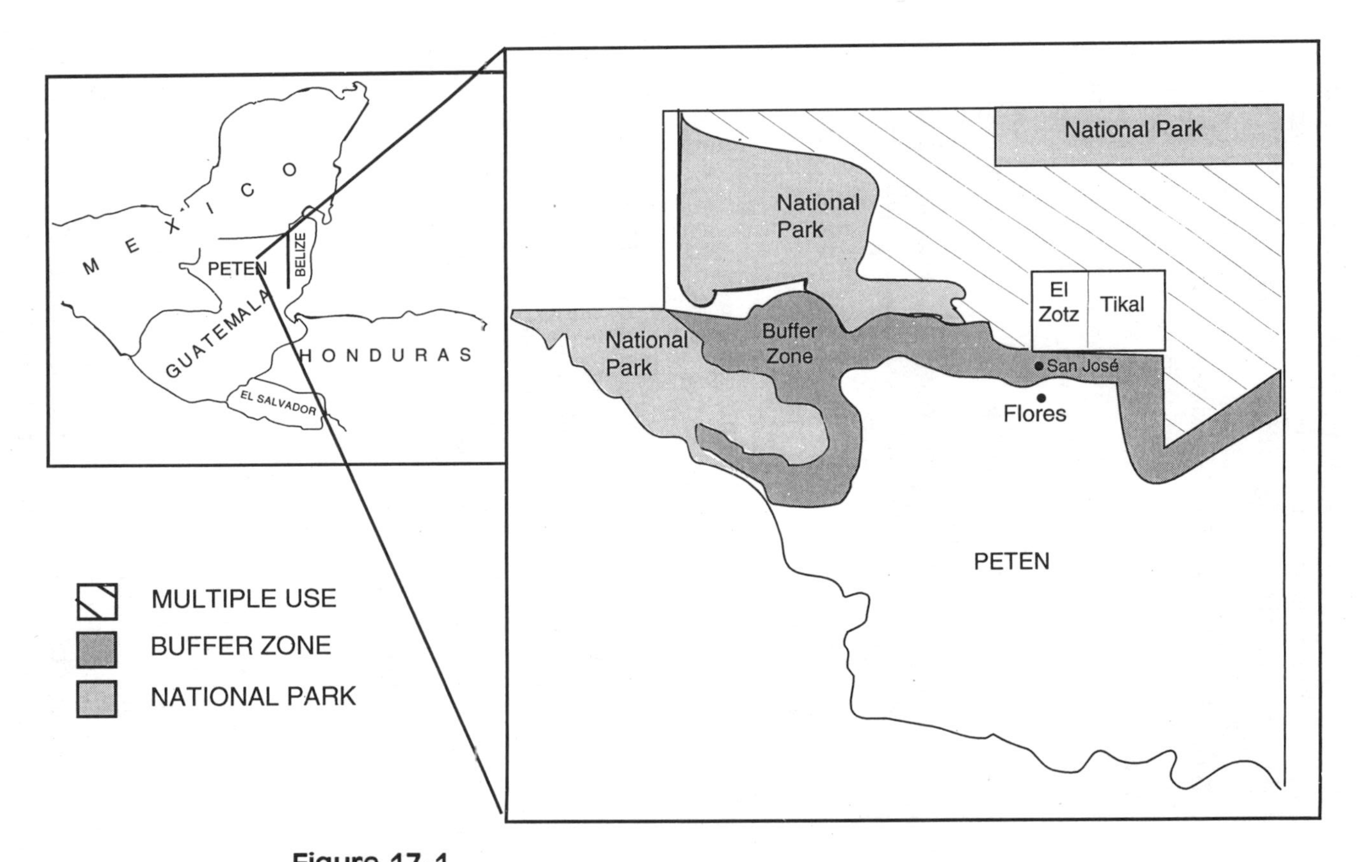

Figure 17-1
Map of the Maya Biosphere Reserve, Petén, Guatemala.

Petén forests have a long history of disturbance, ranging from the Mayan occupation to repeated fires and hurricanes (Cowgill 1961, Snook 1993). Residents have traditionally extracted latex from chicle (*Manilkara zapota*), harvested seeds from allspice (*Pimenta dioica*), and cut leaves from palms for floral arrangements (*Chamaedorea* sp.). Such exploitation is not entirely benign and affects forest composition (Holdridge et al., 1950; Lamb 1966).

Wood extraction in the MBR began with Spanish colonization and accelerated with the entrance of larger companies in 1910. Early efforts focused on individual trees of mahogany and logwood (*Haematoxylon campechianum*) with easy river access, and their extraction resulted in minimal disturbance. Detrimental logging impacts increased with the introduction of mechanized equipment in the 1940s, sawmill construction in 1959, and the granting of industrial forest concessions in the early 1960s (PAFG 1992). Although industrial concessions were canceled in 1989, illegal harvesting continues to degrade the forest. Most of the stated problems affect the Bio-Itzá Reserve to varying degrees.

Natural Forest Management

Human activities have wrought considerable changes in the MBR, and much of the forest within the multiple-use and buffer zones (such as the Bio-Itzá) may be considered a mosaic of patches distinct in terms of age, structure, and species composition. Patches in these zones may be legally harvested for timber and nontimber products through natural forest management (NFM).

In general, NFM includes the harvest of commercial-sized trees and subsequently utilizes protective and silvicultural measures to encourage natural regeneration and growth of well-formed commercial species in the residual stand. Managing natural regeneration is cheaper than reforestation, and it serves to sustain or increase the forest's value without causing unacceptable negative impacts (Buschbacher 1990; Johnson and Cabarle 1993).

NFM is applicable for large or small patches of primary and secondary forests. Wide-scale NFM could aid conservation by linking patches and reducing their isolation, providing travel corridors for wildlife, and minimizing the need for exploitation of large, undisturbed blocks of forest by acting as a buffer (Harris 1984). The inherent mixed-species approach can foster the development of diverse forest-based industries by maintaining a varied portfolio of natural resources and enabling rural dwellers to satisfy economic needs without resorting to more damaging activities.

Obstacles to Natural Forest Management

Despite its benefits, numerous obstacles prevent NFM from achieving greater success in the Biosphere. To date, many foresters still lack the necessary knowledge and experience to successfully implement natural forest management. A growing population, unrealistic conservation strategies, vague forest-use policies, and lack of commitment by donor and governmental agencies are just a few of the sociopolitical impediments to NFM in the Petén (Gretzinger 1994).

Management efforts are also complicated by species heterogeneity and an underdeveloped market for species other than mahogany and spanish cedar. The Petén boasts over 300 tree species (FAO 1968); however, of the 135 species identified by Stanley (1994) in a recent inventory (trees with diameter at breast height [dbh] equal to or larger than 25 centimeters), only 11 were considered as commercial on the current Guatemalan market. The same inventory revealed that despite a total wood volume of 64.2 cubic meter per hectare, only 8.3 cubic meters per hectare (13 percent) pertain to economic species suitable for harvest at present.

Ecology of Commercial Species

Successful natural forest management hinges on adequate understanding of the ecological requirements of tree species. Solar illumination, which is especially critical for the growth of tropical species in dense canopy forests, is the easiest to manipulate (Finegan 1993). Few species prosper under total shade conditions and most respond positively to increased illumination (Denslow 1987; Whitmore 1989). Harvesting and silvicultural treatments alter light availability, thus influencing residual tree growth.

The classification of species into groups (guilds) based on response to light is a common first management step in the tropics (Baur 1964; Hutchinson 1987). However, considerable debate has ensued regarding guild definitions (Denslow 1987; Whitmore 1989). To complicate matters, light requirements are not static and may change over the course of a plant's life history (Clark 1994). Several authors have pointed out the limitations of placing species into discrete guilds and suggest that tropical trees be viewed along a continuum between extreme grades of shade tolerance and intolerance (Swaine and Whitmore 1988).

We agree conceptually with the continuum view. From a field-oriented management perspective, however, it is necessary to classify commercial species into guilds. The chosen system essentially classifies species by their ability to germinate and grow in shade, and assigns them a number along the shade continuum. Guild names have been deliberately avoided in deference to numeric values, which may be more easily adopted by managers with preferences for particular terms. Guild values should be viewed as qualitative descriptors rather than as quantitative measures. The practical utility of this approach will become apparent when we present the silvicultural index model.

Guild class 1 refers to species such as *Pouteria* spp. that can germinate and develop normally in heavy shade and show little growth response to increased light. Guild class 2 species (e.g., *Calophyllum brasiliense*) also germinate and develop in shade but are destined to suboptimal growth rates without increased illumination. Although guild class 3 species such as *Terminalia amazonia* can germinate in partially shaded conditions, they appear to remain in a quiescent state without substantial light increases. Guild class 4 species (e.g., *Pseudobombax ellipticum*) are capable of germinating under partial shade but will eventually die without direct sunlight. Lastly, guild class 5 species (e.g., *S. macrophylla* and *C. odorata*) need high solar illumination to germinate and develop to maturity. If such species germinate in shade, they usually die within the first few seasons.

Seed dispersal vectors are also critical to the design of management strategies. Species in guild classes 3–5 usually have small, light-weight, widely dispersed seeds of highly prolonged viability and the ability to germinate in high-light conditions in disturbed areas (Gómez-Pompa et al., 1991). Common vectors are wind or animals, including birds, bats, and smaller terrestrial mammals (Roth 1987). Species in guild classes 1 and 2 usually have fleshy, colorful seeds prone to distribution by animals. Such seeds are relatively short-lived under high-light conditions and germinate best in low-light conditions (Brokaw 1985).

In most undisturbed tropical forests, animal-dispersed species are more common than wind-dispersed species (Ibarra-Manriquez et al., 1991). Regardless, 80 percent of the most common, commercially acceptable species in the Petén are prone to wind dispersal, thus suggesting past widespread disturbance.

Diagnostic Sampling: A Tool for Managing Forest Patches

Given their small size, fragmented distribution, and often degraded conditions, forest patches do not lend themselves to costly, traditional inventories. Sampling techniques must be simple, rapid, and inexpensive. Originally developed to assess the timber management potential of large tracts of land, diagnostic sampling (DS) can also satisfy the stated requirements for forest patches. The CATIE project has frequently employed DS to assess forest remnants throughout Central America.

This method concentrates on sampling the frequency of selected precommercial trees (less than the minimum-cut diameter for the species). These trees, usually the largest, best-formed individuals with a vigorous crown, are termed *leading desirables* (Hutchinson 1991) and must meet a range of quality criteria that define them as trees most likely to constitute future timber harvests.

The basic premise behind DS is that in humid forests, light distribution is the limiting environmental factor in tree growth. Under normal site conditions, a healthy tree with a fully illuminated crown should grow at its maximum rate. By quantifying the crown illumination of leading desirables, one may calculate the percentage of the stand most likely to benefit from thinning.

Diagnostic sampling uses a simple design for data recording and analysis. Contrary to most conventional forest inventories, computers are unnecessary for analysis, and one tabulation form serves multiple ends. The ease of analysis allows the active involvement of rural communities in forest management planning.

Forest remnants face increasing pressures, and a rapid timber appraisal can help determine whether these areas stay forested or are cleared for cattle or agricultural production. A 200-hectare forest patch can be surveyed by DS within three to four days by a five-person team. The low cost of DS is advantageous for rural communities, government agencies, and NGOs on limited budgets. Diagnostic sampling conducted by CATIE in the Rio San Juan area of Nicaragua cost approximately $5.00 per hectare whereas traditional inventories in the same region cost $17.80 per hectare (Castañeda et al., 1994).

While DS is valuable in forecasting the productive timber potential of forest stands, it cannot fully substitute for traditional inventories. The results of DS are more difficult to analyze statistically (Hutchinson 1991), and neither total stand basal area nor total volume can be calculated. In addition, diagnostic sampling is not suitable for areas such as dry forest ecosystems with relatively open canopy, where solar illumination is not the primary limiting environmental factor.

Methodology

Diagnostic sampling methodology has been described by many authors, including Barnard (1950), Wyatt-Smith (1961), Nicholson (1965), and Hutchinson (1991). In this section, we address only key conceptual points and those that represent modifications on previous work.

Diagnostic sampling was first applied in Malaysia in the 1930s (Browne 1936), when it was realized that the number of commercial seedlings in a stand was less important than their spatial distribution. The technique arranged continuous two by two-meter plots along transects to determine frequency and distribution of seedlings in forests to be logged. If present, one seedling of a commercial species was chosen for each plot. Logging was permitted only when commercial seedlings were found in at least 40 percent of the plots sampled (1,000 seedlings per hectare).

The commonly used plot size has since increased to 10 by 10 meters to sample the stocking and status of advanced regeneration, and to correspond to the area necessary for the crown of a leading desirable to develop unimpeded to commercial maturity (defined as 50 centimeters dbh in Malaysia). This plot size assumes an average 20:1 ratio between crown and bole diameter for all species irrespective of site quality (Dawkins 1963). Maximum stocking of leading desirables would thus be 100 evenly spaced trees per hectare. This value serves as a useful guide for comparing stands, although uneven-aged forests may actually have more than 100 trees per hectare of varying sizes that could constitute future harvests. Certain prerequisites must be met prior to initiating DS in the field. Since only currently commercial species can be chosen as leading desirables, a list of such trees must be compiled. Species with a substantially higher economic value (e.g., *C. odorata* and *S. macrophylla*) should form a separate group and be given priority in selecting leading desirables. We use the following five groups to classify tree species in the Petén according to their economic potential:

1. AAA: high economic value and commercial desirability (mahogany and spanish cedar)
2. ACT: moderate economic value and current market acceptance (13 such species for the Bio-Itzá are listed in table 17-1)

Table 17-1

Number of trees per hectare with a larger minimum-cut diameter in the logged, high-forest strata of the Bio-Itzá Forest Reserve, San José, Petén, Guatemala (1,016 plots of 10 by 10 m).

Species	Log quality	
	Acceptable[a]	Unacceptable[b]
Commercial		
AAA[c]	—	0.2
ACT[d]	6.5	0.3
Noncommercial		
POT[e]	2.9	0.7
NCV[f]	2.5	0.6
PRO[g]	—	—
Total	11.9	1.8

[a]Tree containing at least one sound log larger than 3 meters in length.
[b]Tree not containing one sound log larger than 3 meters in length.
[c]Mahogany and spanish cedar.
[d]Actually commercial species.
[e]Potentially commercial species.
[f]Noncommercial species.
[g]Protected species.

3. POT: currently of little, if any, economic value, but with commercial potential (based on physical properties or markets in other countries)
4. NCV: species of no economic value that are not foreseen to be commercial within the near future
5. PRO: protected tree species that may not legally be harvested for timber (e.g., *Manilkara zapota and Pimenta dioica*)

Since DS principally predicts future production and unmanaged tropical forests typically contain a substantial number of large relic trees, a false impression of stand composition and economic value may be obtained if upper-diameter-size limits are not set for leading desirables (Wyatt-Smith 1961). This upper size limit should coincide with the determined minimum-cut diameter for each commercial species. Based partially on results from research in similar forests in Belize and Mexico, a minimum cut diameter of 60 centimeters was established for mahogany and cedar (AAA); the legal minimum diameter of 45 centimeters has been accepted for other commercial species (ACT). To test the validity of the minimum-cut rules, over 50 growth and yield plots have been established in the Petén by CATIE.

For each 10 by 10-meter plot, DS essentially contains two steps. First, the leading desirable is selected, noting its size class (pole, sapling, or seedling), dbh, crown illumination class (Dawkins 1958), and level of vine infestation. The latter two parameters are taken to ascertain the need for a silvicultural treatment. Second, the dbh, commercial height, and log-quality class of all trees occurring in the plot over the minimum-cut diameter are registered (Stanley 1992). This step is especially important if DS will be the only sampling method utilized, and it allows for the statistical analysis of harvestable wood volume.

In order to gain a reliable estimate of the spatial distribution of leading desirables, the plots should be laid out in continuous transects throughout the sample area, with a 5 to 10 percent sample intensity. By estimating the upper canopy height in each plot and noting evidence of human disturbance, a simple forest strata map can be developed. This feature is advantageous where air photos are either unavailable or out of date.

Results from the Bio-Itzá Reserve

In July 1992, project technicians and community members carried out diagnostic sampling in the Bio-Itzá Reserve. The following forest strata were identified: logged high forest (1,300 hectares), logged high forest with evidence of fire (250 hectares), unlogged high forest (1,760 hectares), and *corozal,* or forest dominated by the *Orbignya cohume* palm (290 hectares). In the following section, we illustrate results from the selectively logged high forest strata since it is most representative of remaining forest patches in the Petén.

Current Commercial Status

As shown in table 17-1, selective logging has resulted in the extraction of all the well-formed, harvestable mahogany and cedar trees (AAA), leaving only 0.2 poor quality individuals per hectare. For commercial species of lower value (ACT), 6.5 trees per hectare are harvestable. However, we recommend harvesting only 5 trees per hectare, with the rest remaining for regeneration, wildlife habitat, and biodiversity purposes.

Financial Viability

Conventional opinion is that management for lesser-known timber species (ACT) without the presence of high-value mahogany or spanish cedar is unprofitable in the Petén. Financial analysis of the DS results reveals that this may not necessarily be true for the logged high forest strata.

A timber harvest on forest similar to the Bio-Itzá revealed that trees in the ACT group yielded a net commercial volume of 1.89 cubic meters per tree (Castiglione 1994). Using a conversion rate (roundwood to sawnwood) of 1 cubic meters equal to 200 board feet (bf), and a harvest of five trees per hectare as mentioned above, we estimate that the logged high forest strata could produce 1,890 bf per hectare. To be conservative, 300 hectares of this strata could remain unharvested to protect critical biological zones. Given a commercial volume of 1,890,000 bf for the 1,000 hectares available for harvest, and utilizing a 25-year cutting cycle, 40 ha could be cut per year to provide a total commercial volume of 75,600 bf annually.

Results from research on logging costs in the Petén (Gretzinger 1995a) estimate a production cost of U.S. $0.52/bf for sawn lumber delivered to Guatemala City. This cost includes diagnostic sampling, management plan preparation, administration, road and site maintenance, technical assistance, felling, extraction, milling, transportation, and taxes (but does not include postharvest activities such as silviculture).

Using Guatemala City prices of approximately U.S. $0.58/bf for lumber of the ACT class (export prices are higher), San José residents could expect net annual earnings of approximately $113.40 per hectare ($4,536 for all 40 hectares) harvesting the degraded high forest. While these returns may not inspire rapid outside investment in the forestry sector, several points put these earnings in perspective.

First, wages paid to community members for participating in certain aspects of the operation were viewed as financial costs in the analysis, although they represent an economic benefit to the community as a whole. Second, no quantification of earnings from nontimber forest products or tourism has been attempted. Third, volume increments over the course of the cutting cycle have not been included. Fourth, demand and prices for lesser-used species rises over time. For example, as recently as 1991, Guatemala exported few such species. Timber for export currently includes *Calophyllum brasiliense, Terminalia amazonia,* and *Pseudobombax ellipticum,* among others. Based on these factors, it appears that integrated forest management could provide attractive financial returns for the San José community.

Percentage Stocking of Precommercial Trees

In uneven-aged management, future harvests will come from the current stocking of trees with dbh greater than 10 centimeters. For tropical forests of North Queensland, Nicholson (1972) considered a stand to be adequately stocked if 40 percent or more of the plots had a leading desirable of this size. Results from San José are comparable: 75 percent of the sample area contained a leading desirable of some size, and 38 percent harbored one with a dbh greater than 10 centimeters (table 17-2).

Future Harvests

Each diameter class in table 17-2 roughly corresponds to a 25-year cutting cycle, assuming an average annual diameter increment of 0.6 centimeters for commercial species (Stanley 1994; Snook 1993). One can use this information to estimate the number of harvestable trees for the next cutting cycle. For example, in the 45 to 59-centimeter-diameter range, a total of three AAA trees were selected as leading desirables in the sample. Dividing this number by the total area of sample plots (2.25 hectares), suggests that 1.3 mahogany or cedar trees per hectare could be available for harvesting within 25 years. However, since diagnostic sampling selects only one individual per plot, this value may underestimate future harvests because other AAA trees in this diameter class may also be present. Accounting for both a possible overestimation and tree mortality, there should be at least one mahogany tree per hectare for harvest in 25 years. Also since subsequent cycles should yield greater numbers of ACT species than presently harvestable (table 17-2), it appears that sustainable timber production is feasible in the logged, high-forest strata throughout repeated cutting cycles.

Necessity of Silvicultural Treatment

By examining Table 17-1, one can determine if the abundance of noncommercial large trees could impede the regeneration of economic species. Many authors state that the removal of these large trees is an essential step in previously unmanaged forests (Baur 1964; Hutchinson

Table 17-2

Total number of leading desirables (LDs) by crown illumination class in the Bio-Itzá Forest Reserve, San José, Petén, Guatemala (diagnostic sample of logged high forest with 225 plots, 10 by 10 m).

Diameter class of the leading desirables	Number of LDs per crown illumination class							
	Abundant illumination		Partial illumination		Deficient illumination			Stocking
	AAA	ACT	AAA	ACT	AAA	ACT	Total	(%)
dbh in cm								
45–59.9	3	—	—	—	—	—	3	1.3
30–44.9	8	11	—	—	—	1	20	8.9
15–29.9	6	20	1	4	1	4	36	16
10–14.9	—	3	3	7	4	11	28	12.4
Subtotal	17	34	4	11	5	16	87	38.6
(%)	(19.5)	(39.1)	(4.6)	(12.6)	(5.7)	(18.3)	(100)	—
Saplings	1	—	2	3	5	9	20	8.9
(%)	(5)	—	(10)	(15)	(25)	(45)	(100)	—
Seedlings	—	—	1	1	4	56	62	27.6
(%)	—	—	(1.6)	(1.6)	(6.4)	(90.3)	(100)	—
No LD found	—	—	—	—	—	56	56	24.9
(%)	—	—	—	—	—	(100)	(100)	—
Total							225	100

1991; Wadsworth 1987). Given changing markets, of the 6.7 noncommercial trees per hectare shown in table 17-1, the well-formed POT trees may become economically valuable and should not be eliminated. Due to their protected status, PRO species cannot be cut legally. Thus, only four noncommercial large trees per hectare could be eliminated.

Since the Maya Biosphere Reserve is managed for multiple objectives, timber production is not the only consideration. For example, large, upper-canopy trees provide important avifaunal habitat by augmenting vertical foliage distribution (Harris 1984). In consideration of such functions, we recommend that in timber production areas, such trees be eliminated only if their total exceeds 10 trees per hectare, or when they directly compete with an economically valuable tree. This guideline is based on the previously mentioned 20 : 1 crown–bole ratio and implies that the 10 or more large trees per hectare would heavily shade over 10 percent of the surface area. In the Bio-Itzá case study, a specific silvicultural treatment to eliminate these large trees is not recommended.

Liberation thinning is a silvicultural treatment specifically oriented for precommercial trees, freeing their crowns from competition. This technique has demonstrated promising results in Malaysia and Costa Rica (Hutchinson 1987, 1993). Diagnostic sampling can be especially helpful in determining the necessity of this treatment. As shown in the crown illumination columns in table 17-2, trees with fully illuminated crowns (58.6 percent) are assumed to be growing at their maximum rate, and a thinning would not significantly augment their growth rates. The remaining percentage of leading desirable trees (41.4 percent) would benefit from liberation thinning. This percentage is large enough to warrant a liberation treatment. The expected reduction in mortality of liberated trees is an added benefit that should also be considered.

Silvicultural Index Model for Long-Term Sustainability

While silvicultural recommendations to this point have focused on increasing the growth rate and decreasing mortality of precommercial-sized trees, they may not satisfy the ecological requirements for regeneration of economic species. For example, it is uncertain whether liberation thinning would foment the regeneration of these species. To date, there has been no easy method to predict which silvicultural system would most likely sustain a range of species over the long term.

To address this issue, we designed an index that utilizes the ecological guild values previously described to determine the most appropriate silvicultural system for 15 commercial species in the Bio-Itzá Reserve. The guild value in the following equation is squared to increase the range between minimum and maximum values and thus corresponds closer to the broad shade continuum described by Swaine and Whitmore (1988). The silvicultural index value (SIV) has a minimum value of 1 and a maximum of 100.

$$\text{Silvicultural Index} = \frac{\sum (\text{Guild}^2 \times \text{SDF} \times \text{Vector} \times \text{Frequency})}{\sum \text{Frequency}}$$

A soil disturbance factor (SDF) from 1 to 2 is used in the index. Species that need exposed soil to successfully germinate (e.g., mahogany) receive an SDF of 2. Also built into the equation is a seed dissemination vector value, with 1 signifying animal or bird dissemination and 2 signifying wind dissemination.

Lastly, the index equation takes into account the frequency of leading desirables, as noted during the diagnostic sampling. A species distributed throughout the sample area should be more heavily weighted than a rare, clumped species. Frequency values are assigned to each species, with a value of 1 equivalent to a frequency of leading desirables 1–5 percent of the plots, 2 for a frequency of 5–10 percent, and 3 for a frequency over 10 percent. Since DS records only the "best" precommercial-sized tree in each plot, the chosen silvicultural system more heavily emphasizes the regeneration of well-formed, evenly spaced species. Therefore, the product of the equation is essentially a weighted average based on ecological site requirements.

When substantial price differences exist between commercial species (as in the Petén), the index should be separately calculated for the different economic groups, as illustrated in table 17-3. The index value for the AAA group is the sum of each species individual index value, divided by the sum of the frequencies (300/3 = 100). For the ACT group, the silvicultural index value is 26 (502/19 = 26.4). Since the AAA and ACT index values are significantly different, two silvicultural systems should be implemented, with priority given to AAA species because of their much greater economic worth. A value of 100 signifies that the AAA group should be treated in an even-aged system that creates large gaps with exposed soil (see figure 17-2).

To achieve successful regeneration for the AAA class, figure 17-2 suggests that approximately 75 percent of the basal area should be eliminated, with the remaining 25 percent comprised mostly of commercial species. Implementing this system is somewhat costly and subjects the forest to drastic changes. Therefore, it should be carried out only where there is a presence of mahogany or cedar that could serve as seed trees. Where there is a predominate wind direction during seed fall, as in the Petén, the mahogany or cedar seed shadows should receive the basal area reduction. *Cedrela odorata* in the Petén exhibits a cone-shaped seed shadow, approximately 0.25 hectares in size (Stanley and Gómez 1993). With an average of one cedar or mahogany per hectare over 25 centimeters dbh (the size considered capable of adequate seed production), this silvicultural system should affect no more than 25 percent of the treatment area.

As shown in figure 17-2, with an SIV of 26 for ACT species, a selection system with no more than 40 percent of the basal area eliminated in harvesting and treatment should be successful in regenerating the

Table 17-3

Silvicultural index values (SIV) for commercial species calculated by ecological guild, soil disturbance factor (SDF), seed dissemination vector, and diagnostic sampling frequency for Petén, Guatemala.

Economic class	Guild	Guild^2	SDF	Vector	SIV	DS freq	Total
AAA							
Cedrela odorata	5	25	2	2	100	1	100
Swietenia macrophylla	5	25	2	2	100	2	200
Total						3	300
Group SIV							100
ACT							
Aspidosperma megalocarpon	2	4	1	2	8	3	24
Aspidosperma stegomeres	2	4	1	2	8	1	8
Astronium graveolens	2	4	1	2	8	2	16
Calophyllum brasiliense	2	4	1	1	4	2	8
Gymnanthes lucida	2	4	1	2	8	1	8
Lonchocarpus castilloi	4	16	2	2	64	1	64
Platimyscium dimorphandrum	4	16	2	2	64	1	64
Pithecellobium arboreum	2	4	1	1	4	1	4
Pseudobombax ellipticum	4	16	2	2	64	1	64
Schizolobium parahybum	5	25	2	2	100	1	100
Terminalia amazonia	3	9	2	2	36	3	108
Vatairea lundelli	3	9	1	2	18	1	18
Zanthoxylum elephatiasis	4	16	1	1	16	1	16
Total						19	502
Group SIV							26.4

Note: Guild classification based on Gretzinger et al. 1993, Augspurger 1984, Barton 1984, Hartshorn 1980, Brokaw 1985, and personal observation.

majority of the important commercial species. Liberation thinning combined with a harvest on average eliminates 30–40 percent of the basal area. However, as shown in table 17-3, this system will not foster regeneration of species with high individual SIVs such as *S. parahybum.* Sampling results indicate that these species are poorly distributed and their low economic value does not justify a separate management scheme. Nevertheless, depending on location, these high-SIV species

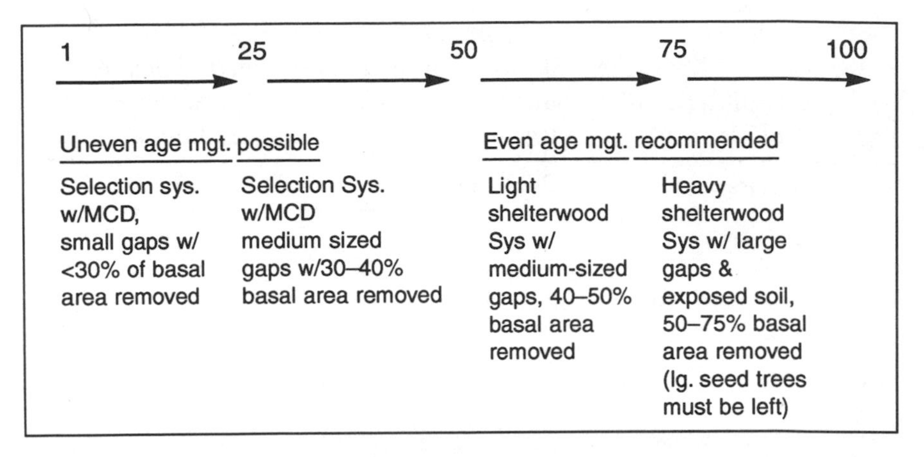

Figure 17-2

Rating scale for interpreting silivicultural index values. (MCD = minimum cut diameter; basa area removal based on all species over 10 cm dbh.)

should be able to take advantage of the large gaps created for mahogany and cedar.

The recommendations for sustainably managing mahogany and cedar versus the ACT species seem to contradict each other; however, the proposed implementation of the two systems within the same stand is simple:

1. Where there is a well-formed mahogany or cedar over 25 centimeters dbh (approximately one tree per hectare), reduce the basal area by 75 percent in a 0.25-hectare, cone-shaped zone on the opposite side of the predominate wind direction. Much of the understory and nondecomposed soil organic layer should be removed, possibly via prescribed burning.
2. In areas unaffected by the shelterwood system, a selection system that uses a minimum-cut diameter should be undertaken. On a research level, thinning should be used to liberate the suppressed crowns of pole-sized trees with the objective of proving that the increase in growth will outweigh the cost incurred.

Conclusions

Given the precarious situation of many tropical forest patches, an immediate conservation action should be the production of forest goods that yield competitive financial returns. If rural dwellers do not receive a monetary benefit from the forest, then intrinsic values (biodiversity, carbon sequestration, wildlife habitat, etc.) will be endangered; farmers will eliminate the forest in favor of traditional agricultural activities.

Diagnostic sampling was shown to be a relatively simple, rapid, and cost- effective field technique that provided basic information for decision making. Results from DS, financial estimates, and the application of the silvicultural index were used to develop long-term management strategies. Despite a degraded condition and absence of commercial-sized, high-value species in the Bio-Itzá Reserve, we have shown that a timber harvest could be profitable and that long-term production is feasible.

The Bio-Itzá Reserve, although atypical in size, represents a somewhat novel combination of proposed management techniques, which are applicable to much smaller patches. As patch size decreases, however, timber management costs become prohibitive; however, this problem can be overcome by grouping individual patches together into one administrative block. For example, in the Rio San Juan area of Nicaragua, where farmers typically own 10–20 hectares of forest, the CATIE project is promoting the formation of an association of farmers that would actively manage these forests. Management costs would be paid for by the association and profits disbursed based on the size of each member's forest.

San Miguel Talamanca in Costa Rica is a working example of a similar organization, where since 1992 an association of 12 farmers has been managing small forest patches. The incorporation of appropriate technology, such as the use of water buffalos for log skidding, minimizes costs. CATIE has provided technical assistance to help the association maximize profits by integrating the harvest of timber and nontimber forest products. This type of integrated use is perhaps the best hope for conserving the remaining forest patches. However, more information is needed to ascertain how logging affects the production of nontimber forest products. Preliminary studies suggest that certain marketable palms and small plants are negatively impacted by the increased light resulting from logging (Gretzinger 1994). Still, specific ecologic requirements of understory plants may be evaluated against requirements of the timber species and weighted by financial or other objectives to determine the best course of action.

Although ecotourism in selected sites has generated significant income, profits have not been evenly distributed and can open a Pandora's box in a rural community, causing, among other problems, inflationary pressure on land prices. Nevertheless, ecotourism can play a role in community forest management but should not be considered a panacea. There are some instances where forest-based ecotourism projects have been successfully incorporated with timber management. For example, in Quintana Roo, Mexico, a number of rural communities augment their income from harvesting timber and nontimber products by offering "ecotreks" (Gretzinger 1995b).

In various forms, the CATIE forest management project applies the methodology presented in this chapter in demonstration sites throughout Central America. These sites are, for the most part, degraded forest remnants that have little or no perceived value by the local populace. By applying an integrated approach and using the philosophy that "seeing is believing," the project hopes to increase the number of hectares under forest management in Central America.

Although, this chapter presents techniques that are subject to modifications with greater use, it represents one of the first attempts to integrate the ecological requirements for species regeneration with the management of forest patches in the Petén to ensure a sustained production of timber. It should be stressed that no recipes exist, and management objectives, which are often subjective and vary from owner to owner, must be taken into account.

Acknowledgments

We thank Phil Cannon, Glenn Galloway, Paul Martins, Francis Putz, and César Saboga for their valuable comments and suggestions, which helped to improve this chapter. We are also grateful to Carlos Gómez and Spencer Ortíz for their dedication in carrying out the inventory fieldwork in the Bio-Itzá Reserve. This work was financially supported by USAID under RENARM project funding.

References

Augspurger, C. K. 1984. "Seedlings survival of tropical tree species: Interactions of dispersal distance, light-gaps, and pathogens." *Ecology* 65(6): 1705–1712.

Barnard, R. C. 1950. "The elements of Malayan silviculture." *The Malayan Forester* 13 (3): 122–136.

Barton, A. M. 1984. "Neotropical pioneer and shade-tolerant tree species: Do they partition tree-fall gaps?" *Journal of Tropical Ecology* 25: 196–202.

Baur, G. N. 1964. "The ecological basis of rainforest management." Sydney, Australia: Forestry Commission of New South Wales.

Brokaw, N. V. L. 1985. "Gap-phase regeneration in a tropical forest." *Ecology* 66: 682–687.

Browne, F. G. 1936. "Milliacre surveys." *The Malayan Forester* 5: 177.

Buschbacher, R. J. 1990. "Natural forest management in the humid tropics: ecological, social, and economic considerations." *Ambio* 19: 253–258.

Castañeda, A., A. Castillo, C. Sabogal, and F. Carrera. 1994 *Aprovechamiento Mejorado en el Bosque Tropical Húmedo: Estudio de Caso en el sitio "Los Filos," Rio San Juan, Nicaragua.* Turrialba, Costa Rica: CATIE.

Castiglione, J. R. 1994. *Informe de Consultoria Realizada para El Proyecto OLAFO/CATIE en San Miguel, Petén, Guatemala.* Turrialba, Costa Rica: CATIE.

Clark, D. 1994. "Plant demography." In *La Selva: Ecology and Natural History of a Neotropical Rainforest,* edited by L. A. McDade, S. B. Kamaljit, H. A. Hespenheide, and G. S. Hartshorn, pp. 90–105. Chicago: University of Chicago Press.

Cowgill, U. M., 1961. "Soil fertility and the early Maya." 1961. *Transactions of the Connecticut Academy of Arts and Sciences,* vol. 42. New Haven, Connecticut.

Dawkins, H. C. 1958. "The management of natural tropical high forest with special reference to Uganda." Imperial Forestry Institute (G.B.) Paper no. 34.

Dawkins, H. C. 1963. "Crown diameters: Their relation to bole diameter in tropical forest trees." *Commonwealth Forestry Review,* 2 (2): 318–333.

Denslow, J. S. 1987. "Tropical rainforest gaps and tree species diversity." *Annual Review of Ecological Systems,* 18: 431–451.

FAO (United Nations Organization for Food and Agriculture). 1968. *Inventario Forestal para el Departamento del Petén.* Guatemala.

Finegan, B. 1993. *Bases Ecológicas para la Silvicultura.* Turrialba, Costa Rica: CATIE.

Gómez-Pompa, A., and A. Kaus. 1990. "Traditional management of tropical forests in Mexico." In *Alternatives to Deforestation: Steps toward Sustainable Use of the Amazon Rainforest,* edited by A. B. Anderson, pp. 45–64. New York: Columbia University Press.

Gómez-Pompa, A., T. C. Whitmore, and M. Hadley. 1991. *Rainforest Regeneration and Management.* Carnforth, UK: Parthenon.

Gretzinger, S. 1994. "Response to disturbance, community associations and successional processes on upland forest in the Maya Biosphere Reserve, Petén, Guatemala." M.S. thesis, North Carolina State University, Raleigh.

Gretzinger, S. P. 1995a. "Manejo sostenible del bosque natural en Bethel, la Reserva de la Biosfera Maya." Turrialba, Costa Rica: CATIE.

Gretzinger, S. 1995b. "El manejo forestal comunitario en la selva Maya: La perspectiva campesina." Turrilba, Costa Rica: CATIE.

Gretzinger, S. P., M. E. Salazar, M. A. Manzanero, E. Moratoya, and H. R. Aguilar. 1993. "Impactos de los aprovechamientos forestales industriales en un bosque primario del Petén, Guatemala." Paper presented at the Third Central American Forestry Congress, Petén, Guatemala.

Guillen, A., C. Gómez, C. Flores, C. Matus, M. Manzanero, and O. Navas. 1993. *Trabajo de Ejercicio: Propuesta de Plan de Manejo de Bosques de la Reserva Bio-Itzá, San José, Petén, Guatemala.* Turrialba, Costa Rica: CATIE.

Harris, L. D. 1984. *The Fragmented Forest: Island Biogeography Theory and the Preservation of Biotic Diversity.* Chicago: University of Chicago Press.

Hartshorn, G. 1980. "Neotropical forest dynamics." *Biotropica* 12(suppl.): 23–30.

Holdridge, L. R., B. F. Lamb, B. Mason Jr. 1950. *Los Bosques de Guatemala.* Turrialba, Costa Rica: CATIE.

Hutchinson, I. D. 1987. "The management of humid tropical forest to produce wood." In *Management of the Forest of Tropical America: Prospects and Technologies,* proceedings of a conference, edited by J. C. Figueroa, F. H. Wadsworth, and S. Branham, pp. 121–155. San Juan, Puerto Rico: USDA Forest Service.

Hutchinson, I. D. 1991. "Diagnostic sampling to orient silviculture and management in natural tropical forest." *Commonwealth Forestry Review* 67(3): 223–230.

Hutchinson, I. D. 1993. "Silvicultura y manejo en un bosque secundario tropical: Caso Pérez Zeledón, Costa Rica." *Revista Forestal Centroamericano* (C.R.), CATIE, 2(2): 13–18.

Ibarra-Manríquez, G. B., B. Sánchez-Garfias, and L. González-Garcia. 1991. "Fenología de lianas y arboles anemocoros en una selva calido-humeda de México." *Biotropica.* 22: 242–254.

Johnson, N., and B. Cabarle. 1993. *Surviving the Cut: Natural Forest Management in the Humid Tropics.* Washington, DC: World Resources Institute.

Lamb, B. F. 1966. *Mahogany of Tropical America: Its Ecology and Management.* Ann Arbor: University of Michigan Press.

Leonard, J. H. 1987. *Natural Resources and Economic Development in Central America.* New Brunswick, NJ: Transaction Books.

Lundell, C. L. 1937. *The Vegetation of Petén.* Washington, DC: Carnegie Institute.

Nations, J. D., and R. B. Nigh. 1980. "The evolutionary potential of Lacandon Maya sustained-yield tropical forest agriculture." *Journal of Anthropological Research* 36: 1–30.

Nicholson, D. I. 1965. "Review of natural regeneration in the dipterocarp forests of Sabah." *The Malayan Forester* 28: 4–24.

Nicholson, D. I. 1972. "Compartment sampling in North Queensland rainforests as a basis for silvicultural treatment." *Commonwealth Forestry Review* 51(4): 314–325.

PAFG (Plan de Accion Forestal-Guatemala). 1992. *Concesiones Forestales a Gran Escala: Caso de Guatemala.* Guatemala: PAFG.

Roth, I. 1987. *Stratification of a Tropical Forest as Seen in Dispersal Types.* Dordrecht, the Netherlands: W. Junk Publishers.

Schwartz, N. B. 1990. *Forest Society: A Social History of Petén, Guatemala.* Philadelphia: University of Pennsylvania Press.

Snook, L. C. 1993. *Stand Dynamics of Mahogany* (Swietenia macrophylla) *and Associated Species after Fire and Hurricane in the Tropical Forests of the Yucatan Peninsula, Mexico.* Ph.D. dissertation, Yale University, New Haven.

Stanley, S. A. 1992. *Analisis del Inventario de la Finca "Istancia," Propiedad de CUDEP, Petén, Guatemala y Recomendaciones para el Manejo Sostenible* Turrialba, Costa Rica: CATIE.

Stanley, S. A. 1994. *Plan de Manejo Forestal, Unidad de Manejo Arroyo Colorado, Petén, Guatemala.* Turrialba, Costa Rica: CATIE.

Stanley, S. A., and C. Gómez. 1993. *El Efecto de un Tratamiento Silvicultural Orientado Conforme al Viento Predominante sobre la Regeneración de Cedrela odorata en El Petén, Guatemala.* Unpublished working document. Turrialba, Costa Rica: CATIE.

Swaine M. D., and T. C. Whitmore. 1988. "On the definition of ecological species groups in tropical rain forests." *Vegetatio* 75: 81–86.

Terborgh, J. 1992. "Maintenance of diversity in tropical forests." *Biotropica.* 24(2b): 283

Wadsworth, F. H. 1987. "A time for secondary forestry in tropical America." In *Management of the Forests of Tropical America: Prospects and Technologies,* edited by J. Figueroa, F. H. Wadsworth, S. Branham, pp. 189–198. Rio Piedras, Puerto Rico: Institute of Tropical Forestry.

Whitmore, T. C. 1989. "Canopy gaps and the two major groups of forest trees." *Ecology* 70: 536–538.

Wyatt-Smith, J. 1961. "A review of Malayan silviculture today." *The Malayan Forester* 24: 5–18.

Chapter 18

Community Restoration of Forests in India

Mark Poffenberger

In the past, most biodiversity studies have focused on areas with large tracts of old growth. Those concerned with biodiversity conservation have often viewed human interaction with the ecosystem negatively. The potential benefits to biodiversity preservation through forest restoration have rarely been discussed. Yet, in recent decades in rural India, an estimated 10,000 villages have initiated protection of several million hectares of degraded forest, allowing it to regenerate and regain some of the biodiversity it had lost (Poffenberger et al. 1995). This has occurred largely without support from forest departments, large donors, or national or international conservation NGOs.

The emergence of a grassroots environmental movement in eastern India suggests that many rural communities are cognizant of the importance of maintaining healthy forest ecosystems and demonstrating that they are the most effective agents in the protection of biological resources. In the past, outsiders have often taken a paternalistic approach to the involvement of local community groups in forest management. While development and conservation rhetoric increasingly supports community "participation," strategies continue to be designed and driven by outside agents. This chapter suggests a reversal in strategy is warranted, whereby outside actors find better ways to support local initiatives in managing forest resources, in part through creating an enabling environment in which community can take action. Such a shift is necessary, since many rural communities have been legally alienated and disempowered as managers of forest resources over the past century.

The Indian National Joint Forest Management Support Group, in conjunction with the Asia Forest Network, has conducted a series of studies over the past five years which show that biodiversity increases rapidly when degraded natural forests are protected by community groups. This chapter reviews some preliminary findings regarding the political contexts in which community forest protection has emerged and the effects of natural regeneration on forest biodiversity and rural incomes. The chapter then explores some of the institutional implications of decentralizing public forestlands management.

Much of India's forestland is fragmented or interspersed with settlements and agricultural lands. In many areas in West Bengal, Orissa, and Bihar, forest patches are no larger than 25 to 1,500 hectares. Most of India's largest national parks are populated by numerous communities. Forest villages are frequently comprised of tribal people, who are heavily dependent on, and highly knowledgeable about, forest resources. Yet, Indian forest policies evolving over the past century have steadily undermined their rights and responsibilities in management. Empowering rural resource users to act as managers of natural ecosystems may be the best and only way to ensure their continued existence.

India's Grassroots Forest Protection Movement

In India, nearly 23 percent of the total land area is officially designated public forestland (Poffenberger 1990). Over the past century, due to disturbances from commercial logging, fires, grazing, and fuelwood collection, healthy forest cover has declined to less than 10 percent of the country. It is estimated that as much as 80 to 90 percent of the subcontinent was under forest cover in prehistoric India. Forest clearing likely began in the Gangetic plains and other riverine areas with alluvial soils. It was not until the end of the eighteenth century that commercial timber extraction began growing in importance to meet the needs of the British navy. During the second half of the nineteenth century, India's forestlands began to be nationalized and a government forest department was established to oversee their use. Demands for logs (sleepers) to build India's extensive railway lines greatly increased felling. As the country industrialized during the early twentieth century, timber exploitation increased correspondingly. Demands of the First and Second World Wars placed further pressure on the forest. Throughout this period India's forest communities, often tribal peoples, saw their forest rights eroded and the natural resource base depleted. After independence, deforestation accelerated as shifting policies and political changes created uncertainties that stimulated unsustainable-use practices. Forest departments found themselves under pressures from politicians to extend logging rights to well-connected commercial interests at highly subsidized rates. Villagers watched as their forests were felled, often feeling they should attempt to capture some of the benefits before others took the tree away. In a response both to a management vacuum and growing market demands, commercial fuelwood cutting by impoverished families became common. State forest departments, with a few thousand, often office-bound staff, were unable to restrain the millions of rural forest users.

During the 1980s, India's forests were disappearing at a rate estimated at 1.48 million hectares annually (*New York Times* 1990). Losing forest produce created severe problems for many rural, subsistence-based communities. In addition, forest clearance was negatively affecting microclimate conditions, well levels, agricultural pest populations, and other environmental factors. In some cases, because of their loss of forest-based products and environmental services, communities were forced to migrate to lowland agricultural areas or urban centers to survive.

Indian tribal resistance to outside interference by the state or other powerful forces is well documented. From the late eighteenth century, with the emergence of the British colonial government, tribal rebellions grew increasingly common.[1] In recent decades too, tribal groups have organized politically and adopted both violent and nonviolent strategies to counter outside threats. In West Bengal, and later in Andhra Pradesh, Naxalite communist organizers have attempted to mobilize tribal communities to resist government forces with violence (Poffenberger and McGean 1995). More recently, tribal communities in eastern

[1]The literature on tribal and peasant resistance to colonial control of the forests in India is extensive. Scholars who have explored this topic include Ramachandra Guha, Madhav Gadgil, Ranajit Guha, Subhachari Dasgupta, and Vandana Shiva. The *Subaltern Studies* series, edited by Ranajit Guha, (New Delhi: Oxford University Press, 1994) is highly recommended. R. C. Verma's *Indian Tribes through the Ages* (New Delhi: Ministry of Information and Broadcasting, Government of India, 1990) also provides a useful summary.

India have created a political coalition to lobby for the establishment of an independent state (*Jharkhand*), comprising parts of Bihar, West Bengal, and Madhya Pradesh (Munda 1988). An overriding objective of these tribal sociopolitical movements has been to gain greater land security and recognition of their ancestral rights to resources. While tribal separatist political movements may not have directly stimulated the emergence of forest protection groups, the growth of tribal political leadership as part of India's decentralized democratization has given new confidence to historically suppressed communities to take direct action to deal with many problems, including those emerging around forest resources.

Community forest management groups are concentrated in the vast forest tracts of the Chotanagpur Plateau, which extends from southwestern Bengal and south Bihar through some districts of Orissa; however, an increasing number of village-based forest protection groups are being identified in Madhya Pradesh, Gujarat, Rajasthan, Karnataka, and other states. Since the 1970s, small community groups have been taking over degraded sal (*Shorea robusta*) forests, often comprised of trees and shrubs so heavily hacked that they are no more than one meter in height, in an attempt to halt degradation and initiate regeneration. In many cases, these natural forests had been logged in the 1950s, 1960s, and 1970s, and were then further degraded through uncontrolled overcutting and grazing by local communities. Concerns over scarcities of fuelwood, fodder, fruits, seeds, and a wide range of forest products have often initiated community discussions regarding ways to resolve local resource shortages.

Local leadership, which generally begins in one or two communities, is a critical element in the initiation of management. After it is shown that forest protection and regeneration are possible, other villages begin to organize similar strategies (Poffenberger 1995). In communities with strong leaders, whether traditional tribal clan elders, youth club members, or a new generation of political organizers, villagers begin to formulate management strategies based on the establishment of tighter protection through patrols and watchers (Poffenberger 1995).

Over the past two decades, community forest protection initiatives in eastern India have occurred without formal assistance from forest departments or nongovernmental organizations. In fact, throughout the 1970s and 1980s, most forest departments have been preoccupied with social forestry projects involving the establishment of eucalyptus and other fast-growing species on community lands, road and railway embankments, and other common property. Concern over deforestation prompted the Forest Act of 1980, which further excluded communities from the utilization of forest resources, with a ban on logging instituted in 1986. Until 1990, there was no government policy to support the devolution of rights and responsibilities to community groups for public forestland management.

There is evidence, however, that at the village level many forest guards and field-level officers have attempted to work with villagers and encourage them to protect local forests. Discussions with some forest protection group leaders in Bihar, Orissa, and West Bengal reveal that individual forest guards and rangers had encouraged them to take action.[2] Consequently, the input of the forest department appears to have been sporadic and based on the personality and interest of individual staff. Some foresters who have attempted to bridge the conflicts between communities and their departments feel they have little support from their own agencies.[3] In other areas, where villagers have seen the forest department sell cutting rights to loggers while beating or arresting them for fuelwood collection, hostilities and antipathy between the sides remained high. In parts of Bihar, villagers so distrusted forest department staff that they warned them they would be beaten if they entered the forest.

Joint Forest Management

The West Bengal Forest Department was one of the first states to begin experimentation with joint forest management. The rise to power of the Communist Party of India Marxist (CPIM) in 1972 was certainly a factor in this shift in orientation. The CPIM's campaign platform dictated greater attention to the needs of rural areas, including an aggressive land-reform campaign. On occasions when the forest department sought police assistance to constrain villagers from cutting timber and fuelwood, they were often met with forcible resistance from community groups. Given the populist sociopolitical environment in West Bengal during this period, many villages began acting on their own, exerting local control over forest access in their area.

At the field level, as early as the 1970s, some forest officers recognized the need to establish joint forest (JFM) management agreements

[2]Field reports by the Joint Forest Management National Support Group, 1993–94.
[3]Discussions with Sarangi range officer, Dhenkanal, Orissa, 1993.

with local villages. In return for the community's commitment to protect degraded forests from fires, grazing, illegal cutting, and agricultural encroachment, forest officers agreed to share forest products with village members. Successful experiences with comanagement from the Arabari research station near Kharagpur in southwest Bengal was particularly influential in showing forest department officers the viability of this new approach. Although Arabari was important in developing an approach to forest management partnerships, until the late 1980s it and other forest department efforts in collaborative management received little attention. Informal agreements between forestry field staff and community groups often collapsed when the officer involved was transferred. It was only with the passage of the first West Bengal Resolution in July 1989, and the National Joint Forest Management Resolution in June 1990, that joint forest management began to gain greater official recognition in one part of India. By late 1994, sixteen Indian states had passed resolutions supporting extending some rights and responsibilities for public forests to village management groups.

Patterns of Community Organizing around Forest Patch Protection

Recent studies have helped illuminate patterns of sociopolitical action leading to the establishment of forest access control systems in eastern India. This social process spreads to neighboring communities allowing effective forest protection to be gradually extended over an increasing number of demarcated forest patches. Forest protection is commonly initiated by a small settlement (*sahi, falia,* etc.) comprised of 10 to 100 households. Once access controls are imposed on a designated area of forest by one settlement, hamlets in the neighborhood that have shared access in the past are usually drawn into management negotiations. In many cases, an increasing number of villages establish access controls until most forestlands have come under the exclusive protection and utilization of specific local communities. Neighboring communities squeezed out of access-controlled forests often begin protection activities in other areas, as a reaction to their loss of access to the initial area closed. During the period in which access controls are being initiated, forest resource extraction pressures (and potential conflict) may increase on neighboring open-access forests, sometimes accelerating their degradation. In other cases, compensation may involve resource substitutions. Villages that are completely excluded from the division of forest protection and use rights either reach agreements as secondary users or poach fuelwood and fodder, thereby remaining in conflict with managing communities. These communities who are left out remain threats to these newly emerging, decentralized forest management systems.

While resource-use pressures appear to shift during the transition from open to controlled access, once most forestlands come under intensified management, greater ecological stability and biomass productivity usually characterizes the affected forests. In the Sarangi Range, in the Dhenkanal District, community forest protection groups have been growing in number for over 40 years. By 1995, virtually all state reserve and protected forests (20,000 hectares) were under the control of over 100 villages. Healthy young secondary forest regeneration is presently evident throughout the range, though a few patches of unprotected forests are still degrading (Poffenberger and McGean 1995).

In some areas, once most of the forest patches have been brought under protection, usually by 5 to 20 adjacent communities, an apex organization may form to represent the villages' interest in dealing with the forest department and other outside organizations. These organizations also function as dispute-resolution mechanisms. In Nayagarh area in Orissa, several hundred villages comprise a large apex group (*Maha Sangha*), which is supported by elected political representatives. Member villages contribute dues toward its operation. Similar apex groups are also being identified in Bihar and West Bengal. In the future, apex bodies could reduce the need for forest departments to work with thousands of individual communities; they may also take on an applied research and extension function.

While the process of community organizing around forest patch protection is becoming clearer, a number of questions remain.

- What are the sizes of forest patches around which the community organizing sequence occurs?
- What factors affect the size of the area, both in terms of the number of communities involved and the spatial area covered?
- What types of communities tend to initiate protection?
- What types of communities are left out of the process?
- What factors influence the rate of expansion (constraints and impetus to replication)?
- How are resource shortages met and compensated in the early phases of forest closure?
- How might or do community clusters coordinate their management activities and decision making to avoid conflicts?

- What size of apex organization is most appropriate for settling disputes and facilitating collaborative management?

Ongoing research in India and other regions of the world should help provide answers to these questions.

Valuing Forest Regeneration

The creation of a network of local community forest management groups has allowed for the establishment of intensified ecosystem protection at a time when growing demographic pressures threatened to destroy India's natural forests. The result has been to facilitate the restoration of degraded forests in many areas in the eastern part of the country. While valuing the benefits accruing from these changes is difficult, it is useful to review some of the biological gains, local economic benefits, and environmental services resulting.

Biological Benefits

In India, an estimated 100 to 150 million hectares, representing between 30 to 50 percent of the nation's land area, is classified as degraded. Much of this area is disturbed forestland, covered with scrub growth often less than one meter in height, and generally possessing a limited range of plant species. Yet, it is estimated that between 30 to 86 million hectares still possess healthy root stock of coppicing tree species. The sal (*Shorea robusta*) forests of the east, the teak (*Tectona grandis*) forests of the west, and the terminallas, acacias, and *Butea monosperma* forests of the semiarid regions all coppice vigorously, generating growth rates of up to one to two meters per year in the first 10 years of protection. Dramatic changes in biodiversity occur as the canopy trees grow and forest cover closes, creating multiple tiers and microniches for shade-tolerant herbs, climbers, and shrub species, especially during the first decade of regeneration. Shade-tolerant species, including mushrooms and medicinal herbs, replace pioneering shrub species.

Local Benefits

The restoration of natural forests also reestablishes their productivity for indigenous communities. In one study of regenerating sal forests in southwest Bengal, researchers found 214 species of flora and fauna after five years of regeneration (Malhotra 1991a). Tribal community members utilized 72 percent of the species present, including 47 medicinal plants, 39 edible species, 6 types of tubers, and 11 varieties of mushrooms, of which 28 species were used for preparing products for sale (Malhotra 1991b). Interestingly, the study also estimated that the value from nontimber forest products exceeded that from sal stems harvested for construction poles by over 300 percent.

In a small remnant teak forest in southeast Gujarat, discussions with communities indicated that, due to logging, grazing, and fires, the useful plant composition declined from 57 species before 1964 to 11 species a few years ago. After several years of protection, however, a number of species once thought to have become extinct are reappearing (Poffenberger et al., 1992). In regenerating forests in the western ghats of Karnataka, Gadgil, and Ravindranath (1991) found 19 tree species in a forest after 12 years of protection, 26 after 14 years of protection, and 55 after 50 years of protection.

Participatory rapid appraisals (PRAs) of changing species composition in regenerating forests have also demonstrated the wealth of botanical knowledge possessed by community groups, especially tribals. During a recent PRA in southeastern Gujarat, in a group discussion lasting approximately one hour, Vasava and Chaudary tribal women identified 128 species of plants, birds, and animals present in a large natural teak forest, many of which were used. Similarly, extensive botanical knowledge is being discovered among tribal communities in other parts of India (Poffenberger et al., 1992).

Forest protection not only supports greater biodiversity, but increases the flow of many forest products, making sustained-yield management of nontimber forest products (NTFPs) increasingly attractive to community groups. By contrast, when sal and teak forests were clean-cut on rotation under earlier working plans, the availability of most NTFPs dropped sharply. Forest products also tend to flow disproportionately to low-income families, due to their lower opportunity costs in collection. In a mixed teak forest in southeastern Gujarat, it was found that the collection of seeds, gums, and leaves generated between Rs. 8 to 12 per day. In contrast to the state wage rate of Rs. 27 per day or income from logging activities, which ranged from Rs. 28 to 40 per day (Poffenberger et al., 1992: 60).

Increased forest product flows can lead to improvements in nutrition by expanding the availability of tubers, fruits, seeds, nuts, mushrooms, and edible leaves. In Dhenkanal District in central Orissa, Saura tribals reported subsisting primarily on forest tubers and other edible plants for six months of the year. Reliance on forest tubers as a staple food in the diet of tribal peoples appears widespread from initial studies in eastern India.

Regenerating forests also create expanded employment opportunities, as the productivity of grasses, leaves, and seeds allow for small cottage industries to develop. In the Jamboni range in southwest Bengal, a recent study of 12 communities indicated mean household income from major NTFP exceeded Rs. 2,300 (approximately U.S. $100) per year (Malhotra 1991b: 11). The greater availability of a range of NTFPs is also allowing some forest communities to reduce their dependence on commercial fuelwood headloading. After six years of protection in Raigarh village in southwest Bengal, men had reduced their labor allocation for fuelwood cutting from 47 to 14 percent, while increasing their labor inputs for NTFP collection from 12 to 41 percent. In this area, the collection of 24 medicinal species was an important source of cash. The most commercially important of these were bel (*Aegele marmelos*) and kalmegh (*Andrographis paniculats*), while the highest-value medicinal was sabai (*Euliopsis binata*) fiber grasses and the collection of kendu leaves (*Diospyros melanoxilon*) for Indian cheroot rolling (*bedi*).

While the shift from fuelwood extraction to the collection of NTFPs, especially medicinals, has increased employment opportunities in Raigarh and reduced general forest disturbance, a number of medicinal species appear to be overexploited (Pal, n.d.). Medicinal plants that require uprooting or debarking for product collection are particularly threatened. To improve income generation for medicinals and other NFTPs, and to ensure that harvesting practices are sustainable, culturing operations, either within the natural forests or on private agricultural lands, need to be explored. These changes allow communities dependent on socially disruptive seasonal agricultural migration to begin to break this pattern by productively managing natural forests, maintaining their communities, and reducing school dropout and medical problems.

Many of the forest ecosystems of India possess a high proportion of tree species capable of regenerating through suckers growing from the root or stump. This allows even badly degraded scrub land rapidly to reestablish a closed-canopy, young secondary forest through coppice growth. Species composition changes as the canopy closes, with sun-loving herbs and shrubs being replaced by shade-tolerant plants. Natural regeneration through voluntary community protection costs between 3 to 5 percent of the amount needed to establish monoculture plantations on the same land area. At the same time, NTFPs flowing from naturally regenerating forests usually yield more income and employment to community members, especially women, when contrasted with monoculture plantations. Under natural regeneration many of the forest ecosystems studied also began yielding NTFPs much faster than degraded lands placed under monoculture tree plantations. Many grasses and leaves are available after three to six months of protection.

Other Environmental Service Benefits

Regenerating natural forests are also being associated with improved hydrological functions by slowing runoff and improving soil moisture and groundwater levels. While it is difficult to measure accurately changes in water availability that can be directly attributed to improved forest cover, community members frequently believe that increasing well levels are taking place. They also note favorable microclimatic shifts that reduce temperatures and increase humidity during the long dry season. Such perceptions encourage further efforts to improve resource management.

Tribal communities observe that many bird species have returned to the regenerating forest, helping to control insect problems on agricultural lands. In some cases, however, the return of large mammals is seen as undesirable. In West Bengal, forest protection committees are experiencing problems with elephant herds now migrating back into their areas as forests regenerate.

Managing Forest Ecosystems

Conventional forest department planning and monitoring systems were designed to address timber stocking levels. They did not consider such issues as the role of communities in the sustainable management of forests and changing biodiversity. In most states the locations of forest protection groups are poorly known. Where inventories or registration have been done, there is little information indicating which FPCs (forest protection committees) are effective in facilitating regeneration and where forests are continuing to degrade and loose biodiversity. Working-plan information on vegetative conditions is usually badly outdated, and, in many areas, management objectives are changing rapidly. In addition, working plans reflect broad management objectives for large areas rather than identifying highly localized needs and conditions required for JFM program management.

Currently, the Indian Forest Service focuses monitoring activities on changes in crown density, with good forests categorized as those with 40 percent or greater canopy closure, moderate forests as those with 10 to 40 percent closure, and poor forests as those with under 10 percent closure. These categories are not sensitive either to gradual or to rapid changes occurring in many forest ecosystems. Much of the degradation occurring in terms of loss of biodiversity, soil erosion, biomass removal, and fires may occur under the canopy. Further, the early phases of natural regeneration may not be reflected in canopy monitoring statistics. Monitoring systems that assess critical changes in vegetative conditions are needed. Such information systems should be able to identify and distinguish between forests that are continuing to degrade ecologically in the absence of access controls and those that are regenerating under community protection.

Perhaps of even greater concern, forestry development projects have increasingly focused the attention of foresters on the administration of activity budgets and the achievement of planting targets. These administrative duties place immense demands on staff time and draw attention away from the more fundamental needs to monitor forest conditions and improve field-level management operations. It is not uncommon for a district forest officer with 100,000 hectares of land to concentrate virtually all of his staffing resources in a small area of a few hundred hectares because a development project has captured the department's attention.

At this early phase in the emergence of community forest management partnerships, there is an urgent need for forest departments to determine where villages are already protecting forests and to identify high-priority areas for establishing community-based forest-access controls. Forest departments need methods to rapidly identify and monitor locations of existing or potential forest protection groups, as well as the condition of local forests. At the present time, few divisional forest officers (DFOs) or range officers regularly use topographic maps of their areas that accurately identify local communities and forest boundaries. Topographic maps on a scale of 1:50,000, however, are easy to obtain and very useful for spatial inventorying of FPC activities and vegetative conditions. These maps can be used to document where access controls are emerging or established, where forests are regenerating or degrading, and priority areas where JFM activities have high potential and require encouragement.

A number of field trials have been conducted with methods to map the location of FPCs and the vegetative conditions of forest tracts at the beat, section, and range level. Most DFOs and range officers who have been exposed to the rapid-mapping-assessment methods feel that they are extremely useful in quickly establishing an overview of JFM activities in their territories. After trials in Sarangi Range in Dhenkanal Division of Orissa, one forest range officer noted: "I am pleased with this map. I can see my entire range—boundaries, locations of active and inactive villages. I can see problem areas; the overall status of the forest composition. I have a holistic view."

The process of mapping is the first step to making foresters cognizant of village management strategies, problems, and needs. As such, it is also a preliminary activity toward acknowledging, legitimizing, and supporting the role of communities in forest protection.

Under both community and joint forest management systems of production, emphasis frequently shifts from optimizing timber output to enhancing the flows of many forest products. To manipulate forests sustainably to enhance the yield of NTFPs, new information regarding patterns of regeneration, changes in forest species composition, density, and shifts in the volume flows of many different products over time is required. The management objectives for newly protected forest areas should be defined jointly by the community with territorial foresters. The manipulation and marketing strategy for each site should also be developed together. Cash-and-kind product-flow requirements will need to be determined through joint consultations. After considering available technical methods, market absorption levels, and price conditions, decisions regarding the best options for manipulation, including clear and selective felling, thinning, pruning, and enrichment planting, need to be determined.

Just as much agricultural research has moved on-farm in recent decades, JFM programs now require research be done "in-community and forest." Community members who now manage the forests need to take the lead in establishing study priorities and collecting data to better determine how different management options perform in real-life situations and under varying social and ecological conditions. Selected community people should become equal members of newly established research teams. Supportive programs could be designed to assist communities to take responsibility for regeneration and manipulation monitoring programs, including keeping records of vegetational changes, product flows, and income generated. Eventually, some emerging apex organizations comprised of numerous FPCs could play a coordinating role, in conjunction with state forest departments, in applied forest management research and extension.

Conclusion

Empowering communities to protect natural forests not only renews the ecosystem but strengthens communal institutions as well. In this rapidly changing world, cultural diversity is at least as threatened as biodiversity. India possesses one-third of the world's tribal peoples, many of whom live in the forest areas in the central part of the subcontinent. Their languages, belief systems, knowledge, and material culture are strongly tied to the forest. As the forests are cleared, the social systems of the tribal people are undermined. Without supplemental forest foods and products, families are forced to spend increasing periods away from the community as migrant laborers, or to become slum dwellers in urban areas. Empowered to facilitate natural regeneration, communities regain their confidence and stability. As one Bhil man stated, "the gods had left us, but now they have returned."

References

Gadgil, M., and N. H. Ravindranath. 1991. "Natural regeneration through community participation in degraded forest lands in the western Ghats." Banglalore, India: Centre for Ecological Sciences, ITS.

Malhotra, K. C. 1991a. "People, biodiversity and regenerating tropical sal forests in West Bengal, India." Paper presented to the International Symposium on Food and Nutrition in the Tropical Forests, September, Paris, France.

Malhotra, K. C. 1991b. "Joint forest management in West Bengal: Study of non-timber forest produce." Indian Institute of Bio-Social Calcutta: Research and Development.

Munda, R. D. 1988. "The Jharkhand movement: Retrospect and prospect." *Social Change* 18: 2.

New York Times. 1990. "Global deforestation," July 6, 1990.

Pal, S. n.d. "Agroecosystem study of Raigarh villages," Rama working paper, Krishna Mission, Calcutta.

Poffenberger, M. 1990. *Joint Management of Forest Lands: Experiences from South Asia.* New Delhi: Ford Foundation.

Poffenberger, M. 1995. "The resurgence of community forest management in eastern India." In *Natural Connections,* edited by D. Western and R. M. Wright, pp. 53–79. Washington, DC: Island Press.

Poffenberger, M., and B. McGean. 1995. *Village Voices: Forest Choices.* New Delhi: Oxford University Press.

Poffenberger, M., B. McGean, and A. Khare. 1995. "Communities sustaining India's forest in the 21st century," In *Village Voices: Forest Choices,* edited by M. Poffenberger and B. McGean. New Delhi: Oxford University Press.

Poffenberger, M., B. McGean, A. Khare, and J. Campbell, eds. 1992. *Field Methods Manual, vol. II. Community Forest Economy and Use Patterns, PRA Methods in South Gujarat, India.* New Delhi: SPWD.

Chapter 19

Challenges in Promoting Forest Patches in Rural Development Efforts

Larry Fisher and Roland Bunch

Introduction

Trees are a central feature of most rural landscapes. In lowland irrigated rice lands and in semiarid regions, trees and forest patches dot the rural landscape; they are apparent in forest galleries, within narrow ravines, along riparian strips, surrounding springs and sacred areas, and, most frequently, in home gardens and within villages. The presence of trees as a central feature of the rural economy is most prominent in communities that have a strong history of forest farming. Forests, trees, and their products provide a wide array of benefits for rural households and communities—in traditional food-security strategies, as well as in the transition from grain-crop systems toward more market-oriented farming systems. As subsistence, food-crop-centered systems evolve into more cash-based economies, tree crops (along with livestock) frequently become the pivotal feature of the system (Hayami 1994).

Nevertheless, agricultural expansion and encroachment, loss of ancestral control of communal lands, the introduction of new technologies, and changing legal and administrative contexts have dramatically reduced the forest area and tree cover in rural areas throughout the world. This has led to increased attention by social activists and environmentalists concerned about the integrity of rural communities and their ecosystems.

This chapter will explore opportunities for reversing these trends. Throughout the world, government agencies, and increasingly nongovernmental organizations (NGOs), have attempted to promote forest conservation, agroforestry, and agricultural improvement programs to protect biodiversity and improve the livelihoods of rural communities. The following analysis is an attempt to review some of this experience and relate it to the promotion of forest patches in the rural landscape.

In this discussion, we will address the practical aspects of how to better integrate forest patches into rural development programs. To answer the question, What does a forest patches program look like? we will draw from a broad spectrum of activities—from direct technical- and extension-training efforts to more indirect attempts to influence conservation-related behavior through policy initiatives. A listing of relevant forest patch programs necessarily includes efforts focused on soil and water conservation, agricultural extension, reforestation, watershed management, buffer zones, protected areas and wildlife management, and integrated rural development programs (Stern et al., 1993).

Program initiatives can be divided into four basic elements: (1) technical, (2) management, (3) organizational, and (4) legal and policy aspects. To be sure, no single agency can effectively implement all of these elements simultaneously. Program leaders should begin with activities that address the most immediate and compelling needs identified with local communities and also work from the comparative strengths of their organizational settings.

At the same time, successful programs and organizations must evolve to meet changing needs and opportunities; thus programs may initially begin with technology development and dissemination, and subsequently move into more policy-reform and advocacy-oriented activities. Given the multifaceted nature of these programs, strong local organizational capacity and strategic organizational alliances are essential features of long-term success. Creating linkages between programs and institutions offers the range and complementarity that cannot be achieved by any individual program.

This chapter will outline some of the key issues within the four general categories of technology development, program management, organizational capacity building, and policy reform. The overall analysis offers a survey of the range of activities and program development approaches that can contribute to successful forest patch efforts.

Technology Development

The technologies chosen within agricultural and forestry development programs frequently have major impacts on the protection and/or generation of forest patches. Program personnel should consider the following factors in the development and promotion of these technologies: agricultural intensification efforts, reduction of migratory agriculture, agroforestry, use of forest products, and off-season labor.

Agricultural Intensification Efforts

If farmers can produce all the subsistence crops needed for the household on one hectare or less, rather than two or three hectares, more land will be available for forest or orchards. Furthermore, even though such intensification frequently requires more labor per hectare initially, the overall labor requirement per unit of food produced can often be decreased. This is clearly a major factor in farmer decision making to intensify cultivation practices.

Agricultural intensification can free up land for low-labor land-use practices such as growing fruit and forest trees. For example, in the village of Las Venturas, San Marin Jilotepeque, Guatemala, many farmers with only two or three hectares learned to intensify their agriculture during the 1970s, using a variety of improved soil, water conservation, and crop production practices. A 1994 study found that these farmers had, in the intervening years and by themselves, developed a system of sustainable forest management. On the one-half to two hectares they no longer farmed, many were maintaining forests by planting out volunteer tree seedlings and harvesting only that quantity of more mature trees that could be done so sustainably (Bunch and Lopez 1994).

Reduction of Migratory Agriculture

Lal (1989) estimates that some 11 million hectares of forest are cut each year to replace degraded agricultural land. There are many effective technologies that can be used to reduce fallow periods and assist farmers in the transition to more sedentary forms of agriculture. Among the strategies being utilized to improve fallow management are assisted, or natural, regeneration, the use of cover crops in a variety of slash-mulch systems, alley farming, and judicious application of fertilizers and other soil amendments. Many of these technologies build upon traditional soil and water management practices used by forest farmers throughout the tropical world. There are numerous examples of how programs have incorporated and advanced local agricultural practices.

In Flores, Indonesia, a traditional fallow system (based on a fast-growing leguminous tree, *Leucaena leucocephala*) has been used in the development and promotion of alley farming hedgerows for soil conservation and fertility management around the world (Metzner 1982; Kang 1989; L. Fisher 1992). In other parts of Indonesia, indigenous use of the intrusive weed *Eupatorium inulifolium* has effectively doubled the cropping period and led to greater conservation of neighboring forested areas (Malcolm Cairns, pers. comm.). In Guatemala, the Centro Maya has been using a velvet bean (*Mucuna pruriens*) and maize intercrop to reduce the practice of slash-and-burn agriculture in the humid tropical Peten region. Some farmers there have now grown maize in the same plots for eight years straight in areas where farmers traditionally farm the land for only three years at a time.

Agroforestry

Agricultural systems that incorporate trees can also reduce the cutting of firewood and other pressures on forest patches, and can even create such patches on agricultural land. Such agroforestry systems include the use of fruit orchards and small, mixed plantations; contour tree barriers (alley farming hedgerows); live fences; multitiered associations of trees with understory crops; and home gardens that include perennials; wood lots; and protected or managed forest areas. On the island of Sumba, Indonesia, adoption of alley farming hedgerows now provides much of the fuelwood needs for families that had previously collected firewood within protected forests. The hedgerows have also been used as shade for transition to economically important tree crops such as coffee, cacao, and cinnamon. In Cebu, the Philippines, livestock raising has stimulated the planting of a range of fodder trees and shrubs, which have expanded forest patches and reduced pressures on existing forest areas (Cerna et al., n.d.). Animal credit schemes have contributed to family incomes and provide economic and food security during the off-season.

Use of Forest Produce

Activities that reinforce the value of forest stewardship can greatly enhance conservation practices. In Xishuangbanna Prefecture, southern China, management of protected forest areas has been strengthened by the use of the medicinal plant Chinese cardamum (*Amomum villosum*) as an important source of income (Eberhardt, forthcoming). The collection of a variety of nontimber forest products such as honey, dyes, Dioscorea tubers, rattan, and medicinal plants has been the most effective food-security strategy during lean periods in semiarid Nusa Tenggara, Indonesia. In a different context, revenues from ecotourism have reinforced conservation of forest habitat for howler monkeys in the case of the Community Baboon Sanctuary in Belize (Lyon and Horwich, chapter 11, this volume). Validating the use of forests that benefit local communities, and developing new forest products and sustainable management practices, can encourage protection of forested areas. It should be noted, however, that the tendency toward commercialization of forest products and services can also lead to overexploitation, resulting in increased degradation of the resource base and increasing income disparities and conflicts within communities (Dove 1994).

Use of Off-Season Labor

The cutting of trees for firewood or charcoal production frequently occurs during the agricultural off-season, when farmers have few sources of gainful employment. If agricultural or other off-season activities can provide a significant demand for high-value labor and offer increased income levels, then off-season exploitation and destruction of forests will eventually decrease. Production of a variety of handicraft products and processing of agricultural produce, as well as certain agricultural activities, can fill this need. For example, most soil conservation work can be done during the off-season; irrigated crops can be grown near water sources; and improved drainage, raised beds, or thick mulches can allow farmers to grow crops when rains are too intense for traditional methods.

In Guinope, Honduras, large numbers of farmers used to cut trees and gather pine resin because of a lack of gainful employment during the dry season. Now, following the implementation of a successful agricultural extension program during the 1980s, resin collection has virtually ended, and firewood cutting has decreased. This change is largely due to the additional soil preparation activities that are now carried out in the dry season, as well as the increase in plots irrigated through microirrigation schemes (Bunch 1988).

Program Management Aspects

The development of programs to encourage forest patches requires some level of organizational and management capability. Rural development experience around the world suggests that programs should begin with preliminary analysis and dialogue with local communities. The development of local leadership and organizational capacity, basic mechanisms for accountability, and reinforcement of learning approaches (i.e., monitoring and evaluation) will ensure continued responsiveness and sustainability. Programs that ignore these rather simple steps risk alienating local communities or losing sight of new opportunities as they emerge.

Rural Survey, Analysis, and Planning

Working with rural communities assumes site-specific analysis and initiative. The use of participatory appraisal methods is a key step in understanding the local setting. These techniques have the added advantage of building a sense of common purpose and collective action. Participatory appraisals should engage wide representation within the community and between the community and external agencies. They can be used to identify critical problem issues, possible solutions, and develop a plan of action. The engagement of local communities in a well-organized assessment phase, and in the articulation of program goals and plans, is often cited as one of the most fundamental factors in determining program success (see, inter alia, Bunch 1982; Seymour 1985; Wells 1994).

In the last two decades, the emergence of innovative, yet simple applied social survey methods have enabled a more manageable, rapid, and vivid process of participatory, community-led assessment and planning. The range of tools developed under the general categories of rapid rural appraisal (RRA), agro-ecosystems analysis (AEA), participatory rural appraisal (PRA), and other applied survey methodologies have led to more substantive exchange and dialogue within local communities and with public service agencies (see, e.g., Chambers 1992; Khan and Suryanata 1994; Khon Kaen University 1987; McCracken et al., 1991). Some of the most effective tools that have been used in the development of reforestation programs include resource mapping, transects, matrix ranking (e.g., for agroforestry species and wealth ranking), seasonal calendars, institutional and social diagrams (Vochten and Mulyana 1994).

These tools and approaches, now widely practiced in the development community, are increasingly documented to be effective as practical, rapid research methods that also empower rural communities to take initiative in developing programs for rural change. Participatory appraisals have also proven to be an effective mechanism for negotiating agreements between outside agencies and local communities (Fisher et al., 1994).

Farmer-Centered Program Development

The staggering physical and cultural diversity of tropical environments suggests that most technical and social solutions will have to be determined locally. If ecological or sustainable agriculture is to have a chance of spreading throughout the tropics, it will require a virtual explosion of new knowledge. While green revolution technologies were believed to have wide relevance to farmers throughout the tropics, it took almost 30 years for chemical agriculture to reach most village farmers. Even so, the application of chemical inputs has been inconsistent and has often been criticized for the lack of adoption by marginal farmers on more degraded lands. In these areas, the use of green manures or cover crops may be more appropriate but will require careful selection of species for different maize or sorghum varieties and adaptation to conditions of altitude, slope and soil type, rainfall pattern, farming systems, and insect populations.

Since these marginal, highly diverse environments are precisely the areas where regeneration and protection of forest patches is more critical, the only way to accomplish success is to engage farmers directly in the process of technology development, adaptation, and dissemination. Effective programs often begin by encouraging villagers to establish very simple experiments and to evaluate and share their results with others. The development of farmer extension networks, facilitated by NGOs or state agencies, is one of the essential features of effective agricultural or agroforestry development programs (Hayami 1994; Seymour 1985), and one that will also be needed for developing new knowledge and management systems for forest patches.

Limiting Technologies and Activities

The most effective approach to agricultural improvement is not trying to change everything farmers do simultaneously, but starting where farmers are, determining basic constraints, and then gradually changing a few critical factors to achieve high levels of success. Using a limited, well-chosen (and widely relevant) intervention can simplify program administration, reduce costs, motivate farmers more efficiently, achieve greater social justice, and keep the management of program activities within the hands of local communities (see Bunch 1982). For example, World Neighbors–assisted programs in eastern Indonesia and in the Philippines have demonstrated the catalytic role of soil and water conservation practices in galvanizing community participation, reinforcing local voluntary working groups, and significantly transforming local farming systems (Cerna et al., n.d.; L. Fisher 1992).

Program Organization and Accountability

Rural development programs must be designed to strengthen local organizational capacity. Accountability is critical: program leaders are best identified and supervised within local settings. Local organizations (user groups, tribal federations, farmer networks, or village-level government) should participate in the selection and orientation of project staff, in the definition of their job descriptions, and in their performance evaluations. Many successful agroforestry programs have relied upon farmer leaders as project staff and extensionists (Bunch 1982; Seymour 1985). Management of program finances should also be positioned as closely as possible to the local level.

The monitoring and evaluation of program activities, essential to program learning and responsive adaptation, should begin with the participatory appraisal approaches described above. Program objectives and indicators should be defined and assessed through local organizational structures—and the procedures for evaluation should be consistent with program-defined needs. Conducted in the field and in office setting, these evaluations, which will be both informal and formal, will use oral and practical methods, as well as written documentation. More rigorous diagnostic research methods can also be employed to offer objective validation for the assessment of program impact.

Integrated Approaches

Experience with rural development programs suggests that it is difficult to dissociate any single activity from the complex needs of communities and the fabric of change taking place within them. Efforts to promote forest patches will likely be viewed by local people within this broader framework. For example, during initial discussions of agroforestry program support in Flores, Indonesia, community leaders insisted that a more immediate priority was the construction of a gravity-flow water system. Because of the great distances to water sources, villagers were too busy collecting water during the dry season. Without accessible water in the community, they argued, it would be difficult to maintain the nurseries at these distant springs; transporting seedlings would also be a major constraint. The completion of the water system led to broad participation in soil and water conservation and tree-planting efforts in these communities.

Agroforestry and forest conservation must therefore be viewed within a wider community development context, and programs should be prepared to extend their vision and activities in addressing rural needs. Forestry-oriented programs must also consider relevant activities in agriculture and livestock, as well as education, health, family planning, and other community concerns (Bunch 1982; Wells 1994). While no one program, agency, or community could possibly implement all these activities, it is important for programs seeking to promote forest patches to take into consideration the wider social, technical, and institutional context and to leverage linkages with other agencies to address these related community needs.

Organizational Development

For forest patch initiatives to succeed, local organizations, both formal and informal, will be a critical factor. Many activities, such as nursery management, land clearing, joint forest conservation and protection, and enforcement, require community participation and some measure of social organization. While new organizational mechanisms may be needed for specific activities or for broader program management, it is important to recognize that organizational capacity for collective action already exists in most rural communities. As Robert Fisher has pointed out in his analysis of Nepal's Community Forestry Program, "failure to understand the sociology of local organizations has been a major contributing factor in the relatively poor performance of community forestry in the past" (R. Fisher 1991). It is therefore important to understand the fundamental structure of a local community, such as its organizations, functions, membership and representation, values, regulations, and leadership, roles (Fisher et al., 1994).

Building upon these local dynamics, most programs exhibit several levels of organizational structure and involve many types of institutions fulfilling a variety of roles. Local community organizations provide the natural framework for much of a program's implementation, NGOs have been effective catalysts for mobilizing community initiative and participation, government regulatory agencies are vested with legal and policy authority, and research institutions can offer technical input for program issues.

Local Organizations and Capacity Building

Strengthening local capacity involves much more than technical training and leadership skill. Individual capacity building is central to development; however, building "social capital"—the collective capacity of local organizations to mobilize resources and implement effective programs—is an equally important challenge for development programs. Assessing and strengthening local organizational capacity necessarily deals with a range of skills and capabilities, which may include the capacity to mobilize resources, manage and monitor development program activities, develop and maintain intercommunity links, articulate vision, and negotiate with outside agencies (Gubbels 1994).

NGOs as Emerging Institutions

NGOs are increasingly viewed as central actors in conservation efforts around the world. In many cases, they have become a critical linking mechanism between government agencies and local groups (Tolisano and Poffenberger 1991). NGOs are often praised for their responsiveness to local needs, for their strong commitment to participatory approaches, and for their innovative approaches to training, advocacy, and local capacity building. At the same time, NGOs often lack technical expertise; have limited, time-bound resources to work with; and may fail to make connections of localized program efforts to the broader institutional and policy context (see Farrington and Bebbington 1994; L. Fisher 1994; Korten 1992; Winder 1992).

As NGOs have emerged as a legitimate alternative to bureaucratic, centralized government institutions, they are increasingly being relied on to implement extension and conservation programs. There are compelling examples of small-scale NGO programs that have served as models for the development of national programs. However, many NGO programs that are most effective at the local level, and have relied on dynamic leadership for their implementation, have stalled as they have been scaled up into larger efforts. There have also been reports that the rapid rise of NGOs have created a climate of unrealistic expectations. In the Philippines, for example, recommendations by NGOs within the Asian Development Bank–funded national reforestation program led to the creation of bogus, fly-by-night organizations that were often established simply to gain access to funding (Korten 1993; Pasicolan and Sajise 1993).

Government Agencies

The government is the abiding institutional setting that offers both the regulatory context and the social services with the greatest influence on conservation activities. Governments often have tremendous material and staff resources, along with an overarching, standardized structure under which programs function. Of course, these strengths are also considered major weaknesses in assessments of governments' limited responsiveness and effectiveness.

Government agencies—such as agriculture, forestry, rural development, and public works—directly influence forest patches. Forest patch programs must be designed to work within what is often an extremely complex, rigid, and glacial bureaucratic framework. Practical difficulties include lack of coordination and continuity, jurisdictional conflicts between line agencies, staff turnover, and frequent shifts in national- or provincial-level policies. Program leadership must develop effective coordination mechanisms and stimulate feedback loops through applied research and accessible documentation. While pilot projects at the local level are essential to provide vivid examples and learning opportunities, they must be leveraged effectively through sustained linkages with relevant government agencies at the district, province, and national level.

New coordinating structures or informal mechanisms will often be needed to mitigate existing administrative divisions. The convening of working groups comprised of representatives of relevant agencies is one way to bring about change within the bureaucracy (Korten 1989). Informal interagency networks, such as NGOs, research institutions, and government agencies, have also proven effective. But a key weakness in these structures can be the emphasis on institutional representation, which results in a lack of voice for rural people who have the most insight and strongest stake in the decision-making process. Program efforts must be cognizant of the need to select effective representative spokespersons from local communities and to find the proper fora in which they can participate in these discussions.

Finally, it is important to note that even as major substantive changes occur at the policy level, there remains the monumental challenge of implementing these changes down through the system. Extensive training will be needed for the reorientation of field extensionists and public officials.

Alliances/Networks/Coalitions

No single agency can implement the range of interventions needed for truly effective and sustainable forest patch programs. Therefore a mix of institutions, activities, and approaches will be needed, and it is important to seek both coordination and complementarity among the various actors that can contribute to these efforts. A variety of networks and coalitions can be established in pursuit of these goals. These may include single agency coalitions—or "horizontal networks" (e.g., among NGOs, farmer organizations), or they may include interagency and/or interdisciplinary coalitions—"vertical networks" (NGOs, researchers, government officials, and farmer leaders). Networks may also form within different geographical units (provincial, national, regional) and may focus on specific or general themes (see Alders et al., 1993).

Informal networks present opportunities to develop linkages between actors separated by distances of geography, discipline, or institution. They can be an effective forum for facilitating cooperative exchange of skills, technical information, and material resources. They can assist in coordinating activities to reduce duplication and are able to address large, complex development problems that cannot necessarily be resolved at any one level (grassroots, national policy). However, networks often suffer from lack of clear objectives, conflicting ideologies or perspectives of participants, and dominion by individuals or agencies with strong agendas (Moeliono and Fisher 1991). Since networks are often informal coalitions, they may lack the resources and/or the enforcement mechanism for effective implementation. For these reasons, it is important to establish clear objectives early on, to focus on concrete activities, and to encourage broad participation of all stakeholders through regular and varied activities. Networks will require the commitment of a core of individuals and, ultimately, material resources for sustained support.

Policy Issues

Many local land-use practices and decisions over resource management are often greatly influenced by a policy context that is determined at great distances from forest areas (Wells 1994). In Vietnam, Byron (1994) has noted that in addition to technological considerations, a combination of three policy-related factors was important in the widespread adoption of tree farming in the central highlands: (1) allocation of degraded public lands to households, (2) economic liberalization and transition to a market economy, and (3) the importance of new roads and access to market centers.

In fact, there are a number of policy instruments and other external factors that can influence the behavior of rural people in conserving forest patches. It is often difficult to isolate any individual element from the complex set of social and political issues which impinge on natural resource management (see, e.g., Hafner 1990).

Experience in the Philippines, for example, suggests that lack of secure land tenure has not precluded farmer interest in tree-planting efforts (Seymour 1985), cash incentives are a useful but inconclusive factor in determining sustainability of reforestation programs (Pasicolan and Sajise 1993), and strong emphasis on local organizations has not necessarily led to greater program success (Korten 1993). In short, there are rarely simple, single-answer approaches to developing forest patch programs; the range of policy instruments to consider is quite extensive. The mix of policy and program initiatives and their appropriateness to the local context will determine in large part whether farmers participate in these programs. Some of the policy options include land tenure, trade and marketing, roads and infrastructure development, subsidies and credit, population, and community education.

Land Tenure

Security of land tenure and usufruct rights, whether de facto or de jure, is a necessary, if not wholly sufficient, element for decision making on land use (Lynch and Alcorn 1994; Seymour 1985). Farmers and communities must have a strong sense of clarity over the values, rights, regulations, and institutions that underlie tenurial systems—both traditional and state-regulated—if they are to consider long-term investments and the maintenance of potential benefits from forested areas. The undermining of traditional and community-based tenure systems has been one of the major contributing factors to deforestation; and expanding state control over forest management is at the core of this problem, particularly where ineffective and/or unjust enforcement mechanisms are the norm (Alcorn 1993).

Local control over tenure regimes, particularly if it is backed up by effective laws, institutions, and enforcement mechanisms, appears to be the best hope for conservation-related efforts. Indigenous reserves in Latin America and communal forest lease agreements in the Philippines are examples of this relatively new approach. Forest conservation programs must seek to strengthen local organizational capacity for the clarification and enforcement of tenure and land use. In addition, these programs must address the policy and institutional framework that provides the context for these community-based systems.

Trade and Marketing

Home gardens, orchards, woodlots, and forest patches have traditionally provided for essential economic and subsistence needs. Given the isolation of many rural communities, and the challenge of accessing markets, successful agroforestry programs have tended to emphasize with high-value-per-unit-weight commodities and products. Many programs have been developed with the objective of improving market access through diversification of products and buyers, development of cooperatives or other locally managed commercial organizations, and greater access to information (Seymour 1985). Forest patch programs have also emphasized the economic imperative of finding useful benefits for private reserves, cottage industries, and other value-added opportunities. However, as mentioned earlier, there is certainly mixed evidence to draw conclusions about the impact of exploiting forest products to encourage conservation.

Roads and Infrastructure Development

Another critical influence on forest management practices is the increased access enabled by road construction. On the one hand, roads and other communication mechanisms provide greater access to markets, information, and other social services. On the other hand, they have proven a destabilizing element on previously isolated communities; road construction itself has led to rapid and extensive deforestation (Hafner 1990). Road construction and infrastructure development remain major priorities for government investment in rural areas. While their effects are often difficult to predict, they are certain to bring about fundamental social and economic transformation. Careful planning and preparatory assessments can often mitigate some of the negative impacts.

Subsidies and Credit

Material incentives for tree planting are hotly debated in practice (L. Fisher 1992). Cash, equipment and materials, or food-for-work incentives, in anticipation of subsequent benefits, can serve to bridge the need for investment capital during the early growth of seedlings. Reforestation programs can provide off-season employment and enable families to stay together during lean periods. However, incentives are often challenged on the basis of limited sustainability. Financial or material support is frequently provided for emergency crisis relief within the framework of short-term projects, creating dependencies that undermine sustained local initiative. In many cases, activities dissipate quickly after the incentives are withdrawn. Material incentives tend to narrow farmer interest to short-term gains and often create strong reasons for sabotaging reforestation projects (i.e., if the project fails, the government will have to provide funding for replanting). Incentives also cause tension in communities because they are typically distributed unequally and often monopolized by wealthy elites, further marginalizing the small holders who are most in need of these programs. If incentives are to be used, they should be done so judiciously, with minor material benefits, geared to local decision making and the multiple objectives of the program (both short- and long-term), and done in a manner that reinforces local management and control.

Population

Demographic aspects of reproduction, migration, and settlement patterns are yet another important influence on the integrity of forest patches. Although population growth and density are certainly key factors in deforestation rates, they may be mitigated by agricultural development and changing land management efforts (Metzner 1982; Tiffen 1994). Of perhaps greater importance are the rapid population shifts taking place in many rural areas, often the result of misguided government policy on resettlement and transmigration. Encouraging lowland populations to migrate into sparsely populated upland areas has led to major deforestation in mainland Southeast Asia. Moving indigenous communities from their traditional lands removes the social buffer and local control, resulting in open-access situations that can lead to rapid forest and land degradation.

Community Education

Cultural values are an essential feature of efforts to promote forest patches. Communal forest management requires shared attitudes about the social, economic, and environmental benefits of these areas. In addition, there is often a strong spiritual and religious dimension of conservation that is frequently overlooked by conservation programs. Sacred groves and sites that are respected for religious or ritual purposes are often the last remaining forested areas in rural communities (Lebbie and Freudenberger, chapter 15, this volume). Program leaders seeking to encourage the conservation of forest patches must therefore understand and respect these beliefs and traditions and, in some cases, work to reinvigorate eroding social traditions or instill new conservation values into training and education activities.

Conclusions

Reinforcing and promoting forest patch conservation requires a combination of initiatives that span technical, program management, institutional, and policy dimensions. Within these general categories, program leaders have to tailor options to address site-specific conditions. Continued monitoring of results is needed to adapt program interventions to the dynamic change that is inherent in all rural development efforts.

Documented experience around the world suggests that there are often chicken-and-egg dynamics inherent in the development of conservation programs, many of which grow out of community-based initiatives that are reinforced by institutional responsiveness and relevant policy changes (Bailey 1994; Seymour 1994; and Poffenberger, chapter 18, this volume). In some cases, programs may not, in fact, have begun as forest patch programs, but moved in this direction as a response to local community interest and creative initiative. In other efforts, forest patch projects have evolved into multifaceted, integrated, rural development programs that address a range of activities and build from the local level to address wider regional or national concerns.

In any case, it is important to understand local contextual factors (cultural aspects, market forces, legal and institutional setting, etc.) that may impact on farmer decision making to protect forest patches or expand tree-planting efforts on private or community land. A good understanding of existing constraints and possible supporting factors will better position any program-related intervention (technical, organizational, legal, or political) designed to stimulate or reinforce the conservation and/or generation of forest patches.

The analysis of factors that contributed to the success of a community forestry project in Vietnam shows the importance of a mix of programmatic approaches:

> *...building on the local traditions of forest protection, the project encouraged communities to explore measures that could result in long term protection and management of the resource. This included extensive consultation with communities, the establishment of local protection teams, the formation of local rules on forest use, training in simple silvicultural techniques, and forest survey to gather information for long term management plans.* (Neave and Quang 1994)

While a general formula for success seems relatively easy to define, examples of truly effective and sustainable forest conservation programs remain painfully rare. Given the widespread failure of reforestation efforts, bold innovation and institutional change will be needed. Community-level organizations and NGOs must be prepared to find constructive ways to work with the government to access their more extensive technical and material resources and to influence the broader policy framework; governments must seek ways to decentralize authority and decision making to the local level to encourage participation and responsive action, as well as address the challenge of diversity (Seymour 1985; Korten 1992; Farrington and Bebbington 1994). And new alliances will be needed to create linkages between local-level initiatives and national policy settings, to enable the important sharing and complementarity between agencies.

References

Alcorn, J. 1993. "Remnant forest use and ownership: Patterns and issues." In *Forest Remnants in the Tropical Landscape: Benefits and Policy Implications,* edited by J. Doyle and J. Schelhas, pp. 41–44. Washington, DC: Smithsonian Migratory Bird Center.

Alders, C., B. Haverkort, and L. van Veldhuizen. 1993. *Linking with Farmers: Networking for Low External Input and Sustainable Agriculture.* London: Intermediate Technology Publications.

Bailey, C. 1994. "Global and regional issues in natural resources degradation." Comments to the International Food Policy Research Institute's International Advisory Committee Meeting, Kampala, Uganda, May 16–17, 1994.

Bunch, R. 1982. *Two Ears of Corn: A Guide to People-Centered Agricultural Improvement.* Oklahoma City: World Neighbors.

Bunch, R. 1988. "Case study 5: Guinope integrated development program, Honduras." In *The Greening of Aid: Sustainable Livelihoods in Practice,* edited by C. Conroy and M. Litvinoff, pp. 40–44. London: Earthscan Publications, Ltd.

Bunch, R., and G. Lopez. 1994. "Soil recuperation in Central America: Measuring the impact 4 to 40 years after intervention." Paper presented to the International Institute of Environment and Development's International Workshop, Bangalore, India, 28 November to 2 December 1994.

Byron, N. 1994. "Tree farming by small farmers on degraded lands in Vietnam." *CIFOR NEWS,* Number 5, December 1994.

Cerna, L., L. Moneva, R, Trangia, M. Tabasa, and R. Noel. n.d. *The Mag-uugmad Foundation: Farmer-Based Approaches to Local Resource Management in the Philippines.* Oklahoma City: World Neighbors.

Chambers, R. 1992. *Rural Appraisal: Rapid, Relaxed, and Participatory.* Sussex, UK: Institute of Development Studies.

Dove, M. R. 1994. "Marketing the rainforest: 'Green' panacea or red herring?" Asia Pacific Issues Series No. 13, East–West Center, Honolulu, Hawaii.

Eberhardt, K. forthcoming. "Social factors influencing land use of upland swidden farmers in Xishuangbanna Prefecture, China," M.S. thesis, Cornell University, Ithaca, NY.

Farrington, J., and A. Bebbington. 1994. "Interactions between NGO's, governments, and international funding agencies in renewable natural resources management." Paper presented to the International Symposium on Rain Forest Management in Asia, Centre for Development and the Environment, University of Oslo, March 23–26, 1994.

Fisher, L. 1992. "From *kebe* and *lamtoronisasi* to farm planning: Experiences promoting alley farming, soil conservation, and uplands development in southeastern Indonesia." Paper presented to the International Conference on Alley Farming, IITA, Ibadan, Nigeria, September 14–18.

Fisher, L. 1994. "Non-governmental organizations, democratization, and the reorientation of professional forestry." Unpublished paper, Department of Natural Resources, Cornell University.

Fisher, L., V. Russell, and T. Robertson. 1994. "Sustaining conservation in protected areas: Strategies for managing conflict." Paper for the Northern Arizona University 1994 Graduate Student Interdisciplinary Symposium: "Exploring the Challenges of Conflict: Issues and Alternatives in the 20th Century," Flagstaff, Arizona, November 4–5.

Fisher, R. J., 1991. "Local organizations in community forestry." FAO Regional Wood Energy Development Program, Regional Expert Consultation on Local Organizations in Forestry Extension, Chiang Mai, Thailand, October 7–12.

Gubbels, P. 1994. "Criteria for assessing local organizational capacity." Internal office memo, World Neighbors West Africa Regional Office, Ougadougou, Burkina Faso.

Hafner, J. 1990. "Forces and policy issues affecting forest use in northeast Thailand, 1900–1985." In *Keepers of the Forest: Land Management Alternatives in Southeast Asia,* edited by M. Poffenberger, pp. 69–94. West Hartford, CT: Kumarian Press.

Hayami, Y. 1994. *In Search of Modern Sustainable Agriculture in Upland Indonesia.* Jakarta, Indonesia: FAO-UNDP/Bureau of Planning Ministry of Agriculture.

Kang, B. T. 1989. "Alley cropping/farming: Background and general research issues." Paper presented to the AFNETA Inaugural Meeting, August 1–3, Ibadan, Nigeria.

Khan, A., and K. Suryanata. 1994. *A Review of Participatory Research Techniques for Natural Resource Management.* Jakarta, Indonesia: Ford Foundation.

Khon Kaen University 1987. *Proceedings of the 1985 International Conference on Rapid Rural Appraisal.* Khon Kaen, Thailand: Rural Systems Research and Farming Systems Research Projects.

Korten, F. 1989. "The working group as a catalyst for organizational change." In *Transforming a Bureaucracy: The Experience of the Philippine National Irrigation Administration,* edited by F. Korten and R. Siy Jr. Quezon City, Philippines: Ateneo de Manila Press.

Korten, F. 1992. "A place at the table: NGO's and the forestry sector." *Unasylva* October 1992.

Korten, F. 1993. "The high cost of environmental loans." Asia Pacific Issues Series No. 7, East–West Center, Honolulu, Hawaii.

Lal, R. 1989. "Conservation tillage for sustainable agriculture: Tropics vs. temperate environments." *Advances in Agronomy* 42: 85–197.

Lynch, O., and J. Alcorn. 1994. "Tenurial rights in community-based conservation." In *Natural Connections: Perspectives in Community-Based Conservation,* edited by R. M. Wright, D. Western, and S. Strum, pp. 372–392. Washington, DC: Island Press.

McCracken, J., J. Pretty, and G. N. Conway. 1991. *An Introduction to Rapid Rural Appraisal for Agricultural Development.* London: IIED.

Metzner, J. 1982. *Agriculture and Population Pressure in Sikka, Isle of Flores.* Canberra, Australia: Australian National University.

Moeliono, I., and L. Fisher. 1991. "Networking for development: Some experiences and observations." Paper presented to the workshop on networking for LEISA, Silang, Cavite, Philippines, March 9–15.

Neave, I., and B. N. Quang, 1994. *What Works: Lessons from a Community Forestry Project in Northern Vietnam.* Monograph Series No. 5, CARE International, Hanoi, Vietnam.

Pasicolan, P. N., and P. E. Sajise, 1993. "Between cash and usufruct rights." *PURC News and Views,* Philippines Upland Resource Center, Ateneo de Manila University, July–September.

Seymour, F. 1985. "Ten lessons of agroforestry projects in the Philippines." Unpublished internal report, United States Agency for International Development, Manila.

Seymour, F. 1994. "Are successful community-based conservation programs designed or discovered?" In *Natural Connections: Perspectives in Community-based Conservation,* edited by R. M. Wright, D. Western, and S. Strum, pp. 472–498. Washington, DC: Island Press.

Stern, M., J. Schelhas, and W. Hartwig. 1993. *Forest Remnants Project Directory.* Washington, DC: Smithsonian Migratory Bird Center.

Tiffen, M. 1994. *More People: Less Erosion: Environmental Recovery in Kenya.* New York: Chichester.

Tolisano, J., and M. Poffenberger. 1991. "Forging an alliance between local users and national forest management agencies: Lessons from developing countries." Paper presented to the Conference on Nonprofit Organizations in National Rural Development: "Steps Toward a New Partnership," Santa Fe, New Mexico, November 19–20.

Vochten, P., and A. Mulyana, 1994. "Reforestation, protection forest, and people: PRA in the planning of an Indonesian government project, a case study." Unpublished report, World Food Program, Kupang, Indonesia.

Wells, M. 1994. "Community-based forestry and biodiversity projects have promised more than they have delivered: Why is this and what can be done?" Paper presented to the International Symposium on Rain Forest Management in Asia, Centre for Development and the Environment, University of Oslo, March 23–26.

Winder, D. 1992. "NGO's and the forestry sector: Case studies from Oaxaca, Mexico." Unpublished report, Synergos Institute, New York.

Contributors

Janis B. Alcorn, Biodiversity Support Program, c/o World Wildlife Fund, 1250 24th Street, NW, Washington, DC 20037

JANIS B. ALCORN is an ethnobotanist with over 20 years experience in international development and research in the area of tropical ecology, indigenous natural resource management systems, and conservation of biodiversity in Asia and the Americas. She received her Ph.D. in botany, with a minor in anthropology, from the University of Texas at Austin in 1982. Prior to assuming her current position as director for Asia and Pacific in the Biodiversity Support Program at the World Wildlife Fund, she taught biology at Tulane University and served as an AAAS (American Association for the Advancement of Science) environment and natural resource advisor to USAID.

Humberto Alvarez-López, Departamento de Biología, Universidad del Valle, Cali, Colombia

HUMBERTO ALVAREZ-LÓPEZ has a Ph.D. from Cornell University and is currently a professor in the Department of Biology at the Universidad del Valle in Cali, Colombia. His research focuses on bird ecology and conservation.

Richard O. Bierregaard Jr., Biodiversity of Forest Fragments Program, National Museum of Natural History, Smithsonian Institution, Washington, DC 20560

RICHARD O. BIERREGAARD JR. graduated from Yale in 1973 and in 1978 received his Ph.D. from the University of Pennsylvania under the supervision of Robert E. Ricklefs. He spent 14 years directing the Biological Dynamics of Forest Fragments Project in Manaus, Brazil. He is currently an adjunct member of the Biology Department of the University of North Carolina at Charlotte. His principal research interests are the conservation and ecology of Neotropical forest birds and in general, birds of prey.

John O. Browder, Urban Affairs and Regional Planning, Virginia Polytechnic Institute and State University, Blacksburg, VA 24601-0113

JOHN O. BROWDER directs the international development concentration of the Urban Affairs and Planning Program at Virginia Polytechnic Institute and State University. Author of numerous articles on Brazilian Amazon development and deforestation, he also serves on the Task Force on Scholarly Relations with the Natural Science Community of the Latin American Studies Association.

Roland Bunch, COSECHA, Apartado 3586, Tegucigalpa, Honduras, C.A.

ROLAND BUNCH has an M.S. in international agricultural development from the California Ploytechnic State University. He has devoted 28 years to agricultural development in Central and South America, focusing on work for World Neighbors. He is the author of *Two Ears of Corn*, a book on people-centered agricultural development, which has been published in nine languages. He is a cofounder of COSECHA and currently serves as head of the Department of Rural Development at the Panamerican Agricultural School in Honduras.

Virginia H. Dale, Environmental Service Division, Oak Ridge National Laboratory, Box 2008, Oak Ridge, TN 37831

VIRGINIA H. DALE is the associate director of the Environmental Sciences Division at Oak Ridge National Laboratory. Her areas of expertise include ecological modeling, land-use change, and landscape ecology.

Larry Fisher, Department of Natural Resources, 10 Fernow Hall, Cornell University, Ithaca, NY 14853-3001

LARRY FISHER has worked with a range of uplands development programs in Southeast Asia, including farmer-based extension in agriculture and agroforestry, social and community forestry programs, agroecosystems research, and biodiversity conservation. He first served with the Indonesian Village Volunteers Program (BUTSI) and has worked with the Ford Foundation, the Asia Foundation, and as the Southeast Asia area representative for World Neighbors. He received a master's in professional studies in agriculture, and is currently pursuing doctoral studies in the Department of Natural Resources at Cornell University.

Mark Schoonmaker Freudenberger, World Wildlife Fund, 1250 24th Street, NW, Washington, DC 20007

MARK SCHOONMAKER FREUDENBERGER is currently the applied social scientist for World Wildlife Fund–U.S., with the responsibilities of the domains of participatory planning, tenure, migration, and conflict management. Formerly with the Land Tenure Center of the University of Wisconsin–Madison, his area of specialty is tenure and resource management in Senegal, The Gambia, and Guinea. His doctoral studies were in the field of regional and rural planning at the University of California at Los Angeles.

Russell Greenberg, Smithsonian Migratory Bird Center, National Zoological Park, Washington, DC 20008

RUSSELL GREENBERG received his Ph.D. from the University of California at Berkeley in 1981. Since then he has been at the Smithsonian Institution as a research fellow and member of the scientific staff. He is currently the director of the Smithsonian Migratory Bird Center, where he pursues his long-term interest in the psychology, ecology, and evolution of migratory birds and the conservation of their tropical ecosystems.

Steven P. Gretzinger, Producción en Bosques Naturales, CATIE, Turrialba, Costa Rica

STEVEN P. GRETZINGER currently lives with his wife and daughter in Costa Rica, where he works with CATIE on the sustainable management of lowland tropical forests. He received a master's degree in forestry from North Carolina State University. He lived in Guatemala for over five years while working on various natural resource issues with the Nature Conservancy, Conservation International, and USAID.

Carlos F. Guindon, School of Environmental Studies, Yale University, New Haven, CT 06511

CARLOS F. GUINDON, a conservation biologist born and raised in Monteverde, Costa Rica, is currently pursuing a doctorate in the School of Forestry and Environmental Studies at Yale University.

Robert H. Horwich, Community Conservation Consultants, Howlers Forever, Route 1, Box 96, Gays Mills, WI 54631

ROBERT H. HORWICH, president of Howlers Forever and director of Community Conservation Consultants, received a Ph.D. from the University of

Maryland. Following postdoctoral work in India with the Smithsonian Institution, he became a researcher for the Brookfield Zoo. He has studied infant development in birds and mammals, developed a method for reintroducing endangered cranes in India and Central America, and worked on community sanctuaries.

Gustavo H. Kattan, Wildlife Conservation Society, Apartado 25527, Cali, Colombia

GUSTAVO H. KATTAN received a Ph.D. from the University of Florida and is currently a research associate for the Wildlife Conservation Society, New York, and the Instituto Vallecaucano de Investigaciones Cientificas, Cali, Colombia. His research interests include bird ecology and conservation, and he is currently working on patterns of biodiversity at the landscape level in the Colombian Andes.

Martin Kellman, Department of Geography, York University, North York, Ontario MSJ 1P3, Canada

MARTIN KELLMAN is professor of geography and biology at York University. His major research interests are in plant geography and ecology and tropical environmental systems. He has conducted tropical field research in northern Australia, the Philippines, Mexico, Central America, and Venezuela.

Aiah Randolph Lebbie, Department of Forestry, 1630 Linden Drive, University of Wisconsin–Madison, Madison, WI 53706

AIAH RANDOLPH LEBBIE is a Fulbright student in the Conservation Biology and Sustainable Development graduate program at the University of Wisconsin–Madison. He was born in Sierra Leone, where he holds a lectureship position in conservation biology at Njala University College. His research interests include the reproductive ecology and population genetics of tropical forest plants, ethnobotany, and conservation development of sacred groves.

Jonathan Lyon, Environmental Resources Research Institute, The Pennsylvania State University, Land and Water Research Building, University Park, PA 16802

JONATHAN LYON, currently completing his doctorate in ecology at the Pennsylvania State University, has been studying the effects of soil acidity on root growth of forest trees in the northeastern United States. He has also researched prairie and wetland restoration in Wisconsin and land use, tree phenology, and rainforest succession following slash-and-burn agriculture in various regions of Belize. He has been working on community sanctuaries since 1985 and is assistant director and research coordinator of Community Conservation Consultants.

Jorge Meave, Laboratorio de Ecología, Facultad de Ciencias, Universidad Nacional Autónoma de México, México DF 04510, México

JORGE MEAVE holds bachelor's and master's degrees from the Universidad Nacional Autonoma de Mexico (UNAM) and a doctorate from York University. He is currently an associate professor in the Laboratoria de Ecologia at UNAM. His research interests are in tropical montane forest ecology.

Paulo Roberto Moutinho, Universidade Federal do Pará, Belém, Pará, 66.040, Brazil

PAULO ROBERTO MOUTINHO received his master's degree in ecology at the Universidade Estadual de Campinas, in São Paulo. He is currently a professor of animal ecology at Universidade Federal do Pará and cofounder of Instituto de Pesquisa Ambiental da Amazonia. Paulo is pursuing a Ph.D. in ecology on the topic of cutter ant influences on forest dynamics.

Carolina Murcia, Wildlife Conservation Society, Apartado Aéreo 25527, Cali, Colombia

CAROLINA MURCIA has a Ph.D. from the University of Florida and is currently a research associate with the Wildlife Conservation Society, New York. Her research interests include plant reproductive biology and the impact of management practices on the composition and structure of Andean forests.

Daniel C. Nepstad, Woods Hole Research Center, Box 296, Woods Hole, MA 02543

DANIEL C. NEPSTAD, an associate scientist at the Woods Hole Research Center and a visiting professor at the Universidade Federal do Pará (UFPa), has studied human impact on Amazonian forests and currently leads a program on the ecology and management of Amazon forests. He recently cofounded the Instituto de Pesquisa Ambiental da Amazonia, a research and education institute based at UFPa. He received his Ph.D. in forest ecology at Yale University.

Christine Padoch, Institute of Economic Botany, The New York Botanical Garden, Bronx, NY 10458

CHRISTINE PADOCH has been conducting research for over 20 years on traditional patterns of agriculture, agroforestry, and forest management in the humid tropics. She has worked extensively in both Latin America and Southeast Asia, particularly in Amazonia and on the Island of Borneo.

Miguel Pinedo-Vasquez, School of Environmental Studies, Yale University, New Haven, CT 06511

MIGUEL PINEDO-VASQUEZ is a forest ecologist specializing in the management of tropical forests and agroforests. He has extensive experience in working with Amazonian communities on the management of their forest resources and is currently conducting research in floodplain forests in Brazil and Peru.

Mark Poffenberger, Program on Environment, East–West Center, 1777 East–West Road, Honolulu, HI 96848

MARK POFFENBERGER is a social scientist specializing in natural resource management in Asia, where he has spent over 20 years as a volunteer, researcher, consultant, and Ford Foundation program officer. He is currently director of the Asia Sustainable Forest Management Network and is affiliated with the East–West Center in Honolulu and the University of California at Berkeley.

Alison G. Power, Section of Ecology and Systematics, Corson Hall, Cornell University, Ithaca, NY 14853

ALISON G. POWER is an associate professor in the Section of Ecology and Systematics and in the Department of Science and Technology Studies at Cornell University. She has a Ph.D. in zoology from the University of Washington. Her research has concentrated on the ecology and epidemiology of insect-transmitted pathogens of plants and on ecological interactions between natural and agricultural systems.

John G. Robinson, Wildlife Conservation Society, 185th and Southern Boulevard, Bronx, NY 10460

JOHN G. ROBINSON has directed the international conservation programs of the Wildlife Conservation Society (formerly the New York Zoological Society) since 1990. Previously, at the University of Florida, he established the Program for Studies in Tropical Conservation, which provides graduate training in conservation to students from tropical countries. His research interests have focused on tropical forest ecology, especially in the Neotropics, and sustainable use of these systems.

John Schelhas, Department of Natural Resources, 10 Fernow Hall, Cornell University, Ithaca, NY 14853-3001

JOHN SCHELHAS is coordinator of the Cornell Program in Ecological and Social Science Challenges of Conservation and a research associate in the Department of Natural Resources at Cornell University. He has a Ph.D. in renewable natural resources, with a minor in anthropology, from the

University of Arizona. He previously worked for the Smithsonian Migratory Bird Center and the U.S. National Park Service. His research interests include land-use and natural resource use choices of small farmers, national park buffer zones, and agricultural intensification.

José Maria Cardosa da Silva, Zoological Museum, University of Copenhagen, Universitetsparken 15, DK-2100 Copenhagen, Denmark

JOSÉ MARIA CARDOSA DA SILVA recently finished his Ph.D. at the University of Copenhagen, where he is currently a postdoctoral fellow. He worked for eight years as an associate researcher at the Museu Paraense Emílio Goeldi, in Belém, and is a specialist in the ecology and phytogeography of Amazonian and Cerrado birds.

Scott A. Stanley, Producción en Bosques Naturales, CATIE, Turrialba., Costa Rica

SCOTT A. STANLEY is a silviculturist with the Production in Natural Forests project in CATIE, Costa Rica. One of his primary responsibilities consists of determining sustainable harvest levels of timber and nontimber forest products for the Maya Biosphere Reserve in Guatemala. He has worked in Central America for over five years.

André A. J. Tabanez, Departamento de Ciencias Florestais, ESALQ/Universidade de Sao Paulo, Piracicaba, Sao Paulo CEP 13418-900, Brazil

ANDRÉ A. J. TABANEZ has an agronomy degree from the University of Sao Paulo and a master's degree in forest sciences at ESALQ (Escola Superior de Agricultura "Luiz de Queiroz"). He is currently researching the biology and management of forest fragments in the Brazilian Atlantic moist forest.

Rosanne Tackaberry, Department of Geography, York University, North York, Ontario MSJ, 1P3, Canada

ROSANNE TACKABERRY is a science advisor to the government of Canada. She holds degrees in geography and biogeography and has research interests in plant ecology and conservation. She has conducted tropical field research in East Africa, Belize, and Venezuela.

Christopher Uhl, 208 Mueller Lab, Pennsylvania State University, State College, PA 16802

CHRISTOPHER UHL is associate professor of ecology in the Department of Biology at the Pennsylvania State University. He began his studies of forest and human ecology of Amazonia in the mid-1970s, and has investigated slash and burn agriculture, the cattle and timber industries, and regional land-use planning. He is cofounder of the Instituto do Homen e do Meio Ambiente, a research center based in Belém, Pará.

Virgílio M. Viana, Departamento de Ciencias Florestais, ESALQ/Universidade de Sao Paulo, Piracicaba, Sao Paulo CEP 13418-900, Brazil

VIRGÍLIO M. VIANA received the degree of forest engineer at the University of São Paulo and a Ph.D. in evolutionary biology at Harvard University. He is currently professor of tropical forestry at the Department of Forest Science of ESALQ/University of São Paulo, where he conducts research on the biology and management of tropical moist forests of Latin America.

Ima Célia Vieira, Museu Paraense Emílio Goeldi, Caixa Postal 399, Bélem, Pará, 66.040, Brazil

IMA CÉLIA VIEIRA is a scientist at the Museu Paraense Emílio Goeldi in Belém, Brazil. She received her master's in plant population biology at the Universidade de São Paulo, Piracicaba, and is currently pursuing her Ph.D. in ecology through Stirling University in Scotland.

Index

Island Press Board of Directors

CHAIR
Susan E. Sechler
Executive Director, Pew Global Stewardship Initiative

VICE-CHAIR
Henry Reath
President, Collector's Reprints, Inc.

SECRETARY
Drummond Pike
President, The Tides Foundation

TREASURER
Robert E. Baensch
Senior Consultant, Baensch International Group Ltd.

Peter R. Borrelli
Executive Director, Center for Coastal Studies

Catherine M. Conover

Lindy Hess
Director, Radcliffe Publishing Program

Gene E. Likens
Director, The Institute of Ecosystem Studies

Jean Richardson
Director, Environmental Programs in Communities (EPIC),
University of Vermont

Charles C. Savitt
President, Center for Resource Economics/Island Press

Peter R. Stein
Managing Partner, Lyme Timber Company

Richard Trudell
Executive Director, American Indian Resources Institute